Groundwater around the World

Groundwater around the World

A Geographic Synopsis

Jean Margat
Jac van der Gun

CRC Press
Taylor & Francis Group
Boca Raton London New York Leiden

CRC Press is an imprint of the
Taylor & Francis Group, an **informa** business

A BALKEMA BOOK

Cover photo:
Karst spring of La Loue in the Jura, France.
Photo: Philippe Chrochet.

*Emerging groundwater is transformed here into the water of a swiftly flowing stream –
an illustration of closely interrelated groundwater and surface water systems.*

CRC Press/Balkema is an imprint of the Taylor & Francis Group, an informa business

© 2013 Taylor & Francis Group, London, UK

Typeset by V Publishing Solutions Pvt Ltd., Chennai, India
Printed and bound in The Netherlands by PrintSupport4U, Meppel

Published by: CRC Press/Balkema
              P.O. Box 11320, 2301 EH Leiden, The Netherlands
              e-mail: Pub.NL@taylorandfrancis.com
              www.crcpress.com – www.taylorandfrancis.com

Library of Congress Cataloging-in-Publication Data

Applied for

ISBN: 978-1-138-00034-6 (Hbk)
ISBN: 978-0-203-77214-0 (eBook)

Disclaimer

The designations employed and the presentation of material in this publication do not imply the expression
of any opinion whatsoever on the part of the authors or the organisations involved concerning the legal
status of any country, territory, city or area or its authorities, or concerning the delimitation of its frontiers
or boundaries. As the global data sets used do not go beyond the year 2010, more recent changes in national
territories are not reflected in the world maps or related tables.

Contents

List of text boxes

List of tables

List of figures

Preface

Groundwater, water below ground surface: a fascinating subject. When rain falls, we see water disappearing into the ground and reappearing somewhere else, creating and feeding the rivers. What happens in between we cannot see, only speculate. Luckily, keen observation and scientific analysis have provided us with an insight into the world of groundwater, as this book vividly illustrates. In particular, the book highlights the kaleidoscopic variation in groundwater around the world: in its appearance and behaviour; in the role it plays in the hydrological cycle; in ecosystems and human life; in the ways it is exploited and used; and in the challenges of its proper and sustainable management. This invisible groundwater remains the most unpredictable and intriguing part of the hydrological cycle.

Lack of visibility is also reflected in the limited attention that groundwater receives in general; 'out of sight, out of mind', one might say. But how important is groundwater to us? About half of the world's population drinks groundwater every day. Groundwater is of vital importance for agriculture and contributes to more than half of the world's production of irrigated crops; it sustains wetlands and rivers, provides stability to the soil and prevents seawater intrusion. At the same time, groundwater is under increased pressure from population growth, climate change and human activities, with a widespread impact in terms of groundwater depletion and pollution. Hence, there are plenty of reasons to pay due attention to the state of groundwater resources as this wonderful book does.

Why 'groundwater around the world'? Why a global look when groundwater is so often depicted as a local resource? There are several reasons. Firstly, groundwater systems can continuously cover thousands to millions of square kilometres, cross the borders of countries and greatly influence the socio-economic development of vast areas. Even when groundwater is relatively localized, its use and protection eventually have a global impact: in a modern world, a flow of goods is also the flow of groundwater used for their production. Secondly, comparison between the many different groundwater systems on Earth may contribute to deepening the knowledge of any particular one: it draws attention to what is most typical and most relevant in our local groundwater system and it can give suggestions on how to use and manage it optimally. Finally, a global look provides a worldwide overview of the opportunities, threats and problems related to groundwater and thus may help identify priorities for development and management action.

Assessing the groundwater resources at the global scale is a very demanding and laborious process. Data required for the assessment is often non-existent, poorly accessible, contradictory, unreliable and difficult or even impossible to check. Current monitoring of the global state of groundwater is far from adequate and, in comparison with surface water, groundwater has only benefited to a limited extent from remote sensing. Consequently, global hydrological models employ simplistic characterisations of groundwater systems, together with numerous assumptions and poorly verifiable approximations of groundwater characteristics and behaviour. Nevertheless, both authors have not been put off by all these complicating factors. Driven by their fascination for groundwater and guided by their more than one hundred years of combined professional experience, they have patiently and meticulously collected, processed, reviewed, analysed and compared enormous quantities of information, in order to derive – by convergence of evidence – the most plausible picture of the many facets of the world's groundwater. They present this picture in such a way in their book that it is not only understandable for hydrogeologists and other groundwater specialists, but also for a much wider group of potential readers.

The subjects covered in this book are without exception relevant and essential for all themes of the Seventh Phase of UNESCO's International Hydrological Programme, entitled 'Water dependencies: Systems under stress and societal responses' (IHP-VII, 2008–2013). The information presented reflects what the International Groundwater Resources Assessment Centre (IGRAC) aims for in more detail in its Global Groundwater Information System. These were reasons for UNESCO-IHP and IGRAC to encourage the authors for preparing this book and to give them the support they needed.

The rapidly increasing worldwide attention to groundwater governance illustrates the need for a book like this. Good governance of our groundwater resources assumes the active participation of all relevant stakeholders, varying from mandated government institutions to end-users of groundwater and those who value groundwater-related ecosystems. The persons involved have different professional backgrounds and knowledge, but a basic common understanding of groundwater and the opportunities and problems it offers is a prerequisite for fruitful communication that will lead to proper decisions and action. This book, with its systematic presentation of groundwater in a broad hydrological and societal context, and with its many maps, tables and examples, provides an easily accessible introduction and a unique geographically oriented reference which we can warmly recommend to anybody interested in groundwater.

Alice Aureli,
Chief Groundwater Resources and Aquifer Systems Section
International Hydrological Programme (IHP)
UNESCO – Division of Water Sciences

Neno Kukuric,
Director
International Groundwater Resources Assessment Centre (IGRAC)

Acknowledgments

Les eaux souterraines dans le monde, written by Jean Margat, was published jointly by UNESCO and BRGM in 2008. The author undertook this work on the initiative of UNESCO's Division of Water Sciences. He received support and advice from this Division, in particular from Alice Aureli and José Luis Martín, and also from Didier Pennequin and Dominique Poitrinal of the *Service EAU* of the French *Bureau de Recherches Géologiques et Minières (BRGM)*. A global database was developed and maps were prepared at BRGM by Guérin Nicolas, with the assistance of Vincent Mardhel, Annabel Gravier and Jean-Jacques Séguin. The book also benefited from the documentation provided by IGRAC and from the useful and valued comments and suggestions made by Vazken Andreassian, Emilio Custodio and Wilhelm Struckmeier.

After the book had been published and disseminated, UNESCO's Division of Water Sciences soon started thinking about an English version, in order to reach a much wider readership around the world. Alice Aureli and José Luis Martín invited and encouraged Jac van der Gun to prepare such a version in English. Because some years had passed before this work started, it was decided not just to translate the original book, but to seize the opportunity to update and expand the information presented and to add new elements as deemed useful. All this was subsequently carried out, in close consultation and co-operation with the senior author, and it has resulted in this present book.

This book, therefore, is a significantly revised second edition of the original book, presented in a different language. It capitalises on the original edition, but also received additional support and contributions by several persons and institutions. In particular, UNESCO-IGRAC took responsibility for preparing a completely new set of thematic world maps (which was skilfully done by Lena Heinrich), because these had to be modified as a result of updating time-dependent variables. On occasion other colleagues were also consulted to seek their advice on specific questions or subjects. These colleagues include Stefano Burchi, Jake Burke, John Chilton, Steven Foster, Todd Jarvis, Leonard Konikow, Jan Nonner, Mario Sophocleus and Raya Stephan. A number of colleagues made a special contribution by reading the draft of the book or part of it and gave useful comments and suggestions for improvement: Emilio Custodio, Lena Heinrich, Todd Jarvis, Wilhelm Struckmeier and Frank van Weert. After the book had been amended accordingly, Claire Taylor carried out the language editing and made valuable suggestions for correcting and polishing the English. Alice Aureli and José Luis Martín of UNESCO's Division of Water Sciences have been a continuous and valuable support throughout the entire project. Finally, this book could not have been written without the efforts of the numerous individuals around the entire world – most of them unnamed – who have been involved in collecting, processing and interpreting data related to groundwater and who have shared this data and information through databases, reports or publications.

To all these persons and institutions, we extend our sincere and warm gratitude.

Jean Margat
Jac van der Gun

Abbreviations and acronyms

AAC	Association Africaine de Cartographie (African Cartographic Association)
ACSAD	Arab Center for the Studies of Arid Zones and Dry Lands
ALHSUD	Asociación Latinoamericano de Hidrología Subterránea para el Desarrollo (Latin-American Groundwater Hydrology Association for the Development)
AMOR	Aquifer management organization
AQUASTAT	FAO's Global Information System on Water and Agriculture
ASCE	American Society of Civil Engineers
BGR	Bundesanstalt für Geowissenschaften und Rohstoffe (German Federal Institute for Geosciences and Natural Resources)
BGS	British Geological Survey
BRGM	Bureau de Recherches Géologiques et Minières (French Institute for Geological and Mining Investigations)
CBLT	Commission du Bassin du Lac Tchad (Lake Chad Basin Commission)
CEDARE	Centre for Environment and Development for the Arab Region
CEE	Central and Eastern Europe
CCS	Carbon Capture and Sequestration
CGMW	Commission for the Geological Map of the World
CGWB	Central Ground Water Board (India)
CIHEAM	Centre International de Hautes Études Agronomiques Méditerranéennes (International Institute for Advanced Mediterranean Agronomic Studies)
CME	Conseil Mondial de l'Eau (= WWC)
COTAS	Comunidades de Usuarios de Aguas Subterráneas (Communities of Groundwater Users, Mexico)
DDT	Dichlorodiphenyltrichloroethane (insecticide)
DPSIR	Framework of analysis interrelating driving forces, pressures, state, impacts and responses
ECE	Economic Commission for Europe
EDC	Endocrine disruptive compound
EMWIS	Euro-Mediterranean Information System on know-how in the Water sector
EPA	Environmental Protection Agency, USA

ERWR	External Renewable Water Resources
ESCAP	Economic and Social Commission for Asia and the Pacific
ESCWA	United Nations Economic and Social Commission for Western Asia
EU	European Union
EUROSTAT	European Union Directorate responsible for detailed statistics on EU member and candidate countries (its database bears the same name)
FAO	Food and Agricultural Organization of the United Nations
GDS	Groundwater Development Stress indicator
GEF	Global Environmental Facility
GEI	Groundwater Exploitation Index
GGIS	Global Groundwater Information System (developed by IGRAC)
GRAPES	Groundwater and River Resources Action Programme on a European Scale
GRAPHIC	Groundwater Resources Assessment under the Pressures of Humanity and Climate Change (UNESCO project)
GWA	Gender and Water Alliance
GWES	Groundwater for Emergency Situations
GWP	Global Water Partnership
GW-MATE	Groundwater Management Team (World Bank)
IAEA	International Atomic Energy Agency
IAH	International Association of Hydrogeologists
IAHS	International Association of Hydrological Sciences
IAS	Iullemeden Aquifer System
IAHR	International Association of Hydro-Environment Engineering and Research
IBRD	International Bank for Reconstruction and Development
ICID	International Commission on Irrigation and Drainage
ICQHS	International Center on Qanats and Historic Hydraulic Structures (Yazd, Iran)
IFEN	Institut Français de l'Environnement (French Institute for the Environment)
IGME	Instituto Geológico y Minero de España (Geological Survey of Spain)
IGRAC	International Groundwater Resources Assessment Centre (since 2011 also called UNESCO-IGRAC)
IHD	UNESCO's International Hydrological Decade
IHP	International Hydrological Programme (UNESCO)
INBO	International Network of Basin Organizations
INSEE	Institut National de la Statistique et des Études Économiques (French National Institute for Statistics and Economic Studies)
INWEB	International Network of Water-Environment Centres for the Balkans
IPCC	Intergovernmental Panel on Climate Change
IRWR	Internal Renewable Water Resources

ISARM	Internationally Shared Aquifer Resources Management
ISMAR	International Symposium on Managed Aquifer Recharge
ISO	International Organization for Standardization
ITGE	Instituto Tecnológico Geominero de España (predecessor of IGME)
IUGS	International Union of Geological Sciences
IWMI	International Water Management Institute
IWRA	International Water Resources Association
IWRM	Integrated Water Resources Management
IWRS	Indian Water Resources Society
IYPE	International Year of Planet Earth
MAP	UNEP's Mediterranean Action Plan
MAR	Managed Aquifer Recharge
MOPU	Ministerio de Obras Públicas y Urbanismo (Ministry of Public Works and Urban Affairs, Spain)
MRMWR	Ministry of Rural Municipalities and Water Resources, Oman
NAS	Nubian Aquifer System
NWSAS	North Western Sahara Aquifer System
OACT	Organisation Africaine de Cartographie et de Télédétection (African Organisation for Cartography and Remote Sensing, Algiers)
OEA	Organisación de los Estados Americanos /Organisation des États Américains (=OAS)
OAS	Organization of American States
OECD	Organisation for Economic Co-operation and Development
OSS	Observatoire du Sahara et du Sahel (Sahara and Sahel Observatory, Tunis)
PAM	Plan d'Action pour la Mediterranée du PNEU (UNEP's Mediterranean Action Plan = MAP)
PAN	Pesticide Action Network Europe
PCR-GLOBWB	Global Water Balance Model (Utrecht University)
PHI	Programme Hydrologique International/ Programa Hidrológico Internacional (=IHP)
Plan Bleu	Mechanism for environmental regional co-operation between the 21 states bordering on the Mediterranean (under UNEP/MAP)
PNUD	Programme des Nations Unies pour le Développement/ Programa de las Nacionas Unidas para el Desarrollo (=UNDP)
PNEU	Programme des Nations Unies pour l'Environment (=UNEP)
PPCPs	Pharmaceuticals and personal care products
PWN	Provinciale Waterleiding Maatschappij Noord-Holland (Water Supply Company of the Province of North Holland, The Netherlands)
RIVM	Rijksinstituut voor Volksgezondheid en Milieu (Netherlands Institute for Public Health and the Environment)
RIZA	Rijksinstituut voor Integraal Zoetwaterbeheer en Afvalwater-behandeling (Netherlands Institute for Inland Water Management and Waste water Treatment)

SADC	Southern African Development Community
SEMIDE	Système Euro-Méditerranéen d'Information sur les savoir-faire dans le Domaine de l'Eau (=EMWIS)
SOFRETEN	Société Française d'Etudes et d'Engineering
TDS	Total dissolved solids (concentration of dissolved salts in water)
TNO	Nederlandse Organisatie voor Toegepast Natuurwetenschappelijk Onderzoek (Netherlands Organisation of Applied Scientific Research)
TRWR	Total Renewable Water Resources
UN	United Nations
UNDESA	United Nations Department of Economic and Social Affairs
UNDP	United Nations Development Programme
UNDTCD	United Nations Division of Technical Cooperation for Development (also: UN-DTCD; predecessor of UNDESA)
UN-ECAFE	United Nations Economic Commission for Asia and the Far East
UN-ECE	United Nations Economic Commission for Europe (also: UNECE)
UNEP	United Nations Environmental Programme
UN-ESCAP	United Nations Economic and Social Commission for Asia and the Pacific
UNESCO	United Nations Educational, Scientific and Cultural Organization
UNESCO-IGRAC	International Groundwater Resources Assessment Centre (also know as IGRAC)
UN-ILC	United Nations International Law Commission
USGS	United States Geological Survey
USWRC	United States Water Resources Council
VLOM	Village Level Operation and Maintenance
WaterGap GHM	WaterGap Global Hydrological Model (Kassel and Frankfurt Universities)
WFD	Water Framework Directive of the European Union
WHYMAP	World-wide Hydrogeological Mapping and Assessment Programme
WHO	World Health Organization
WMO	World Meteorological Organization
WRI	Water Resources Institute, Washington
WWAP	World Water Assessment Programme
WWC	World Water Council
WWDR	World Water Development Report (prepared by WWAP)

About the authors

Jean Margat

Jean Margat is a hydrogeologist. After fifteen years at the *Service Géologique* of Morocco at the beginning of his career, he moved to the *Bureau de Recherches Géologiques et Minières* (BRGM) at Orléans, France, for almost another twenty-five years. There he initiated, carried out and – later on – supervised groundwater investigations. During this period he was also the Director General's personal advisor on water resources. His professional experience has taken him to many areas in France and abroad, first and foremost in arid regions, in particular in Africa and the Middle East.

In addition to his professional activities at BRGM, Jean Margat was Vice-President of the *International Association of Hydrogeologists* (IAH) and afterwards President of the French national chapter of the Association.

Until recently, he was Vice-President of the Blue Plan for the Mediterranean. He is still consulted as an expert by international organisations such as FAO, UNESCO, World Bank and UNDP.

He is the author of a large number of publications related to water resources assessment and management, mapping, water resources terminology and water economics. In 2008 he received the International Hydrology Prize of the IAHS, UNESCO and WMO.

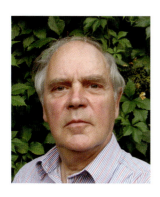

Jac van der Gun

Jac van der Gun is a groundwater hydrologist. After graduating at Wageningen University he worked for one year at a water supply company in The Netherlands and then for four years in Bolivia, participating in water resources assessment activities initiated by UNDTCD. After that he joined the *Institute of Applied Geoscience* of the R&D organisation *TNO (Applied Scientific Research)* in The Netherlands, where he remained until retirement.

At TNO, Jac van der Gun participated in the Groundwater Reconnaissance of The Netherlands, and after a few years took over the overall responsibility for this programme. Soon he was also involved in the international water resources assessment and management projects of the institute. Among others, he was the resident manager of water resources assessment projects in Yemen and Paraguay, and carried out numerous short missions in Asia, Latin America, Africa and Europe for various international and national organisations, providing scientific-technical input, supervising projects, and formulating or evaluating projects and programmes. He has also lectured at UNESCO-IHE on groundwater for more than thirty years.

Jac van der Gun was actively involved in establishing the International Groundwater Resources Assessment Centre (IGRAC) and in 2003 became its first director. He is still active in several groundwater-related projects of international organisations, mostly as a consultant to UNESCO or IGRAC.

Chapter 1

Introduction

Water not only covers three quarters of the Earth's surface: it is also present almost everywhere below ground surface, down to considerable depths and in continuous motion. Groundwater – as we can refer to the vast majority of this subsurface water[1] – is an invisible component of the hydrosphere, representing a hidden part of the water cycle. The general public is not unaware of groundwater, but usually knows hardly anything about it.

Since time immemorial people all over the world have been able to find and exploit subsurface water to supply themselves, or they have managed to drain subsurface water to facilitate mining activities and land use. Over time, they have invented various techniques to capture groundwater and bring it to the surface, making use of different sources of energy, ranging from gravity and muscular energy of humans and animals to wind energy, fossil fuel, electricity or solar pumps. Today, the world's population abstracts two hundred times more water than oil, in terms of volume per year, from the subsurface. Groundwater is widely used in many countries. It is often the primary source of drinking water (supplying half of the world's population) and contributes significantly to irrigation, hence to food security in arid and semi-arid regions. It therefore represents an important component of the water economy. However, for a long time there has been a general misunderstanding about where groundwater comes from and where it is going to. Likewise, proper and accurate knowledge on the place and role of groundwater in the water cycle has been lacking. Groundwater has been the subject of numerous myths rather than of scientific knowledge. Even today, it is a poorly understood resource. Too many simplistic and often erroneous ideas do not do justice to the diversity and complexity of groundwater around the world. For a long time, fundamental misconceptions on groundwater have persisted, as is illustrated in Box 1.1. This suggests that other conceptual flaws in understanding groundwater may still exist, without the professional community being aware of them.

---

[1] Strictly speaking, subsurface water is divided into groundwater and water in the unsaturated zone. The latter is present above the groundwater table and has a much smaller volume than groundwater.

Box 1.1: Where does groundwater come from?

One of the reasons why a scientific notion of the water cycle appeared so late comes from the lack of understanding of groundwater movement. Perceptions related to groundwater were often primarily based on the imagination, or even on mythology, rather than on scientific observation.

From ancient times and up until the seventeenth century, from Aristotle to Descartes, the prevailing belief was curiously enough that groundwater would move in a direction opposite to what we know now, thus from the bottom of the sea to the mountains, from where springs would emerge. This movement was supposed to take place through mysterious channels underground (on its way inexplicably losing salinity ...), as depicted by Father Kircher in 1665 (Figure 1.1).

Only a few brilliant men, in particular the Greek philosopher Anaxagoras (fifth century BC), the Roman architect Vitruvius (first century BC) and Leonardo da Vinci (sixteenth century), had the intuition to attribute the origin of groundwater to the infiltration of rain.

Our current understanding of the subsurface component of terrestrial waters has made it possible to establish the overall direction of groundwater movement in conformance with observations in the field. This corrected view is reflected in pictures designed to illustrate the concept of the hydrological cycle and to educate the general public on it (see next chapter, Figure 2.3). Knowledge of the origin and the final destination of groundwater has become one of the pillars on which the science of hydrogeology was founded. However, only in recent decades has attention been paid to the role that groundwater plays in the water cycle on global and national scales.

Groundwater is an ordinary component of the water cycle, not independent of the other components. A special approach and effort to understand it are nevertheless appropriate, because what most people know about groundwater is not in proportion to the use they make of it. Groundwater resources are often still poorly assessed and hardly managed, even excessively exploited and wasted, but elsewhere underestimated and unused; a wide diversity of conditions can be observed across the globe.

During the twentieth century, exploitation and pollution put unprecedented stress on groundwater in various regions – in both developed and developing countries. This stress was caused by numerous uncoordinated and unplanned individual activities. In other regions, however, people seemed to remain unaware of the potential benefits of this resource and to make no or very little use of it.

This is a book on the geography of the world's groundwater, with emphasis on its physical aspects, but to some extent also dealing with related economic aspects and management. It is somewhere in between a book for scientific and technical readers – such as 'Groundwater resources of the world and their use' by Zektser

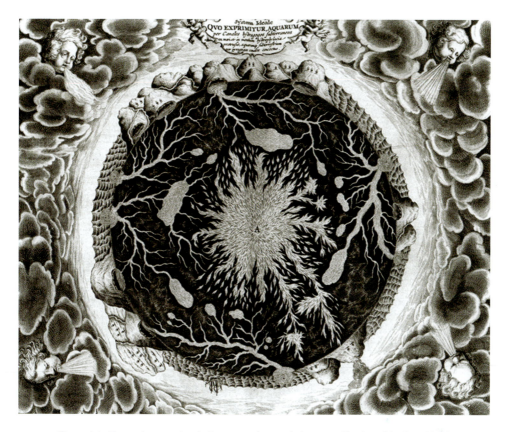

Figure 1.1 Groundwater circulation according to Athanasius Kircher. (Kircher, 1665)

and Everett (2004) – and a text to inform the general public about groundwater or hydrogeology. Its purpose is to provide an overview of the groundwater resources and their current uses around the world, thereby focusing on the extreme diversity of their occurrence, behaviour and potential – which leads to differences in exploration, exploitation and management – rather than on their shared features and theoretical principles. The latter can be found in a variety of textbooks on groundwater or web-based introductions to groundwater; a selection of these is presented in Appendix 6.1.

The book is intended for water resources managers, those in charge of managing the subsurface, hydrogeologists, hydrologists, geographers, engineers and other professionals involved in water, as well as for students, teachers, communication professionals and anyone else interested in the world's groundwater. For those not familiar with the terminology in hydrogeology and related disciplines, a glossary is provided in Appendix 1.

A few words of caution

This book contains a lot of data aggregated by country or by continent, or even over the entire world. Most of this aggregated data has been taken from national and international databases or publications. Assessing the quality of this data is extremely difficult and was considered not feasible in preparing this compilation. Data quality is undoubtedly highly variable, given the large differences in data acquisition practices from country to country – even inside countries – and the varying percentage of areas for which no real data but only rough estimates are available. It is also not uncommon to find different versions of the same statistics in different databases. Reproducing alternative versions (with references to the information sources) has not been avoided in this text, as long as this data was considered plausible. The reader should be alert to the many possible flaws in the individual data and instead look at the global or regional variations or patterns defined by the entire set of data.

Maps are very effective for showing variations and patterns across the globe. Therefore, many world maps are included in the following chapters. The maps that show data by country – corresponding to the conventional political geography – are mainly designed to visualise statistical tables, based usually on national sources. However, variations between countries in size and demography affect the impression that these maps may give on the state of the world and inevitably produce a bias in the perception of the information presented. The effect of country or population size has been eliminated in some of the maps by expressing the displayed variable in units per square kilometre or per inhabitant (per capita). Some maps that are not country-oriented may also reduce the described inconvenience.

A related consequence of using mainly national data to develop a global picture is that it does not sufficiently highlight the regional variations inside countries that are either very large or rich in contrasts.

REFERENCES

Kircher, A., 1665. *Mundus subterraneus*, Liber IV. Joannes Janssonius & Elizeus Weyerstraten, Amsterdam *(between pages 173 and 174)*.

Zektser, I.S., & L.G. Everett (eds), 2004. *Groundwater resources of the world and their use.* UNESCO, IHP-VI, Series on Groundwater No 6. Paris, UNESCO, 346 p.

Chapter 2

Groundwater in the global water cycle

- *Quantities of fresh and saline water inside the Earth's crust*
- *Fresh groundwater fluxes and their geographic distribution*
- *Groundwater in the global water cycle: groundwater recharge and discharge*
- *Links with surface water*
- *Groundwater in arid and semi-arid regions*

2.1 WATER BELOW THE GROUND SURFACE

Water is beyond any doubt the most widespread liquid inside the earth's crust. Most of the permeable porous or fissured rocks – the environment in which aquifers[1] are located –, and even many low-permeability rocks such as clays, contain water down to considerable depths (several kilometres, occasionally more than ten). But the volumes of water and their contents of dissolved solids are highly variable: the volume of invisible water present below one hectare of land may vary from a few thousand to several million cubic metres (m³). It is therefore difficult to calculate the volume of groundwater in a region or country – and calculating the volume of groundwater stored in the entire world is even more difficult.

Various researchers have nevertheless attempted to calculate the total volume of groundwater on Earth (see Table 2.1), in order to compare it with the volumes of water present in other compartments of the terrestrial hydrosphere, and even to explore how it is distributed across the continents (Figure 2.1). An estimated eight to ten million cubic kilometres (km³) of fresh groundwater represents the lion's share of all liquid freshwater on Earth, or to be more precise: about 98 to 99% of it (see Figure 2.2). In comparison, the global volume of freshwater in lakes is less than 1% of the total fresh groundwater volume. However, the volume of freshwater stored in the Earth's crust represents only one per cent of the total volume of water in the

---

1 Hydraulically continuous bodies of relatively permeable porous or fissured rocks containing groundwater are called '*aquifers*', whereas groundwater-filled bodies of poorly permeable formations are called '*aquitards*' or '*aquicludes*', depending on whether or not minor groundwater seepage rates are still possible. Aquifers are the domains where the bulk of all groundwater flow takes place, because their comparatively high permeability favours flow. However, aquitards may hold large volumes of groundwater that can be exchanged with the aquifers through the large surface of contact they share.

Table 2.1 Groundwater stock and flow on the continents. Some global estimates.

| Source | Stock in thousands of km³ | | | Flow |
	Freshwater	Brackish or saline water	Total	Mean groundwater flux in km³/year
Nace 1964, 1968	8 336[1]			
USGS/Nace 1967	8 350			
L'Vovich 1967, 1974 (English translation 1979)	4 000[2]		60 000[3]	12 000
Korzun 1974/USSR Com. UNESCO's IHD 1974	10 530[4]	12 870	23 400[5]	13 320[6]
UNESCO 1978				13 200
UNESCO 1990 and Shiklomanov 1992, 2003	9 800	13 600	23 400[5]	
Baumgarter & Reichel 1975	8 062[7]			
Mather 1984	8 800			
Bureau des Longitudes (France) 1984	8 600[8]			
National Council on Scientific Research USA 1986			15 300	
UNESCO 1988	8 500			
World Resources Institute 1991	8 200			
Döll et al. 2002				14 000[9]
FAO 2003				11 284[10]
Döll and Flörke 2005				12 882[11]
Döll and Fiedler 2008				12 700[11]
Wada et al. 2010				15 200[12]
AQUASTAT 2011				11 968[13]

Notes:
1 Converted from 2×10^6 cubic miles, of which 50% at less than 800 m depth and 50% at greater depths.
2 Groundwater in active water exchange zone, including 85 as soil moisture.
3 After Vernadsky and Makarenko.
4 Includes 16.5×10^{12} m³ of soil moisture and 300×10^{12} m³ of subsurface permafrost ice.
5 Excluding Antarctica: $\sim 2\ 000 \times 10^{12}$ m³ of freshwater ice.
6 Groundwater outflow into streams – annual naturally renewed groundwater resources: calculation by hydrograph separation, excluding an estimated 2 200 km³/year of direct outflow into the oceans.
7 Including 61.3×10^{12} m³ of soil moisture; $3\ 550 \times 10^3$ km³ corresponds to depths less than 800 m.
8 Down to 4 000 m of depth, of which $4\ 000 \times 10^3$ km³ corresponds to depths less than 800 m.
9 Groundwater recharge calculated with the Water GAP-2 model.
10 Sum of country data on groundwater produced internally (national sources compiled for the FAO database, 2003).
11 Groundwater recharge calculated with the WGHM model and mesh width of 0.5°, 1961–1990 average.
12 Groundwater recharge calculated with the PCR-GLOBWB model and mesh width of 0.5°, 1961–1990 average.
13 Sum of country data on groundwater produced internally (compiled by AQUASTAT from national sources, latest update). Data is missing for 3.35% of the territory of the continents; hence the incomplete sum has been corrected for this percentage.

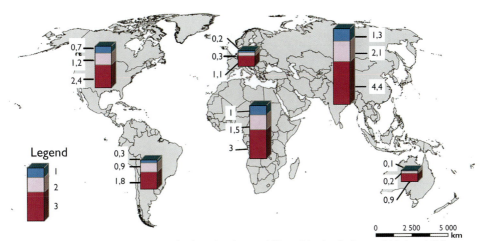

1 *Zone of active exchange, phreatic or confined aquifers down to 100 m of depth – freshwater (global total: 3.6).*
2 *Zone of moderately active exchange, most often confined aquifers, between 100 and 200 m of depth – predominantly freshwater (global total: 6.2).*
3 *Zone of little active exchange, down to 2 000 m of depth, predominantly saline water (global total: 13.6).*

Figure 2.1   Order of magnitude of the volumes of groundwater stored, by continent (in millions of cubic kilometres, or 10^{15} m³). [*Data sources*: Korzun (1974); Shiklomanov (1990); Shiklomanov and Rodda (2003).]

hydrosphere, including the oceans. Estimating volumes of groundwater stored is most appropriate at regional scales, particularly in relation to large aquifers (see Chapter 3) and is rarely done at country level. A few examples of the latter category are presented in Table 2.2.

These astronomic numbers, of which only the order of magnitude is meaningful, are mainly of academic interest and should not mislead us. In fact, unlike most mineral resources, groundwater is in most cases a renewable resource, which allows groundwater to be developed sustainably, making use of the dynamics of groundwater in the present-day water cycle. Only a minor part of the enormous groundwater volumes or 'reserves' is dynamic, which means: more or less regularly replenished by recharge[2] and sufficiently mobile to play the role of a natural regulator or buffer. Although the dynamic reserves probably account for only a few per cent of the total volume ('*stock*' or '*storage*') of groundwater, they are still considerable. The majority of these dynamic groundwater resources is found in phreatic aquifers[3]. Globally aggregated, the seasonal or multiannual variation

---

2 Groundwater recharge, renewal or replenishment is the inflow of water into a groundwater system.

3 A phreatic aquifer or 'water-table aquifer' is an aquifer in which the upper boundary of the groundwater mass forms a surface (water table) that is in direct contact with the atmosphere through the soil pores. This condition favours the aquifer being actively involved in the water cycle. The upper limit of a confined aquifer, on the contrary, is the bottom of an overlying poorly permeable formation

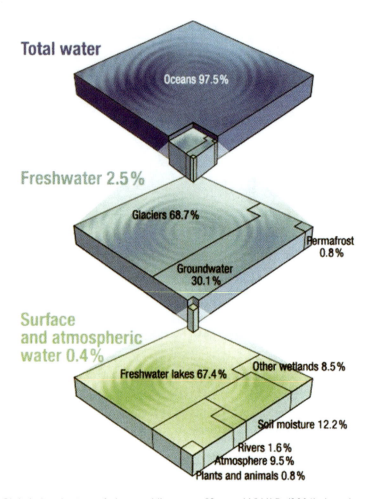

Figure 2.2 Global distribution of the world's water. [*Source:* WWAP (2006), based on data from Shiklomanov and Rodda (2003)]. *Freshwater has a global volume of 35.2 million cubic kilometres.*

in the groundwater volume stored in this category of aquifers is of the order of 10 000 billion m³.

Stocks of water in the earth's crust should therefore certainly not be used as the sole basis for assessing the quantity of the groundwater resources, despite the perhaps misleading impression given by different literature sources[4]. On an aggregated level,

---

(confining bed) that prevents the aquifer from interacting directly with the atmosphere and with surface water bodies.

4 For example, Allègre (1993) writes: "La majorité des ressources en eau sont souterraines: cinq à dix mille fois supérieures aux eaux de surface" (*The majority of the water resources is underground: five to ten thousand times more than surface water*), which is reiterated by Diop and Recacewicz (2003). Similarly,

Table 2.2 Examples of total groundwater reserves calculated by country.

Country	Estimated total volume of groundwater (in km³)	Comments	References
Saudi Arabia	2 185	Down to 300 m of depth	Abdurrahman 2002
Australia	20 000	Great Artesian Basin only	Habermehl 1980, 2001
Brazil	111 661		Rebouças 1988
Egypt	150 000	Nubian Sandstone Aquifer System only	Khater 2005
France	2 000		BRGM 1986
Libya	35 000	Sirte, Kufra and Murzuk Basins only	El Gheriani 2002
USA	125 040	Of which 50% in shallow aquifers and 50% in deep aquifers	USGS 1975

the quantity of a natural resource can be characterised by two key variables: stock (quantity present, volume stored) and flow (rate of growth, rate of renewal). The latter allows sustainable exploitation to take place without depleting the resource in the longer term, while the former buffers variations in inflow over time. Compared to surface water, atmospheric water and most other components of the water cycle, groundwater tends to have a high stock/flow ratio. However, it is due to flow that groundwater participates in the planet's water cycle and that groundwater can, in most cases, be considered a renewable resource.

Table 2.1 suggests that the mean globally aggregated flow – represented by the mean groundwater recharge – is between 11 000 and 15 000 km³ per year for climatic conditions prevailing during the second half of the 20th century.

For most aquifers, flow is more relevant than stock for characterising the groundwater resources quantity. So-called 'fossil groundwater'[5] is an exception: the corresponding water resources are expressed as a volume, like in the case of minerals, but these non-renewable resources include only the extractable fraction of the stock (see Chapter 4).

Zaporozec (2002) states: "Groundwater (…) represents some 98 per cent of the planet's freshwater resources". Shiklomanov and Rodda (2003) also refer to groundwater reserves calculated by continent as 'resources', while SEMIDE's bulletin of June 2005 repeats: "Groundwater represents around 97% of the available water resources on the planet".

5   'Fossil groundwater' or 'non-renewable groundwater' refers to groundwater bodies formed in the geological past that do not receive significant replenishment under current climatic conditions (see Chapter 4).

2.2 HOW MUCH GROUNDWATER PARTICIPATES IN THE WATER CYCLE?

2.2.1 General concepts and features

A schematic picture of the water cycle and some groundwater-related details in this cycle are presented in Figure 2.3. Numerous schematic pictures of the water cycle, often in a landscape setting, exist and have been disseminated in numerous publications. However, they frequently highlight groundwater outflow into the sea, although this is not the most important form of groundwater discharge on a global level, as will be clarified in Section 2.3. Furthermore, they often fail to show the links between aquifers and streams. Because it is difficult to combine local features and the regional water cycle in one single picture, they are shown separately in Figure 2.3. Important groundwater-related aspects of the water cycle are illustrated in a number of cross sections.

Groundwater is not stagnant, although its movement is very slow compared to that of water in streams. According to the specific aquifer conditions, the time it takes groundwater to travel one kilometre may vary from one day to five centuries or more. However, large volumes of water moving slowly through an aquifer may produce a total flux equivalent to that of a stream, in which water does indeed flow at much higher velocities, but through much smaller cross sections perpendicular to the flow direction. In this chapter we use the term *'flux'* to indicate the numerical value of the total flow through an aquifer (or a complex of aquifers and aquitards, often called an 'aquifer system'). Since flux is the rate of flow across a given surface, it may vary according to where such a surface is located inside the aquifer. To avoid this problem, we use flux here as a macroscopic variable averaged over time, equal to the average of the aquifer's long-term recharge and discharge. Under natural conditions, the latter two balance each other out.

Like surface water, groundwater moves in spatially structured systems. For surface water these systems consist of a network of water courses in a *catchment area* or *river basin*, while groundwater flows in *aquifers*. Aquifer systems are less visible than catchment areas: their lateral extent, volume, three-dimensional structure and boundaries have to be inferred mainly from geological information.

An aquifer – literally, 'water-carrier' – is both a reservoir and a transport channel. Groundwater flow in an aquifer is governed by the aquifer's intrinsic characteristics (size, permeability, etc.) but also by its recharge, which – on a global level – is mainly produced by infiltration of precipitation. Most of the groundwater flow eventually ends up in springs and streams. In coastal areas, however, there may be significant subsurface outflow into the sea, while abstraction by wells has become a dominant discharge mechanism in some intensely exploited areas in dry climates. However, on a global level, these are minor components of groundwater discharge compared to the outflow into streams and the discharge via springs. Groundwater renewal (recharge) and groundwater discharge are thus the links in a two-directional exchange between groundwater and other components of the water cycle (Figure 2.3). A correct understanding and identification of all interactions between groundwater and other components of the water cycle is the first step towards assessing groundwater inflows, outflows and change in groundwater storage for a given groundwater system during a specified period (groundwater budget).

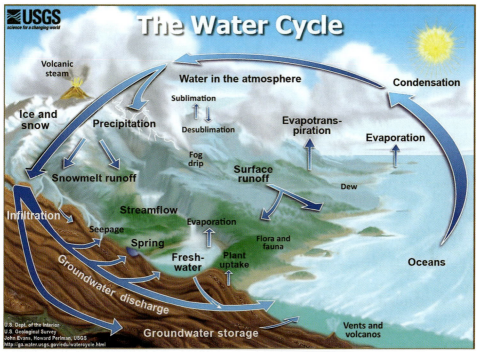

(a) Schematic picture of the natural water cycle. [*Source*: USGS (2012)]

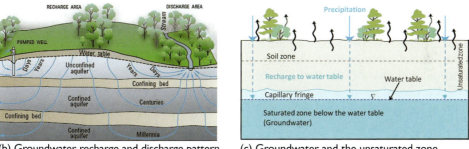

(b) Groundwater recharge and discharge pattern. [*Source*: USGS (2012)]

(c) Groundwater and the unsaturated zone. [*Source*: USGS (2012)]

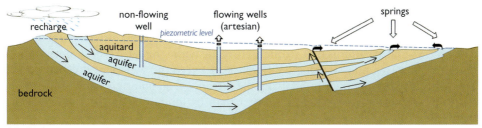

(d) Groundwater discharge by springs and flowing wells.

Figure 2.3 The water cycle and some examples of its groundwater-related details.

Groundwater flow is generally less variable in time than flow in rivers, but – unlike in the case of stream flow – the corresponding rates cannot be directly measured. Assessing the groundwater flux has to be done by indirect methods, e.g. by determining what is entering or what is leaving the aquifers (see Box 2.1). Other commonly used methods are based on the observed hydraulic gradient and the aquifer's hydraulic conductivity[6], or on observed groundwater level variations over a long period of time. Numerical simulation models integrate all these approaches.

Groundwater should not be studied, exploited or managed in isolation from the overall water resources setting of the area concerned. Therefore, before focusing on groundwater fluxes, attention is drawn to the Figures 2.4 and 2.5. The global pattern of mean annual precipitation in Figure 2.4 shows considerable contrasts and this is mirrored in the total renewable water resources (surface water and groundwater combined, or 'blue water'). The latter is depicted in Figure 2.5, although with less detail because of a higher degree of spatial aggregation. However, only some 22% of global precipitation is converted into 'blue water'. More than three quarters of the mean annual global precipitation is transformed into soil moisture and is eventually returned to the atmosphere via evaporation and evapotranspiration ('green water'). The latter sustains crops and natural vegetation.

2.2.2 Global variation in groundwater flux

By applying any of the methods outlined in Box 2.1, maps can be developed that give an impression of the geographical variation in the groundwater flux. For this purpose, the estimated groundwater inflows or outflows have to be expressed as fluxes per unit of area within the corresponding groundwater systems or river basins (*flux density*), followed by further smoothing and generalisation to the extent required by the envisaged scale and resolution of the map. This is how the small-scale world maps shown in the Figures 2.6 and 2.7 have been developed. Figure 2.6 follows the upflow side approach and shows mean annual groundwater recharge by grid cell as generated by the *WaterGAP Global Hydrological Model* (Döll *et al.*, 2002, 2005, 2008). Figure 2.7, produced by L'Vovich (1974), is based on the downflow side approach and shows mean annual groundwater outflow. On a more detailed scale, such 'average long-term values of groundwater discharge' are shown on the *World Map of Hydrogeological Conditions and Groundwater Flow* by Dzhamalov and Zektser (1999). The range of variation in groundwater flux density across the world is extremely large: at the resolution of the above maps it varies by at least three orders of magnitude, from less than one thousand to more than one million cubic metre per square kilometre (equivalent to an annual equivalent water depth of less than one millimetre to more than one thousand millimetres).

Note that the two maps, in spite of different resolutions and different sets of data used, show similar patterns that strongly correlate with long-term world precipitation patterns. However, non-systematic differences between the two maps do occur; these are caused by shortcomings in applying the methods (e.g. ignoring some of the in- or outflow components – see Box 2.1) rather than by limitations in the validity of the data used.

---

6 These hydraulic terms are defined in the Glossary (Appendix 1).

Box 2.1: How to calculate the total groundwater flux through an aquifer

Any aquifer, whatever its size, is an *open system*. In certain parts of its area (recharge zones – mainly where a water table is present) the aquifer from time to time receives water and elsewhere – in its discharge zones – it continuously loses water. Temporary differences between inflows and outflows are balanced by changes in stock, like in a bank account.

To estimate the average flux that links inflows to outflows, two approaches are in common use:

• *The upflow side approach: what the aquifers receive (groundwater recharge)*
Estimates of how much infiltrated atmospheric water is replenishing groundwater per unit area and per unit of time are made on the basis of hydrometeorological variables and soil characteristics. The estimates have to cover a sufficiently long period of time and are sometimes supported by observed groundwater level variations, in the case of an unconfined aquifer. Other contributions have to be estimated as well, such as inflows from streams ('losses' of stream flow), from neighbouring aquifers or from overlying irrigated soils ('percolation losses'). All identified inflow components and their estimated mean annual values are added together to derive an estimate of total flux through the aquifer.

• *The downflow side approach: what the aquifers lose (groundwater discharge)*
This approach is based on the analysis of available hydrological data (in particular, stream flow and spring discharge data). This includes an identification of the baseflow of streams, i.e. defining the share of stream flow originating from groundwater. All the identified out-flow components and their estimated mean annual values are added together to derive an estimate of total flow through the aquifer.

For a long time this has been the most commonly practised approach, but the required hydrological data is rarely available in sufficient detail (both in time and in space) over large areas to allow for a comprehensive estimate without resorting too much to risky extrapolations. In addition, the groundwater flux may be underestimated if it is based exclusively on the analysis of the stream flow, when other potential outflows from the aquifer considered are not negligible. Such outflows include invisible groundwater outflow into the sea or into lakes without an outlet, outflow into neighbouring aquifers, discharge by evaporation or evapotranspiration and – in addition – discharge produced by humans (pumping and drainage activities) at the cost of natural outflow. In such cases, not all the groundwater flow is collected by streams.

Under natural conditions and over a sufficiently long period of time, the two approaches should in principle converge, if all the respective inputs and outputs are taken into account. However in practice the latter is not always properly done.

Due to the diversity of climatic and physiographic conditions and due to differences in the availability of data, the two approaches are usually not equally feasible or convenient for different areas in the world: sometimes the first approach, sometimes the other one is better suited. These differences may also affect the outcomes.

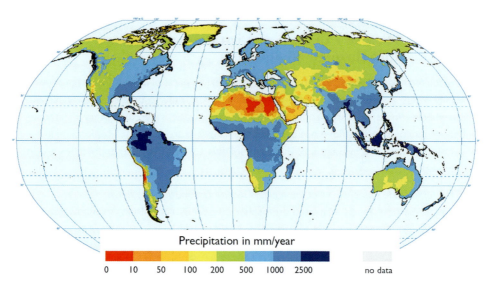

Figure 2.4 Mean annual precipitation in mm/year (period 1961–1990). [*Source:* Insert map of WHYMAP (2008), produced by BGR on the basis of a gridded precipitation normals set prepared by the Global Precipitation and Climate Centre, Offenbach.]

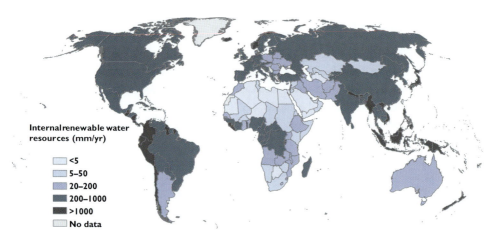

Figure 2.5 Classification of countries according to their mean annual internal renewable water resources (IRWR) per unit area, in mm/year. (*1 mm/year = 1 000 m³/year per km²*). [*Data source:* AQUASTAT (2011). Total water produced internally.]

In general, global and continental estimates of groundwater fluxes, produced since the 1970s by geographers on the basis of these approaches, look credible. On the global level, the total flux of groundwater on the continents is likely to be around 12 000 to 13 000 km³/year, while the breakdown by continent (Table 2.3) shows a degree of contrast similar to that of the stocks (Figure 2.1). At this still very macroscopic scale the groundwater flux is always about a quarter to one third of the mean total flux of 'blue water',

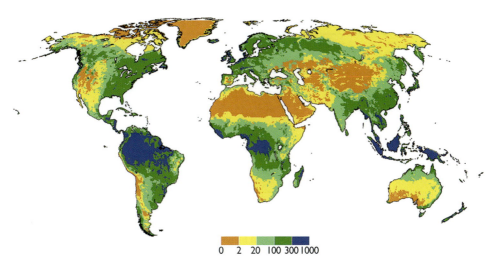

0   2   20   100   300 1000

Figure 2.6 Global distribution of mean annual diffuse groundwater recharge in mm/year (period 1961–1990). (*1 mm/year = 1 000 m³/year per km²*). [*Source*: Döll & Fiedler (2008), calculated by 0.5 degree grid cells.]

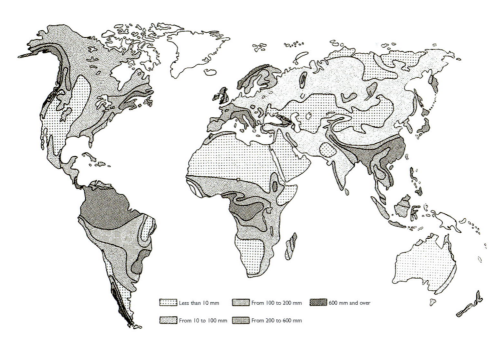

Less than 10 mm   From 100 to 200 mm   600 mm and over

From 10 to 100 mm   From 200 to 600 mm

Figure 2.7 Depth of mean annual groundwater discharge, in mm/year. (*1 mm/year = 1 000 m³/year per km²*). [*Source*: After L'Vovich (1974) and WRI (1991).]

Table 2.3 Distribution of groundwater fluxes over the continents.

Continent	Groundwater discharge		
	(a) Source: L'Vovich (1974)		(b) Source: IHD – USSR (1974)
	Mean depth in mm/year (= 1 000 m³/year per km²)	Mean annual flux in km³/year	Mean annual flux in km³/year
Europe	109	1 065	1 120
Asia	76	3 410	3 750
Africa	48	1 465	1 600
North America	84	1 740	2 160
South America	210	3 740	4 120
Australasia	54	465	757
TOTAL	90	11 885	13 325

Source: After estimates by L'Vovich (1974) and the USSR Committee of UNESCO's International Hydrological Decade (1974).

Note: *The difference between (a) and (b) seems mainly due to the fact that only the estimates under (b) include the subsurface outflow into the sea.*

i.e. of the surface water and groundwater fluxes combined. The breakdown by major climate zones (Table 2.4) shows more variation and is much more meaningful: scarcely 2% of the global groundwater flux corresponds to the arid and semi-arid zones, while the bulk of the flux is shared between cool-temperate and equatorial-tropical humid zones, in almost equal portions. Nevertheless, global aggregation of this type of data on the land phase of the water cycle is not of much practical use, even for educational purposes. This is because aggregation smoothens spatial variability, which is a fundamental aspect of the real world. Even estimates by continent, a little more spatially differentiated (Table 2.3), are still too general for practical purposes. Due to the simplifications introduced at these levels of aggregation, the numbers and maps obtained do not sufficiently reflect local hydrogeological conditions (see Chapter 3).

2.2.3   Groundwater flux by country or by river basin

Estimates of the groundwater flux by country, attempted first by L'Vovich (1974), have been irregularly updated and data sets are still far from being homogeneous. Since 1993 these and other national water resources statistics have been systematically collected and periodically updated in FAO's AQUASTAT database on water resources. In order to minimise confusion on the exact meaning of relevant water resources variables, AQUASTAT has developed its own definitions, briefly summarised in Box 2.2.

Table 2.5, based on AQUASTAT, presents internal groundwater flux data compiled for 101 countries that together cover 92% of the earth's land mass, excluding Antarctica. In order to avoid double- counting, subsurface inflows from neighbouring countries are not included, hence the values represent the mean annual groundwater recharge as occurring within each country's own territory ('*groundwater produced*

Table 2.4 Breakdown of mean groundwater discharge according to climate zones.

Climate zone	Groundwater discharge		
	Mean depth in mm/year (= 1 000 m³/year per km²)	Derived flux (rounded off) in km³/year	% of global total
Subarctic	40	~6 000	50
Humid temperate	30 to 150		
Arid	0 to 10	~200	2
Dry tropical (semi-arid)	20 to 30		
Humid tropical	200 to 320	~5 800	48
Equatorial	600		
Equatorial and mountainous	700		
TOTAL	90	~12 000	100

Source: After estimates by L'Vovich (1974).

internally'). In Table 2.5 both the flux and the flux density is specified; the latter for comparison between countries regardless of their size. The results of this inventory show very large variations, reflecting not only the climatic conditions and the size of each country but also the prevailing hydrogeological setting. The mean annual internally-produced groundwater fluxes for the countries listed range from 0.1 to almost 1 900 km³/year. The equivalent country-averaged flux densities vary from less than 1 to almost 700 mm/year. Figure 2.8 shows the global pattern of these groundwater flux densities by country. Totalling the groundwater fluxes of the countries listed in Table 2.5 and extrapolating this sum to 100% of the continents' area yields a global total of some 12 000 km³/year, which is comparable with the numbers shown in Table 2.3. It is close to the estimate of global diffuse groundwater recharge of 12 700 km³/year by Döll and Fiedler (2008), but on the country level there are considerable differences between the estimates of these authors – based on modelling – and those in AQUASTAT. This may be partly due to conceptual differences, but the main reason for the discrepancies is the large uncertainty inherent to any recharge estimation, regardless of what method is used.

Estimates by river basin of any size are more meaningful, even though the corresponding areas – specially the larger ones – may also be heterogeneous from a hydrogeological point of view. Table 2.6 presents a selection of examples, according to L'Vovich (1974). These examples illustrate the large variation in mean flux densities and in the share of groundwater in the total runoff, which is caused by differences in both climatic and hydrogeological conditions. Showing the huge contrasts across the globe is the main purpose of presenting these and other numbers in this chapter. The numbers themselves are of limited use and certainly unsuitable for addressing local problems.

Box 2.2: Renewable water resources terminology as used in FAO's AQUASTAT

In AQUASTAT, *renewable water resources* are defined as the long-term average annual flow of rivers (surface water) and the recharge of aquifers (groundwater) generated from precipitation. They are computed on the basis of the water cycle. It should be noted that this definition reduces renewable water resources to 'blue water' only (water in streams and aquifers – the conventional sources of water supply) and does not include 'green water' (soil moisture intercepted from rainfall and returned to the atmosphere by evapotranspiration or evaporation), nor any unconventional water resource (e.g. wastewater, recycled water).

A distinction is made between *natural renewable water resources*, corresponding to a situation without human influences, and *actual renewable water resources*, where such influences are taken into account. Another distinction is made between *internal renewable water resources* (IRWR), referring to the water resources generated from rainfall within the country (or other spatial unit) considered, and *external renewable water resources* (ERWR), representing the part of the country's water resources that enters from upstream countries through rivers (external surface water) or aquifers (external groundwater). The external renewable water resources include transboundary inflow from neighbouring countries and a part of the resources of shared lakes or border rivers. In assessing the external inflow into a country, the natural inflow is either considered without human influence (natural ERWR) or modified by the in- and outflows secured by treaties or agreements and by possible water abstraction in upstream countries (actual ERWR).

In summary, the following key variables are distinguished:

Internal renewable water resources (IRWR)	=	Surface water produced internally + Groundwater produced internally – Overlap
Surface water produced internally	=	Long-term average annual volume of surface water generated by direct runoff from endogenous precipitation (surface runoff and groundwater contributions)
Groundwater produced internally	=	Long-term annual average groundwater recharge, generated from precipitation within the boundaries of the country
Overlap	=	$Q_{out} - Q_{in}$ = groundwater outflow into streams minus seepage from streams into aquifers (usually this means: net contribution of groundwater to streams)
Natural external renewable water resources (natural ERWR)	=	Average annual inflow into a country in natural conditions, i.e. without human influence

Actual external renewable water resources (actual ERWR)	=	Average annual inflow into a country, taking into account that part of the flow secured through treaties (in upstream and downstream countries) and possible water abstraction in upstream countries
Natural total renewable water resources (TRWR_natural)	=	IRWR + natural ERWR
Actual total renewable water resources (TRWR_actual)	=	IRWR + actual ERWR
Exploitable water resources (manageable water resources or water development potential)	=	Part of the water resources that is considered available for development, taking into account technical, economic, financial, social, environmental and legal feasibilities and constraints

References: FAO, 2003; Margat et al., 2005

On a global scale, the groundwater flux equals approximately 28% of the long-term mean flux of total natural renewable water resources (blue water). However, the share of groundwater shows large geographical variations, so the global average does not mean very much. The comparison of groundwater fluxes and total internal flux (IRWR), calculated by country on the basis of the AQUASTAT database, shows this clearly (Figure 2.9). The groundwater share in the total internal flow ranges from 2% (Finland) to more than 90% (Saudi Arabia, Oman, Qatar, Hungary). It is more than 50% in 40 of the 156 countries with sufficient data to calculate it. However, the calculated percentage may give the wrong impression in cases where recharge from allochtonous rivers[7] forms a substantial part – if not most – of the estimated groundwater flux, because then the groundwater flux is partly of indirect external origin and thus should not be compared with the total internal flux. Examples are Egypt (Nile), Botswana (Okavango) and Pakistan (Indus).

2.2.4   Other cycles

Groundwater is not only a component in the hydrological cycle; it is also involved in several other cycles. In the geochemical cycle it plays a prominent role as the 'lifeblood of the Earth', because it is the primary agent for geochemical exchanges between the hydrosphere and lithosphere (see Box 2.3). Furthermore, groundwater is affected by climate change caused by changes in the global carbon cycle and it also contributes to food chains around the world.

---

7 Rivers originating in a neigbouring country.

Table 2.5 Groundwater produced internally: estimated mean annual fluxes by country.

Country	Groundwater flux km³/year	mm/year	Country	Groundwater flux km³/year	mm/year
AFRICA			Ukraine	20	33
Algeria	1.5	0.6	United Kingdom	9.8	40
Angola	58	47	ASIA		
Botswana	1.7	2.8	Armenia	4.3	145
Burkina Faso	9.5	35	Azerbaijan	6.5	75
Cameroon	100	210	Bangladesh	21	147
Central African Rep.	56	90	China	829	86
			India	432	131
Chad	11.5	9.0	Indonesia	457	238
Congo	122	357	Iran	49	30
Dem. Rep. Congo	421	180	Iraq	3.2	7.3
Ethiopia	20	18	Israel	0.5	24
Gabon	62	232	Japan	27	71
Ivory Coast	38	120	Kazakhstan	6.1	2.2
Kenya	3.5	6.0	Lao PDR	38	160
Liberia	45	404	Lebanon	3.2	308
Libya	0.5	0.3	Malaysia	64	194
Madagascar	55	94	Mongolia	6.1	3.9
Mali	20	16	Myanmar	454	669
Morocco	10	22	Oman	1.3	6.2
Mozambique	17	21	Pakistan	55	68
Namibia	2.1	2.5	Philippines	180	600
Niger	2.5	2.0	Saudi Arabia	2.2	1.1
Nigeria	87	94	Syria	4.8	26
Senegal	3.5	18	Thailand	42	82
Somalia	3.3	5.1	Turkey	69	88
South Africa	4.8	3.9	Turkmenistan	0.4	0.7
Sudan	7.0	2.8	United Arab Emirates	0.1	1.5
Zambia	47	62.5			
Zimbabwe	6.0	15.4	Uzbekistan	8.8	20
EUROPE			Viet Nam	71	217
Albania	6.2	216	Yemen	1.5	2.8
Austria	6.0	71	THE AMERICAS		
Belarus	18	87	Argentina	128	46
Croatia	11	195	Bolivia	131	119
Czech Republic	1.4	18	Brazil	1874	220
Denmark	4.3	100	Canada	370	37
Finland	2.2	6.5	Chile	140	185
France	120	219	Colombia	510	448
Germany	46	128	Cuba	6.5	59
Greece	10	78	Ecuador	134	473
Hungary	6.0	64	Guyana	103	479
Iceland	24	233	Haiti	2.2	78
Ireland	11	154	Honduras	39	348
Italy	43	143	Mexico	139	71
The Netherlands	4.5	108	Nicaragua	59	456
Norway	96	296	Paraguay	41	101
Poland	12	40	Peru	303	236
Portugal	4.0	43	Surinam	80	490
Romania	8.3	35	Uruguay	23	130
Russia	788	46	USA	1384	144
Spain	30	59	Venezuela	227	249
Sweden	20	44	OCEANIA		
Switzerland	2.5	61	Australia	72	9.4

Source: Selected data, based on AQUASTAT (2011).

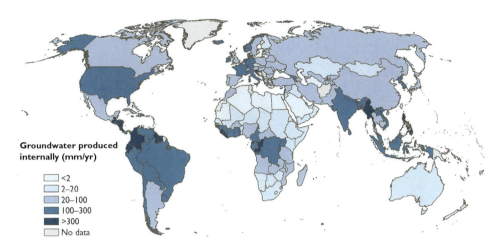

Groundwater produced
internally (mm/yr)

- <2
- 2–20
- 20–100
- 100–300
- >300
- No data

Figure 2.8 Classification of countries according to their mean annual internal groundwater flux per unit area, in mm/year. (1 mm/year = 1 000 m³/year per km²). [*Data source:* AQUASTAT (2011). '*Groundwater produced internally*'.]

Table 2.6 Examples of groundwater runoff estimates by river basin.

River basin	Mean annual precipitation (mm/year)	Groundwater runoff	
		Mean annual depth (mm/year)	Share of total runoff (%)
Agusan at Talakagon (Philippines)	4 100	1 700	65
Bueno at Rio Bueno (Chile)	3 250	1 682	59
Rhine at Lustenau	1 505	565	47
Mekong at Pnom-Penh	1 460	160	25
Murray at Tintaldra (Australia)	1 290	139	42
Ganges at Ramganga (India)	1 182	104	23
Niger at Koulikoro	1 550	87	21
Oder at Slubice	637	78	50
Luaba (Upper Congo) at Bukama	1 250	72	52
Colorado at Kamso	450	70	40
Seine at Paris	632	65	33
Euphrates at Hit	300	40	35
Mahanadi at Sambalpur (India)	1 424	35	5
San Pedro de Atacama at Cuchabarache (Chile)	175	28	93
Krishna at Vijayavada (India)	832	20	9
Orange at Prieska	503	7	23
Los Angeles at Long Beach (USA, California)	350	3	6

Source: After L'Vovich (1974). *The basins are sorted in order of decreasing groundwater runoff depth.*

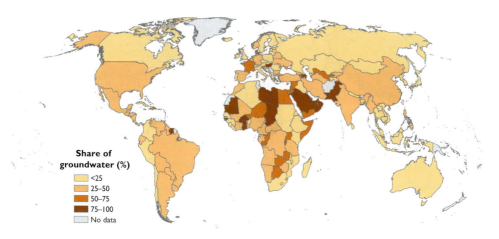

Figure 2.9 Share of groundwater in the long-term average of the total internal water resources flux, estimated by country. [*Data source*: AQUASTAT (2011).]

Box 2.3: The lifeblood of the Earth

Groundwater plays a crucial role in geochemical exchanges between terrestrial water and the lithosphere. This is due to its generally long residence time underground and the usually very large contact surface between water and the solid matrix below the ground surface.

Subsurface water is responsible for dissolution and hydrolysis – particularly in soils and in the unsaturated zone between the soil and the unconfined aquifers – and it also takes care of the subsequent transport and deposition of dissolved matter. These processes are reflected in the concentration of dissolved solids and their variation inside each aquifer, at different depths, all the way along groundwater flow paths. Therefore, these processes leave their marks on the quality of groundwater as a resource (see also Chapter 4).

Disregarding the detail and just taking a global view, we may assume that a major part of around 2 500 million tons of dissolved solids transported annually by inland waters (UNESCO, 1978) moves with flowing groundwater. A large share of the dissolved matter transported by rivers can be attributed to the groundwater discharge they collect.

According to Zektser and Dzhamalov (1998), groundwater of coastal aquifers discharging directly into the oceans – at a global rate of 2 382 km^3/year – transports 1 293 million tons of dissolved solids to the sea annually. Of this global quantity, each year 470 million tons enter the Atlantic Ocean, 521 million the Pacific Ocean, 296 million the Indian Ocean and 7 million the Arctic Ocean. These inflows of dissolved matter are equivalent to 52% of those discharged by streams into the oceans (Zektser and Everett, 2004).

2.3 GROUNDWATER AND SURFACE WATER: HOW ARE THEY RELATED?

The distinction between surface and groundwater is made on the basis of where it flows or where it is located: surface water is found in streams and lakes, while groundwater is present in aquifers and other subsurface domains. This distinction between visible and invisible water is clear at a certain location and at a given moment. However, on a regional scale and over a longer lapse of time (e.g. one year), exchange of water between water courses and aquifers – in both directions – is the rule rather than the exception. The larger the area, the greater generally the share of mobile water that migrates between surface water and groundwater systems. On a river basin or a country scale, the proportions in which the so-called 'effective' rainfall is split into runoff and groundwater recharge immediately after reaching land surface are usually not preserved in the surface water to groundwater flow ratios downflow. Often a major part of the groundwater flow – if not all – emerges, especially in the form of springs, and joins streams where it forms the most stable and permanent flow component, the '*baseflow*' (see Figure 2.10). Under special conditions, exchanges in the opposite direction also occur: in karst areas where surface waters partly or completely disappear, in piedmont areas where streams descending from the mountains lose part of their flow in the alluvial plains (see Figure 2.11), or in arid or semi-arid zones where aquifers below the river beds are recharged by infrequent floods.

Figure 2.10 Baseflow in a small stream in arid northern Yemen: a contribution from groundwater [*Photo:* Jac van der Gun].

Figure 2.11 Storm runoff travelling down towards the Tihama coastal plain (Yemen), where it replenishes the thick unconsolidated aquifer. [*Photo*: Jac van der Gun].

The described migration of water between aquifers and streams prompts two questions related to the relationship between the two main terrestrial water flow components, hence to the involvement of groundwater in the hydrological cycle:

– What share of the stream flow as measured at the outlet of river basins comes from groundwater?
– What percentage of the groundwater recharge is finally captured by streams?

Table 2.7 Contribution of surface water to groundwater fluxes in selected countries in the arid to semi-arid zone.

Country	Mean groundwater flux (km³/year)	Contribution by surface water (mainly floods) (km³/year)	Percentage contributed by surface water (%)
Iran	49.3	12.7	26
Mexico	67	19	28
China:			
– North China Plain aquifer	35.9	7.6	21
– Xinjiang Plains aquifers	–	–	30
United Arab Emirates	0.12	~0.10	83
Saudi Arabia	2.2	~2.0	90
Cyprus	0.41	0.14	34

Referring to the first question, the contribution of groundwater to stream flow at the outlet of river basins may be high or even dominant in areas fully underlain by an aquifer and where a low relief prevents direct runoff from occurring, or in arid regions without functional surface drainage systems. It tends to be low in areas where the total stream flow is abundant or in arid or semi-arid zones without significant aquifers.

Attempts have been made to map the variations of the contribution of groundwater to stream flow, in spite of the difficulties inherent to the noticeable spatial discontinuities of this variable. 'Groundwater runoff coefficients' are defined by relating the share of groundwater (groundwater runoff) to the total catchment runoff, rather than to the average rainfall (which is common practice for calculating 'runoff coefficients'). The relationship between rainfall and groundwater runoff is non-linear and far too complex to be expressed by a simple ratio. Such mapping was done in some countries, in particular in the former USSR, where it was done at the scale 1:5 million in 1965 and later in 1975 at 1:2.5 million. The mapping efforts were expanded to the whole world by L'Vovich (1974), showing groundwater runoff coefficients ranging from 0.01 to 0.60. However, the very apparent correlation between the groundwater runoff coefficient and average rainfall depth, with no visible influence by the hydrogeological conditions, raises questions about the validity of this approach, even on a very small scale. Another attempt, more cautious, mentions only ranges by area for this coefficient (Dzhamalov and Zektser, 1999). Average groundwater runoff coefficients have also been calculated by country, especially for several countries in Central and Eastern Europe:[8] Bulgaria 36%, Hungary 49%, Poland 55% and Romania 28% (Zektser and Everett, 2004).

With regard to the exchange of water in the opposite direction, surface water contributing to groundwater recharge has been less subject to regional estimates. However, the quantity is not negligible, particularly not in arid and semi-arid environments. Table 2.7 shows some examples.

---

8 The discrepancies between these percentages and those derived from AQUASTAT data (see Figure 2.9) may be due to different approaches to estimating the subsurface fluxes (estimating groundwater recharge or discharge), aggregated by country.

Box 2.4: Estimating the overlap between groundwater and surface water in order to calculate their combined natural flux (in km³/year) without double-counting

Most national statistics on natural water resources include separate values for surface water fluxes and groundwater fluxes. Surface water flux is usually considered to be equivalent to measured or calculated discharge at the area's outlet (outflow), while groundwater flux is defined by aquifer recharge (inflow). Since streams and aquifers exchange water in often considerable quantities, it follows that the sum of the independently assessed surface water and groundwater fluxes is affected by double-counting. In order not to overestimate the total renewable water resources (expressed as a rate of renewal), the flux of exchanged water – i.e. the 'overlap' between surface and groundwater fluxes – has to be deducted from this sum.

There are two main approaches to estimating the discharge of groundwater into surface water systems:

1   Direct calculation of the groundwater inflow into the streams by analysing stream flow hydrographs and estimating their baseflow component, which is assumed to correspond with groundwater runoff. A simplified approach (by default) assumes that baseflow rates are equal to dry season stream flow rates.

2   Calculation by difference: overlap $= a - b$, where:

a $=$ natural groundwater recharge, independently estimated (e.g. by modelling) for the area concerned

b $=$ groundwater flux not drained by the area's streams, such as direct groundwater outflow into the sea, groundwater discharge by evaporation or evapotranspiration, or subsurface flow to adjacent areas.

Exchanges in the opposite direction, from surface water systems to aquifers, can be estimated either by measuring the 'losses' of surface water along the thalweg of streams (differential stream gauging) or by calculating the corresponding aquifer recharge, in particular by modelling.

The overlap tends to reach 100% of the internal groundwater flux for landlocked countries (no access to the sea), except when transboundary subsurface flows occur or when significant groundwater discharge takes place by evaporation (as observed in closed basins, in the arid zone).

Double-counting may also occur when calculating the natural rate of renewal of the water resources of a large area (e.g. a continent) as the sum of the values defined for the smaller sub-areas (e.g. countries) of which it is composed. This is because outflows from one sub-area may be inflows to a neighbouring sub-area. For that reason, the concept of 'internal renewable water resources' has been developed (see Box 2.2), which includes only the surface water and groundwater resources generated internally within the area or sub-area concerned.

As a result of these exchanges of water in both directions, there are overlapping quantities in surface water and groundwater budgets that are separately established for one and the same area. In water budget assessments these are indicated as *overlap, repeated account, superposición* or *partie commune,* depending on the country. The mean annual fluxes corresponding to this overlap should be estimated if national or regional *natural water resources* statistics have to be produced without double-counting (see Box 2.4 and Chapter 4). Preparing such estimates is now common practice in different countries where the issue is well understood. Selected examples are shown in Table 2.8.

FAO has worldwide coverage of data on this overlap in its AQUASTAT database, which allows the percentage of overlap included in the internal groundwater flux of almost all the countries in the world to be mapped (Figure 2.12). The estimated values vary over a wide range (from 0% to 100%), perhaps partly due to differences in the methodology used, but are above 75% for the largest number of countries. According to the AQUASTAT statistics, the total overlap on the global scale is around 10 690 km^3/year, or 89% of the total terrestrial groundwater flux (Figure 2.13).

The lion's share of the world's groundwater flux ends up in streams or springs. Only a relatively small portion discharges either directly into the sea or is lost by evaporation (especially in closed basins). Estimates of the global submarine outflow of groundwater are 2 382 km^3/year (Zektser & Dzhamalov, 1988), 2 200 km^3/year (Shiklomanov and Rodda, 2003) and 2 400 km^3/year (Zektser and Everett, 2004), which is equivalent to around 20% of the global groundwater flux. This percentage

Table 2.8 Selected estimates of the overlap between groundwater and surface water fluxes in different countries.

Country	Total internal groundwater flux* (groundwater recharge) km^3/year	Overlap with surface water flux* km^3/year	Overlap as a percentage %	Reference
Australia	72	20	28	AQUASTAT 2012
China	828.8	727.9	88	Wang et al. 2000
D.R. Congo	421	420	99.8	AQUASTAT 2012
France	100	98	98	BRGM/Margat 2001
India	432	390	90	AQUASTAT 2012
Iran	49.3	18.1	37	AQUASTAT 2012
Italy	43	31	72	Plan Bleu 2001
Japan	27	17	63	AQUASTAT 2012
Mexico	139	91	65	AQUASTAT 2012
Morocco	10	3	30	Dir. Hydraulique 1997
Russia	788	512	65	AQUASTAT 2012
South Africa	4.8	3	63	AQUASTAT 2012
Spain	29.9	28.2	94	MOPU 1980
Turkey	69	28	41	Ozcan 2000/DSI 2001
USA (all states)	1383	1227	88.7	AQUASTAT 2012
USA (contiguous states)	1300	1162	89.4	USWRC 1978

Sources: See the cited references.
* Annual means.

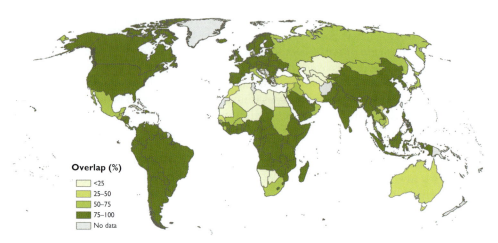

Overlap (%)

- <25
- 25–50
- 50–75
- 75–100
- No data

Figure 2.12 Overlap: percentage of internally produced groundwater overlapping with surface water, estimated by country. [*Data source*: AQUASTAT (2011). Ratio *Overlap[9]/Groundwater produced internally*, in %.]

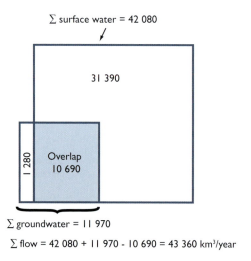

Σ surface water = 42 080

31 390

1 280

Overlap
10 690

Σ groundwater = 11 970

Σ flow = 42 080 + 11 970 - 10 690 = 43 360 km^3/year

Figure 2.13 Global fluxes of terrestrial surface water and groundwater, and their overlap (modified after Shiklomanov, 1998).

seems at first glance to conflict with the estimated total global overlap (89%). However, apart from the error margin in the estimates, one should take into account that not all river basin outlets considered for assessing surface water fluxes are located at the coast and that downstream of some of these outlets the rivers may lose part or all of their flow to groundwater systems. Such conditions can be observed in particular in arid coastal zones and in endorheic basins.

---

9 Flux of water exchanged between streams and aquifers.

2.4 GROUNDWATER IS PROMINENT IN THE WATER CYCLE OF DRY REGIONS

In arid and semi-arid regions, which cover around 30% of the Earth's land surface (Figure 2.14), the renewal of groundwater – like that of surface water – suffers from low and scarce rainfall. The effective contribution of rainfall to recharge is generally less than 10 mm/year, on average. It may even reach zero in some years, and water shortages are aggravated by large variations from one year to another. Groundwater renewal (recharge) is also much less widespread in arid and semi-arid regions than elsewhere and often very localised (see e.g. Simmers, 1997).

Moreover, migration of water between the groundwater and surface water systems is often in the opposite direction compared to what is observed in humid zones. Aquifer recharge is produced less by direct infiltration of precipitation than by infiltration of local runoff and water from allochtonous streams. Sometimes it is enhanced by percolating irrigation water, such as in the valleys of the Indus and the Nile, or in California.

On the other hand, groundwater in regions with a dry climate contributes less to maintaining permanent stream flow than in humid climates. Stream flow in arid zones is often ephemeral and erratic. It is not uncommon for a major part of the groundwater flow to be discharged by evaporation in shallow-water-table zones found in the centre of closed depressions, such as the *chotts* or *sebkhas* of the Sahara and the Arabian Peninsula, and the *salares* of South America and the Iberian Peninsula.

In large parts of the arid and semi-arid zones, both aquifers and river basins are '*endorheic*', which means that their water is not discharged into the open sea and

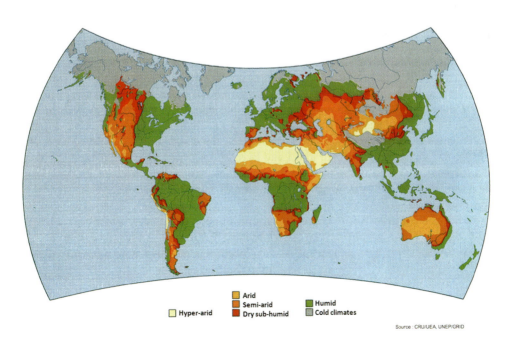

	Arid	
	Semi-arid	Humid
Hyper-arid	Dry sub-humid	Cold climates

Source : CRU/UEA, UNEP/GRID

Figure 2.14 Distribution of the world's arid and semi-arid regions. [*Source:* UNESCO (1984).]

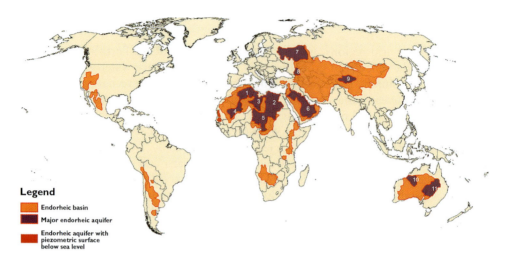

Legend

■ Endorheic basin

■ Major endorheic aquifer

■ Endorheic aquifer with piezometric surface below sea level

Figure 2.15 Large endorheic aquifer systems. [*Source*: UNESCO (1984).]

1 North-Western Sahara Aquifer System
2 Nubian Aquifer System
3 Murzuk Basin
4 Taoudeni Basin
5 Lake Chad basin
6 Arabian Aquifer System

7 Russian Platform
8 North Caucasus Basin
9 Tarim Basin
10 Canning Basin
11 Australian Great Artesian Basin

oceans. The vast endorheic areas (areas of internal drainage) cover more than 10% of the Earth's land area, mainly in arid and semi-arid regions of Eurasia (basins of the Caspian Sea and Aral Sea, Central Asian deserts), the Arabian Peninsula, the Sahara (including Lake Chad) and Australia. The aquifers there are either drained by streams in endorheic river basins or by evaporation in closed depressions. The first category includes the basins of the Volga River (Caspian Sea), the tributaries of the Aral Sea and Lake Balkhash, and many other closed basins of Central Asia, as well as the Jordan Basin (Dead Sea) and the Chari Basin (Lake Chad). Examples of closed depressions with discharge mainly by evaporation are the chotts of the Lower Sahara, the Al Qattara depression in Egypt and Lake Eyre in Australia, all of them largely below sea level. Thus, all Saharan aquifers are endorheic, as well as those of Central Asia, including the Tarim Basin, and also the Canning Basin and most of the Great Artesian Basin in Australia.

Endorheic aquifer systems are also found outside the strictly endorheic zones in arid regions. For example in West Africa, where the Trarza aquifer (Mauritania) and the Maastrichtian aquifer (Senegal) show closed depressions of the piezometric surface, descending even below sea level (Figure 2.15).

In arid and semi-arid regions, groundwater is usually the only permanent source of water. This highlights its value as a resource, in spite of often observed water quality restrictions, such as widespread salinity. In some regions, especially in the Sahara and on the Arabian Peninsula, groundwater reserves in deep aquifers constitute very significant resources – non-renewable but exploitable (see Chapter 4).

REFERENCES

Abdurrahman, W., 2002. *Development and management of groundwater in Saudi Arabia.* GW-MATE/UNESCO Expert Group Meeting, Socially sustainable management of groundwater mining from aquifer storage. Paris, 6 p.

Allègre, C., 1993. *Écologie des villes, écologie des champs.* ePub Format, Paris, Fayard.

AQUASTAT, 2011, 2012. *FAO's Global Information System on Water and Agriculture.* Accessed in 2011 and 2012, respectively (updates refer in all cases to the year 2010).

Baumgartner, A. & E. Reichel, 1975. *The world water balance.* Amsterdam, Elsevier, 179 p + 31 maps.

Bond, R.G., & C.P. Straub (eds), 1975. *Handbook Environmental Control.* CRC Press, Cleveland, Ohio.

BRGM, 1986. Le compte des eaux continentals. In: *"Les comptes du patrimoine naturel".* INSEE, Collection No 535–536, Series D, pp. 137–138.

BRGM/J. Margat, 2001. Contribution to chapter "Eaux continentales". In: *Rapport sur l'État de l'Environnement en France,* Institut Français de l'Environnement.

Bureau des Longitudes, 1984. *La terre, les eaux, l'atmosphère.* Paris, Gauthier-Villars.

Diop, S., & Ph. Recacewicz, 2003. *Atlas mondial de l'eau.* Paris, Autrement.

Dir. Hydraulique Maroc, 1997. *Développement des ressources en eau du Maroc.* First World Water Forum, Marrakech, 20–26 March 1997.

Döll, P., B. Lehner & F. Kaspar, 2002. Global modeling of groundwater recharge. In: G.H. Schmitz (ed.) *Proceedings of Third International Conference on Water Resources and the Environment Research,* Vol. I, pp. 27–31. Dresden, Technical University of Dresden.

Döll, P., & M. Flörke, 2005. *Global-scale estimation of diffuse groundwater recharge.* Institute of Physical Geography, Frankfurt am Main, Frankfurt Hydrology Paper 03.

Döll, P., & K. Fiedler, 2008. Global-scale modelling of groundwater recharge. *Hydrol. Earth Syst. Sci.,* 12, pp. 863–885.

Dzhamalov, R.G., & I.S. Zektser, 1999. *World Map of Hydrogeological Conditions and Groundwater Flow 1: 10 M.* Compiled under supervision of UNESCO by the Water Problems Institute, Russian Academy of Science.

El Gheriani, A., 2002. *The Great Man Made River project.* Colloquium "Eau et Économie", Sept. 2002, Soc. Hydrot. France, Paris.

FAO, 2003. *Review of world water resources by country.* Water Report 23. Rome, FAO, 112 p. Available from: ftp://ftp.fao.org/agl/aglw/docs/wr23e.pdf.

Habermehl, M.A., 1980. The Great Artesian Basin, Australia. *BMR Journal of Australian Geology & Geophysics,* 5, pp. 9–38.

Habermehl, M.A., 2001. Hydrogeology and environmental geology of the Great Artesian Basin, Australia. Chapter 11 in: V.A. Gostin (ed.), 2001, *Gondwana to Greenhouse – Australian Environmental Science.* Geological Society of Australia Inc., Special Publication 21, pp. 127–143, 344–346.

Khater, A.R., 2005. *Groundwater resources in Egypt: development and protection measures.* CIHEAM, International Conference on Water, Land and Food Security in Arid and Semiarid regions, 6–11 Sept. 2005, M.A.I., Bari.

Korzun, V.I. (ed.), 1974. *World water balance and water resources of the Earth.* Leningrad, Hydrometeoizdat.

L'Vovich, M.I., 1967. *World water resources and their future.* Izv. AN SSSR, ser. Geogr. No 6 (in Russian).

L'Vovich, M.I., 1974. *World water resources and their future.* Mysl' P.H. Moscow. English translation A.G.U., Washington, 1979, 415 p.

Margat, J., K. Frenken & J.-M. Faurès, 2005. *Key water statistics in AQUASTAT.* IWG-Env. International Work Session on Water Stastistics, Vienna, June 20–22, 2005. Available from: ftp://ftp.fao.org/agl/aglw/docs/ PaperVienna2005.pdf.

Mather, J.R., 1984. *Water resources: Distribution, use and management*. New York, Wiley, xv + 439 p.

MOPU, 1980. *El Agua en España*. Ministerio de Obras Públicas y Urbanismo, Centro de Estudios Hidrográficos, Madrid.

Nace, R.L., 1964. Water of the World. In: *Natural History*, January 1964.

Nace, R.L., 1968. *Water of the World*. US Department of the Interior, Geological Survey, July 1968.

National Council on Scientific Research of the USA, 1986. Cited by I.A. Shiklomanov in *Assessment of Water Resources and Water Availability in the World*, published by the World Meteorological Organization and the Stockholm Environment Institute, in the series Comprehensive Assessment of the Freshwater Resources of the World, 2nd World Water Forum, The Hague.

Ozcan, A.D., 2000. Communication on the Plan Bleu. Source: *Present Status of Water Resources Assessment in Turkey*. Water and Statistics, State Institute of Statistics, Ankara.

Plan Bleu, 2001. *Unpublished documentation*.

Rebouças, A., 1988. Groundwater in Brazil. *Episodes*, Vol. 11, no 3, pp. 209–214.

Shiklomanov, I.A., 1990. *Global water resources*. UNESCO, Nature and Resources, Vol. 26, No. 3.

Shiklomanov, I.A., 1992. World fresh water resources. Chapter 2 in: P.H. Gleick (ed.): *Water in Crisis – a Guide to the World's Fresh Water Resources*, Oxford University Press, 1993.

Shiklomanov, I.A., 1998. *World water resources: a new appraisal and assessment for the 21st century*. UNESCO, Paris.

Shiklomanov, I.A., & J. Rodda, 2003. *World water resources at the beginning of the twenty-first century*. Cambridge U.P./UNESCO.

Simmers, I. (ed.), 1997. Recharge *of aquifers in (semi-)arid areas*. International Contributions to Hydrogeology, Vol. 19, Balkema, Rotterdam, 277 p.

UNESCO, 1978. *See*: USSR Committee of UNESCO's International Hydrological Decade, 1974.

UNESCO, 1984. *Map of the world distribution of arid regions*. UNESCO, Paris, France.

UNESCO, 1988. *See*: Zektser, I.S., & R.G. Dzhamalov, 1988.

UNESCO, 1990. *See*: Shiklomanov, I.A., 1990.

UNESCO, CGMW, IAH, IAEA & BGR, 2008. *Groundwater Resources of the World, 1: 25 M*. WHYMAP. UNESCO & BGR, Paris, Hannover.

USGS/R. Nace, 1967. *Are we running out of water?* US Geological Survey, Circular No 536, Washington D.C.

USGS, 1975. *Handbook Environmental Control*.

USGS, 2012. USGS web page accessed in January 2012 (http://ga.water.usgs.gov/edu/watercycle.html).

USSR Committee of UNESCO's International Hydrological Decade, 1974. *World water balance and water resources of the Earth*. English translation UNESCO, 1978.

USWRC, 1978. *The nation's water resources 1975–2000. Second water assessment*. Washington DC, Water Resources Council, US Gov. Print Office.

Wada, Y., L. van Beek, C. van Kempen, J. Reckman, S. Vasak & M. Bierkens, 2010. Global depletion of groundwater resources. *Geophysical Research Letters*, Vol. 37, L20402, available from: doi:1029/2010GL044571.

Wang, R., H. Ren & Z. Ouyang, 2000. *China Water Vision: the ecosphere of water, life, environment and development*. Edited by Rusong Wang, China Meteorological Press, Beijing, 178 p.

WRI, 1991. *World Resources 1990–1991*. World Resources Institute, Washington DC.

WHYMAP, 2008. *Groundwater Resources of the World*. Map at scale 1: 25 M. BGR & UNESCO, Hannover, Paris.

WWAP, 2006. *World Water Development Report No 2.*

Zaporozec, A., 2002. *Groundwater contamination inventory.* UNESCO-IHP VI, Series on Groundwater No 2, Paris.

Zektser, I.S., & R.G. Dzhamalov, 1988. *Role of groundwater in the hydrological cycle and continental water balance.* UNESCO-IHP, Paris, 133 p.

Zektser, I.S., & L.G. Everett (eds), 2004. *Groundwater resources of the world and their use.* UNESCO, IHP-VI, Series on Groundwater No 6, Paris, UNESCO, 346 p.

Zektser, I.S. & H. Loaiciga, 1993. Groundwater fluxes in the global hydrologic cycle: Past, present, and future. *Journal of Hydrology,* Vol. 144, pp. 405–427.

Chapter 3

Geography of the world's groundwater systems

- *In what types of geological formation is groundwater present and moving? Down to what depth? In what spatial patterns?*
- *What can and should a geographic overview of groundwater systems present – and how?*
- *Mega aquifer systems*
- *Distribution of the main aquifer types around the world*
- *Groundwater provinces and global groundwater regions*

Groundwater is not only a hidden element of the hydrosphere and an invisible component in the terrestrial segment of the global water cycle, which can be included in a macro-hydrological picture as presented in the previous chapter. It is also an essential and dynamic constituent of the lithosphere, and hence of geoscientific interest and addressed by the discipline of *hydrogeology*.

Groundwater plays an active role in the internal geodynamics of the earth's crust. It does so by transferring pressure and by transporting matter involved in the mechanical and geochemical evolution of the subsoil, and even in volcanism. Studying and understanding this role of groundwater is what Lamarck had in mind when he introduced the term 'hydrogeology' (Lamarck, 1802).

On the other hand, the composition and structure of the subsoil have a strong influence on groundwater. In the first place, on its distribution, movement and chemical composition, as well as on its interactions at or near the surface in the form of in- and outflows and interconnections with streams. In addition, they also determine whether groundwater is accessible, how easily it may be abstracted, or how sensitive groundwater is to various natural and human influences. The properties of the subsoil thus are among the main factors defining the water resources offered by groundwater, as well as the constraints and obstacles that groundwater may present to various activities underground (e.g. mining, tunnelling, use of underground space).

These hydrogeological conditions and an outline of how they vary geographically across the globe are the subject of this chapter. The chosen approach results in defining groundwater systems mainly on the basis of geological characteristics.

Other approaches to defining groundwater systems or for classifying them do exist, e.g. the flow systems approach (Tóth, 1963), the use of groundwater bodies as basic management units (European Commission, 2008) or classifications on the basis of hydraulic state (confined, semi-confined, unconfined). However, these are less suitable for the spatial scales considered here.

3.1   TWO KEY ASPECTS OF THE SUBSOIL: COMPOSITION AND STRUCTURE

Just like in biology where the distinction is made between tissues and organs, both of equal importance, so the hydrogeological setting of the occurrence and movement of groundwater can be characterised by two key aspects:

*   *the composition of the subsoil in terms of rock properties* (in particular porosity, fissuring, permeability, solubility): these properties define the local capacity to store and conduct groundwater (see Figure 3.1).
*   *the regional structures*: these delineate and define the geometry (thickness, lateral extent, depth) of aquifers and other hydrogeological units, their interconnections and external links (e.g. recharge zones, links with surface water); thus the physical framework inside which the dynamics of groundwater takes place (see Figure 3.2).

These two aspects are essential for describing and analysing the hydrogeology of an area. Hydrogeological maps are the most suitable tool to present the corresponding information in a geographical context (see Box 3.1). For each of these aspects dedicated cartographic methodologies and techniques exist that are reflected in the map legend. Hydro-lithological classifications – quantitative or not – address the 'composition' aspect and reveal relevant properties of the subsurface formations, especially those related to storage and movement of groundwater. Usually the maps focus mainly on the zone immediately below the soil – in other words: on the formation outcrops –, but such classifications are also suitable for deeper aquifers or for cross sections in a multi-layer setting. For the identification and presentation of structure, different types and degrees of schematisation may be applied, in particular for delineating aquifer systems. Such a schematisation is based on a typology of 'boundary conditions' of these systems, or, for small-scale maps, on a classification of the degree of structural complexity of the regions.

Each of the two aspects yields useful information. The first one allows for assessing the potential productivity of groundwater abstraction works (e.g. wells) and for characterising the properties and behaviour of the aquifer medium (see Figure 3.3). The second one enables the aquifer systems to be analysed and modelled, in order to evaluate and manage their groundwater resources.

The assessment of local and regional hydrogeological conditions and the conversion of the assessment outputs into useful knowledge and information should combine these two aspects as much as possible. Their joint presentation, however, can be hampered by the diversity and complexity of the data and by the fact that meaningful features may be related to different scale levels. Therefore, available monographs, reports and maps usually focus on either of these two aspects.

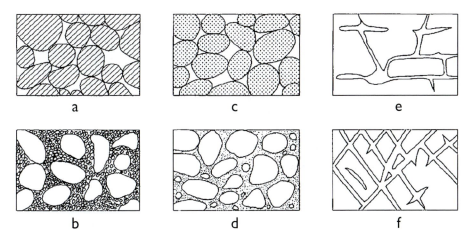

Figure 3.1 Different types of rock interstices, relevant for storage and movement of groundwater. [*Source*: After Meinzer (1923)].

(a) *Well-sorted sedimentary deposit having high porosity;*

(b) *poorly sorted sedimentary deposit having low porosity;*

(c) *well-sorted sedimentary deposit consisting of pebbles that are themselves porous, so that the deposit as a whole has very high porosity;*

(d) *well-sorted sedimentary deposit whose porosity has been diminished by the deposition of mineral matter in the interstices;*

(e) *rock rendered porous by solution; and*

(f) *rock rendered porous by fracturing.*

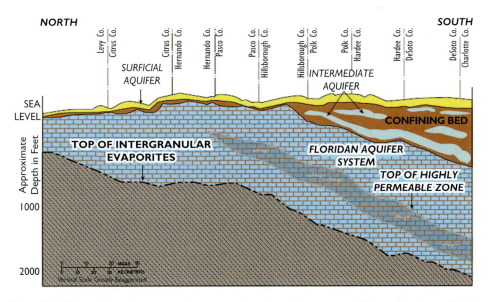

Figure 3.2 Hydrogeological cross section as a tool to show the structure. [*Source*: Southwest Florida Water Management District (2011)].

(a) Poorly sorted unconsolidated alluvial sand and
gravel formation, Yemen (*Photo*: Jac van der Gun)

(b) Cross-bedded friable Wajid sandstone, Yemen
(*Photo*: Ronnie van Overmeeren)

(c) Fractured and karstic Tertiary limestone, Oman
(*Photo*: Jac van der Gun)

(d) Jointed basalts near Lake Baringo, Rift Valley, Kenya
(*Photo*: Slavek Vasak)

Figure 3.3 Close-ups of selected outcropping aquifer rocks. (*Images cover 5 to 25 m² each*).

A brief comment on the spatial scale level may help the understanding of how the hydraulic parameters that are commonly used in hydrogeology and groundwater hydraulics are related to the composition of the subsoil as presented here. The distribution and geometry of the interstices in subsurface formations, and also the actual flow paths of individual particles of water through these interstices are fundamentally defined at a microscopic-scale level. However, in practice this scale level is usually unsuitable for observation and analysis. The disciplines of hydrogeology and groundwater hydraulics therefore have adopted a macroscopic approach to defining groundwater movement and hydraulic parameters of subsurface formations. In this approach, like in many branches of physics, the microscopic reality is replaced by a representative macroscopic '*continuum*' of comparable overall behaviour. The macroscopic physical laws (e.g. Darcy's Law) and the related macroscopic parameters assigned to the continuum (e.g. porosity and hydraulic conductivity) are valid only beyond a certain minimum volumetric size of the porous medium: the so-called '*representative elementary volume*'. Obviously, a representative elementary volume should include a sufficient number of interstices to yield a meaningful statistical average as required in the continuum approach (Bear, 1979; Freeze and Cherry, 1979). In this book, composition and structure of the subsoil are viewed only from a macroscopic perspective.

Box 3.1: Hydrogeological maps

Derived from geological maps, hydrogeological maps have the objective of showing the composition and structure of the subsoil in relation to the occurrence and movement of groundwater. They do so by combining data on the container (aquifer) and the content (groundwater). Like conventional geological maps, they give preference to depicting outcrops, thus the first aquifer below the ground surface. To varying degrees and depending on the purpose rather than on the scale, they try to combine the following:

- a classification of formations in relation to the potential productivity of groundwater abstraction works, or sometimes in relation to the infiltration capacity of water-table aquifers, using an *ad hoc* typology;
- data on groundwater dynamics (piezometric levels, potential field and outflow in discharge zones on a given date) and the relationship between groundwater and surface water;
- the presentation of observed or inferred structural elements at depth (possibly supplemented by cross sections, sketches or three-dimensional drawings), in particular those of delineated aquifer systems (see Box 3.2) which form the framework for assessing and managing the groundwater resources.

Information on groundwater recharge by infiltration of excess rain water, on water quality and on abstraction works can be added to these basic elements, depending on the state of knowledge.

Ever since initial attempts in the nineteenth century, such maps have been developed in many countries. Map scales vary from large (1:200 000 to 1:25 000) for regular coverage to small (1:500 000 to 1:5 000 000) for national, regional or continental summary maps. The maps use rather different legends, but an international standardisation effort, initiated by the International Association of Hydrogeologists (IAH) and UNESCO, aims to promote a uniform mapping methodology (Struckmeier and Margat, 1995). A database on national hydrogeological maps prepared around the world can be consulted at the following web address: http://www.bgr.de/app/fishy/whymap/.

3.2   GLOBAL HYDROGEOLOGICAL PANORAMA

It is impossible to present or even summarise the world's hydrogeology in a few pages, in spite of and also because of the abundance of available monographs in many countries; an additional complicating factor is that the available hydrogeological information varies highly between regions. Beyond the level of detail of the synthesis by Zektser and Everett (2004), a global project initiated by the United Nations Division of Technical Cooperation for Development (UNDTCD), carried out between 1976 and 1990 and coordinated by R. Dijon, has produced a unique series of national hydrogeological summaries for the different regions of the world (UN, 1976, 1982, 1983, 1986, 1987 and 1990). Repeating such an exercise to obtain an updated and

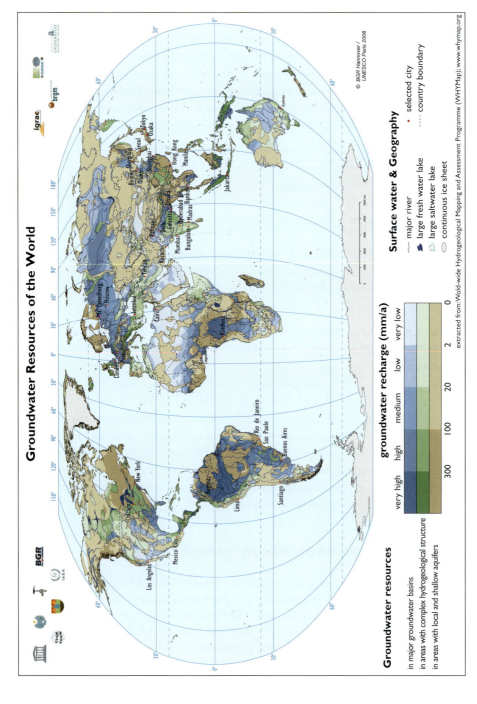

Figure 3.4 Simplified version of WHYMAP's Hydrogeological World Map. [*Source: UNESCO et al.,* (2008); see also: www.whymap.org].

Box 3.2: What is an aquifer system?[1]

Groundwater movement and transfer of water pressures are two distinct phenomena in an aquifer environment. The conceptual model of an aquifer system results from analysing the subsurface in the light of these two dynamic factors.

An aquifer system is a three-dimensional continuous subsurface domain that serves as both a reservoir for groundwater and a preferential natural conduit for groundwater flow ('subsurface highway'). Along its boundaries it has externally imposed hydrodynamic conditions (imposed in- or outflow rates or water levels) that link it with surface water and soil systems. The dynamic responses to various natural or artificial external impulses (e.g. recharge, withdrawal) can spread freely through the aquifer system.

The structure of an aquifer system may be simple (monolayer) or more complex (multilayer configuration). It may be composed of different aquifers and aquitards, between which the reservoir capacity is divided as well as the conductive properties that allow groundwater to flow from recharge areas to one or more discharge zones.

Subsystems of an aquifer system are hydraulically interconnected, thus they are not independent. A hydrodynamic model can be used to produce a schematic representation of the interconnections. Aquifer systems of all sizes exist in nature: their extent ranges from a few to more than one million square kilometres, while thickness can vary from a few metres to several kilometres.

Aquifer systems are the appropriate spatial units for assessing and managing the groundwater resources.

more homogeneous picture would require a large team of specialists, collectively familiar with the hydrogeology of all the regions in the world.

Our purpose here is therefore limited to outlining the hydrogeological basics and the major features of a world geography of groundwater. This geography is full of contrasts and much of it is still unknown. The outline will facilitate a better understanding of what is presented in the next chapters.

The two main variables defining the hydrogeological setting of a continent or region are its geological conditions and climate. Small-scale hydrogeological maps exist for all continents (see Appendix 6.4 and the web address given in Box 3.1) and together they yield a fairly complete global picture, although with emphasis on near-surface conditions. The WHYMAP project, an initiative of UNESCO and IAH, has in recent years developed a small-scale map that presents a global synthesis at the scale 1: 25 000 000 (analogous to the Geological Map of the World). Its consolidated version was published in 2008 under the title 'Groundwater Resources of the World'.

This map, reproduced in reduced and simplified form in Figure 3.4, was designed as a synopsis of available continental or regional hydrogeological maps, after introducing significant but necessary simplifications. Nevertheless, it offers a fairly homogeneous

---

1 See also Section 2.1.

picture of the prevailing type of hydrogeological setting in all the regions of the world. The map legend uses three colours to indicate the three main classes of hydrogeological setting that have been adopted. The classification was inspired by the three basic classes of the international standard legend for hydrogeological maps promoted by IAH and UNESCO (Struckmeier and Margat, 1995) and combines the dominant type of the aquifer medium with the degree of structural complexity. In addition, variations in colour intensity are used to differentiate according to the intensity of the estimated mean annual groundwater recharge.

The three main classes shown on the map can be described as follows:

Class a (blue)   Major groundwater basins. The aquifers they include may range from shallow unconfined aquifers to deep confined aquifers. The major groundwater basins have significant volumes of groundwater in storage, they contain highly productive aquifers and may include artesian zones (flowing wells).

Class b (green)   Areas of complex hydrogeological structure. They include rather productive local aquifers, in particular karst or volcanic aquifers, shallow or deep, and with significant storage.

Class c (brown)   Areas with only local and shallow aquifers. This class includes alluvial aquifers and aquifers in weathered or fissured rock. Stored groundwater volumes are small.

Superimposed mean recharge intensity intervals are shown by differences in colour tone varying from pale (very low recharge, i.e. less than 2 mm/year) to dark (very high recharge; i.e. more than 300 mm/year). The intervals in between are: low (2–20 mm/year), medium (20–100 mm/year) and high rates of mean recharge (100–300 mm/year).

The percentages of the total land area corresponding to each of the three main classes on WHYMAP's hydrogeological world map are shown in Table 3.1, with a breakdown by continent.

All three classes, including the most favourable first one, are present on each continent, although in varying proportions. However, this is not necessarily the case in each individual country, especially in small or medium sized countries (see Figure 3.4). Class 'a' is predominant in Western, Central and Eastern Europe, including Russia,

Table 3.1 Distribution of the three main WHYMAP classes on each of the continents.

	Class a (%)	Class b (%)	Class c (%)
Africa	45	11	44
Asia	32	17	51
Australia–Oceania	33	36	32
Europe	53	19	28
North America	15	27	58
Central and South America	45	11	44
Entire world	35	18	47

in a large part of the Arab countries, in the Northern part of South America and in Southern Africa. It is somewhat less prominent in West and Central Africa, the United States and Australia. Class 'a' is rare in Central and Northern Asia (except in western Siberia) and nearly absent in Scandinavia, Mediterranean Europe, in several African countries (along the Gulf of Guinea, as well as in Namibia, Botswana and Zimbabwe), Canada, Central America, the Andean region and Japan.

Among the different aquifer types there are large differences in structure, hydraulic behaviour and water resources development potential. In order to better understand the geography of groundwater in the world, it is therefore useful to look at each major aquifer type separately in a global perspective. The following main aquifer types will be discussed in turn:

- Sand and gravel aquifers
- Sandstone aquifers
- Karst aquifers
- Volcanic aquifers
- Basement aquifers

Before addressing these main aquifer types (see Section 3.4), the focus will first be on the Earth's largest aquifer systems ('*mega aquifer systems*'), most of which are located in sedimentary basins.

At the end of the chapter, finally, some attention will be paid to groundwater provinces and global groundwater regions, as a point of departure for extending the range of approaches to organising, analysing and presenting information on regional and global variations regarding groundwater.

3.3   THE EARTH'S MEGA AQUIFER SYSTEMS

Surface water flow is organised in river basins of all sizes and a large part of all surface water on the continents is discharged in only a few giant river basins. Similarly, there is also a huge variation in size among the world's aquifer systems and a limited number of very large aquifer systems contains the major share of the world's groundwater volume in storage. The geographical location of these *mega aquifer systems* is less observable in the field than that of large river basins and has to be revealed by hydrogeological studies.

All these large aquifer systems are composed of sedimentary rocks and fall into the category *major groundwater basins* (class 'a' – blue colour) on WHYMAP's hydrogeological world map (Figure 3.4). Thirty-seven very large aquifers have been selected to form the set of the Earth's mega aquifer systems. Their location is shown in Figure 3.5, while their main characteristics are summarised in Table 3.2 and Appendix 3.

Three structural types can be distinguished among these mega aquifer systems:

(1)   Sedimentary basins
These form more or less complex multi-layer aquifer systems, depending on the configuration of the basin floor and on stratigraphy, which may range from Pre-Cambrian to Quaternary. Their thickness may reach thousands of metres (up to 20 000 m on the Russian Platform).

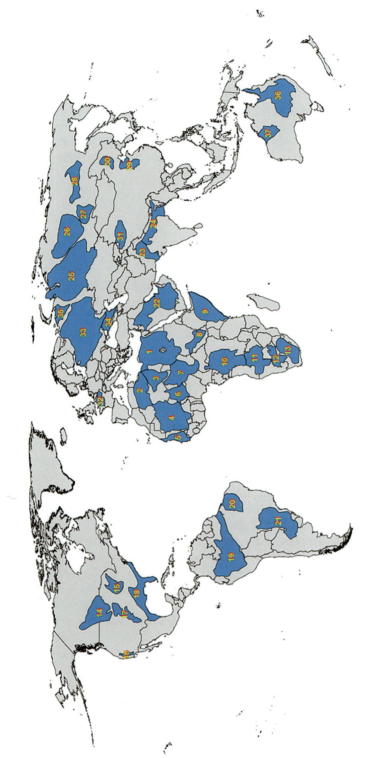

Figure 3.5 The Earth's mega aquifer systems (*For aquifer names and additional information see Table 3.2 and Appendices 3 and 6.3*). [*Source:* J. Margat, October 2005–January 2006 (unpublished)].

Table 3.2 The Earth's mega aquifer systems *(see Figure 3.5).*

	Name	Area (x 1 000 km²)	Maximum thickness (m)	Countries involved (ISO-3 alpha code)
	AFRICA			
1	Nubian Aquifer System (NAS) (Nubian Sandstone and Post-Nubian)	2 199	3 500	EGY, LBY, SUD, TCD
2	North Western Sahara Aquifer System (NWSAS)	1 019	1 600	DZA, LBY, TUN
3	Murzuk–Djado Basin	450	2 500	DZA, LBY, NER
4	Taoudeni–Tanezrouft Basin	2 000	4 000	DZA, MRT, MLI
5	Senegalo–Mauritanian Basin	300	500	MRT, SEN, GMB, GNB
6	Iullemeden–Irhazer Aquifer System	635	1 500	NER, DZA, MLI, NGA
7	Lake Chad Basin	1 917	7 000	NER, NGA, TCD, CMR, CAF
8	Sudd Basin (Umm Ruwaba Aquifer)	365	3 000	SDN, ETH
9	Ogaden–Juba Basin	~1 000	12 000	ETH, SOM, KEN
10	Congo Basin	750	3 500	COG, COD, AGO, RAF, GAB
11	Upper Kalahari–Cuvelai-Upper Zambezi Basin	~700		AGO, BWA, NAM, ZMB, ZWE
12	Lower Kalahari–Stampriet Basin	~350		ZAF, BWA, NAM
13	Karoo Basin	600	7 000	ZAF
	NORTH AMERICA			
14	Northern Great Plains Aquifer	~2 000		CAN, USA
15	Cambrian–Ordovician Aquifer System	250		USA
16	Californian Central Valley Aquifer System	80	600	USA
17	Ogallala Aquifer (High Plains)	450	150	USA
18	Atlantic and Gulf Coastal Plains Aquifer	1 150	12 000	USA, MEX
	SOUTH AMERICA			
19	Amazon Basin	1 500	2 000	BRA, COL, PER, BOL
20	Maranhão Basin	700	3 000	BRA
21	Guaraní Aquifer System	1 195	800	BRA, ARG, PRY, URY
	ASIA			
22	Arabian Aquifer System	>1 485	6 500	SAU, JOR, KWT, BHR, QTR
23	Indus Basin	~320	300	PAK
24	Indus–Ganges–Brahmaputra Basin	~600	600	IND, NPL, BGD
25	West Siberian Basin	3 200	6 000	RUS
26	Tunguss Basin	1 000	4 000	RUS
27	Angara–Lena Basin	600	3 000	RUS
28	Yakut Basin	720	1 200	RUS
29	North China Aquifer System (Huang Huai Hai Plain)	320	1 000	CHN
30	Song–Liao Plain	311	300	CHN
31	Tarim Basin	520	1 200	CHN
	EUROPE			
32	Paris Basin	190	3 200	FRA
33	Russian Platform Basins	~3 100	20 000	RUS
34	North Caucasus Basin	230	10 000	RUS
35	Pechora Basin	350	3 000	RUS
	AUSTRALIA			
36	Great Artesian Basin	1 700	3 000	AUS
37	Canning Basin	430	1 000	AUS

For more details and references: see Appendix 3 and Appendix 6.3.

The basin fill consists of highly permeable layers (such as sand, sandstone or even carbonate rock), alternating with layers of lower permeability (clay, marl, etc.) that allow vertical exchange of water (seepage), sometimes even with impermeable layers that impede seepage (rock salt). Lateral changes of sediment facies add complexity to this vertically layered structure. These basins usually include one or more unconfined aquifers and a variable number of partially interconnected confined or semi-confined aquifers. Deep down, where groundwater renewal is limited, groundwater is often brackish or even saline.

There are numerous examples of this type of aquifer system: the Great Artesian Basin of Australia, the basins of the Sahara (e.g. the North Western Sahara Aquifer System and the Nubian Aquifer System), the Northern Great Plains Aquifer of North America (shared by Canada and the USA), the West Siberian Basin, etc. Two structural variants can be distinguished:

- In the so-called *'Paris type'* (of which the Paris Basin is the prototype), sedimentary aquifers that are deep and confined in the centre of the basin – sometimes artesian – emerge laterally towards the peripheral zone of the basin, where they become phreatic and produce discharge through springs.
- In the *'Pannonian type'* (of which the Basin of Hungary is an example), the upper layers completely cover the deep aquifers. These aquifers can be explored only by drilling and it is therefore more difficult to identify them. Discharge of groundwater is mostly by seepage through the covering layers.

Both types of basins are often 'artesian', especially when they include confined zones over large areas. This turns them into significant energy storage reservoirs (see Chapter 4, Figure 4.16).

(2)   *'Graben' or rift valley systems*

Rift valleys are trough-like crustal depressions limited by faults, usually filled with alluvial deposits. These may form water-table aquifers, but multi-layer systems are observed as well, with confined aquifers either in the lower part of the graben fill or in permeable formations of the subsided crust. Artesian conditions may be present. Examples: Central Valley of California, the North China Plains, the Indus-Ganges-Brahmaputra basin in India and the Rhine Graben in Europe.

(3)   *Piedmont alluvial plains*

Aquifers in these accumulations of clastic material at the foot of mountain ranges are usually unconfined. Examples: the High Plains Aquifer (Ogallala) in the United States and many aquifers scattered along the flanks of the Andes in South America.

Although they have been specified as a single system, several of the world's mega aquifer systems are in reality complexes of adjacent and similar but hydraulically independent aquifers. This is particularly true for the Ogallala Aquifer of the High Plains in the United States or the basins of the Russian Platform.

The dominant characteristics of the world's mega aquifer systems include:

- *Large area*: the largest ones measure a few million square kilometres and therefore are comparable in size to the world's major river basins.

- *Huge reserves*: mostly in confined conditions and without much variation over time, largely to be considered as non-renewable water resources (see Chapters 4 and 5). The identified mega aquifer systems together appear to contain most of the world's groundwater reserves, but they undoubtedly have only a minor share in the global groundwater flux. The bulk of the global flux is produced instead by countless small and medium-sized local aquifers that are predominantly unconfined and interact actively with surface water systems and the atmosphere.
- A *marked continental location*: only a few of them have a large coastline.
- Most of them are *transboundary*, which means that they are crossed by international boundaries and consequently are shared between neighbouring countries.

Depending on climate, the mega aquifers of the world show significant variation in dynamic behaviour:

- In humid areas, they are mainly drained by streams and only a minor proportion have any discharge directly into the sea (submarine outflow). Discharge of deep confined aquifers is above all by upward seepage through aquitards into near-surface layers, or by artesian springs. Groundwater flow in such aquifers tends to converge to a central zone of piezometric minimum, where groundwater becomes virtually stagnant.
- In arid areas, mega aquifers are predominantly endorheic (see Figure 2.15). This means that their natural discharge is either by outflow into streams that have no outlet to the sea, or by evaporation in closed basins, where piezometric levels are sometimes below sea level. Examples are the Nubian Aquifer in Egypt, the aquifers of the Lake Chad Basin, other Saharan aquifers and most of the Great Artesian Basin in Australia.

Comparing the boundaries of the thirty-seven mega aquifers considered with the boundaries of the world's major river basins, it can be concluded that these are rather independent from each other. Only nine of these aquifers are embedded within a single major river basin.

3.4 MAIN AQUIFER TYPES

3.4.1 Sand and gravel aquifers

Unconsolidated sand and gravel deposits of Quaternary age or slightly older (Tertiary) form the most widespread and most exploited aquifer systems in the world. These sediments find their origin in the erosion of rocks: the erosion products are deposited after having been transported by streams (alluvial or fluvial deposits), ice (glacial and fluvio-glacial deposits), wind (aeolian deposits) or the sea (marine deposits). The sand and gravel deposits have high porosity and permeability, are often intercalated with sediments of finer texture (silts and clays) and occasionally include components of large size (boulders and blocks).

Aquifers of fluvial or fluvio-glacial origin – sometimes called alluvial aquifers – are the most widespread type of sand and gravel aquifers. Depending on size and setting, two categories can be distinguished:

- *Local aquifers*, present in the valleys of virtually all streams. These aquifers – often referred to as *alluvial valley aquifers* – usually have limited width and their thickness

Figure 3.6 Valle Central de Cochabamba, Bolivia: a deep intermountain graben filled with unconsolidated Quaternary aquiferous deposits (*Photo*: Jac van der Gun).

rarely exceeds a few tens of meters, except in valleys coinciding with subsidence zones, such as the Rhine graben in Europe or the Santa Clara Valley in California. These aquifers usually have strong hydraulic links with streams and exchange water with them on a regular basis and in two directions: the aquifers are recharged during periods of high stream discharge and drained by the streams during low flow periods. Alternatively, the aquifers may be recharged along one part of the river and drained along another part. Often these local aquifers are an intermediary between surface water systems and regional aquifers – by transmitting water to the latter and by draining them. That explains why they are easier to exploit than more isolated groundwater systems and why it is more difficult to quantify their resources.

- *Extensive aquifers*, located in major flood plains, such as the Indus Valley in Pakistan, the Huang Huai Hai Plain in North China and the Mesopotamia Plain in Argentina. They are located above all in major deltas, such as the Ganges delta in Bangladesh and the Nile delta in Egypt, or in piedmont belts bordering mountain ranges, such as the Po Valley in Italy, the Indus–Ganges–Brahmaputra basin at the foot of the Himalayas in India, the High Plains (Ogallala Aquifer) in the central part of the United States, or in deposits along the Atlantic slopes of the Andes, from Argentina to Venezuela. The larger of these aquifers belong to the category of mega aquifers and have been mentioned before (Section 3.3, Table 3.2). Their thickness may be considerable in areas of recent subsidence, particularly in the

Huang Huai Hai Plain of North China where the Quaternary multilayer aquifer system may reach 1 000 m, as well as in the Po Valley in Italy. These aquifers can contain substantial groundwater reserves, e.g. an estimated 1 000 km^3 in the Central Valley of California and 15 000 km^3 in the Ogallala Aquifer.

Most sand and gravel aquifers are unconfined, with relatively shallow water tables (a few metres to a few tens of metres deep). However, confined aquifers are also found locally, and confined aquifer beds are common in multilayer aquifer systems, e.g. in the North China Plain or in the Nile delta. Sometimes even artesian conditions are observed, such as in the Santa Clara Valley in California or in the Po Valley in Italy.

These aquifers are primarily recharged by infiltration of excess rainwater, in humid zones at mean annual rates that may exceed 200 mm/year. In their upflow zones they may also receive contributions from streams and inflows from hidden limestone or volcanic aquifers at their borders (as is frequently the case in the Mediterranean region). In arid and semi-arid regions, alluvial aquifers in the valleys of seasonal or ephemeral streams often have permanent subsurface flows. They may also receive additional recharge by infiltration of excess irrigation water, sometimes the lion's share of their recharge. This is the case for example in the Indus Plain in Pakistan or in the Nile valley and Nile Delta in Egypt, where these contributions to the alluvial aquifer were estimated to be 7.5 km^3/year.

Although they are usually drained by streams in a way that is fairly unseen, under certain favourable conditions (such as local depression, low permeability cover) alluvial aquifers may feed concentrated springs with occasionally high flow rates. A typical example is formed by the line of *fontanili* in the Po Valley in Italy, where the Silo spring (Veneto) breaks all records with an average flow of 50 m^3/s. In addition, sand and gravel aquifers contribute a significant share of the direct outflow of groundwater into the oceans. Interfaces between fresh and saline groundwater are very common in unconsolidated coastal aquifer environments. Depending on local conditions, such interfaces may be rather sharp or consist of thick transition zones between saline and fresh groundwater (mixing zones).

The geographical distribution of the unconsolidated local or alluvial valley aquifers (the first category described above) largely follows the branches of the stream networks. Because of their small size it is difficult to map them individually at a very small scale; in WHYMAP's map of Groundwater Resources of the World (Figure 3.4) they are therefore not shown explicitly, although included in all three main legend classes. The extensive sand and gravel aquifers (second category of the sand and gravel aquifers addressed here) are included in one single category of this map – the major groundwater basins (shown in blue) –, but they are not delineated separately either.

Unconsolidated sand and gravel aquifers are the world's most easily accessible and most widely tapped aquifers. Their variable thickness and degree of homogeneity result in a wide range of well yields, but often these are high (hundreds of m^3 per hour). Since these aquifers mostly coincide with areas of intensive human occupation (e.g. urbanisation, intensive agriculture), they also suffer under the highest pressures of exploitation and pollutants. The pollution risk is not only high due to the presence of pollutants, but also because these aquifers are often highly vulnerable to pollution (see Section 4.8). In the valleys of permanent and abundant streams, the exploitation of alluvial aquifers often triggers recharge from river water – which then becomes indirectly captured (induced recharge).

3.4.2 Sandstone aquifers

Sandstone is the consolidated version of a sandy deposit. Its age (mostly ranging from Palaeozoic to Mesozoic) has provided sufficient time for consolidation to take place. The individual grains are glued together by a cement of calcite, silica, clay or other minerals, while the weight of overlying younger strata may have produced compaction. Just as sand and gravel beds often alternate with clay and silt layers, so sandstones are commonly interbedded with siltstones or shales. Cementation and compaction have significantly reduced the original primary pore space (*primary porosity*), but external pressures have often produced secondary openings in the solid rock, commonly known as fissures (*secondary porosity*). As long as consolidation is still moderate (friable sandstones), sandstones may have an effective double porosity, which means that groundwater storage and flow are facilitated by both primary and secondary porosity. In highly consolidated sandstones, virtually all groundwater is stored inside fissures and almost all flow is through fissures; hence secondary porosity there is dominant.

Porosity and hydraulic conductivity of sandstone aquifers are lower than those of unconsolidated sand and gravel aquifers, but because sandstones extend over large areas and are often thick, they form an important category of aquifers in a global context. Well-drilling success rates are usually high, like in unconsolidated clastic aquifers, due to the relatively uniform aquifer properties. Several of the mega aquifer systems presented in Section 3.3 consist largely of sandstone, e.g. the Cambrian-Ordovician Aquifer, the Northern Great Plains Aquifer System, the Guaraní Aquifer System and

Figure 3.7 Outcrops of Tawilah Sandstone, main aquifer formation in the Sana'a area, Yemen (*Photo*: Charles Dufour).

the Maranhão Basin in the Americas, the Nubian Sandstone Aquifer System and the Karoo Basin in Africa, as well as the Canning Basin and the Great Artesian Basin in Australia. An additional number of mega aquifer systems includes sandstone as one of the main water-bearing rock units, e.g. the Amazon basin, the North Western Sahara Aquifer System, the Taoudeni basin, the Lake Chad Basin, the Upper and Lower Kalahari Basins and the Congo Basin. Nevertheless, there are many sandstone aquifers around the world that are much smaller in extent, but very important as local sources of water supply. Examples are the Tawilah Sandstone aquifer in the Sana'a basin in Yemen and the Patiño aquifer in central Paraguay.

Like unconsolidated sand and gravel aquifers, large sandstone aquifers are included in the category 'major groundwater basins' of WHYMAP's map of Groundwater Resources of the World.

3.4.3   Karst aquifers

The outcrops of carbonate rocks – mainly limestones and dolomites – cover almost a quarter of the world's land surface and are especially abundant in the Northern hemisphere. Most carbonate rocks originate as sedimentary deposits in marine environments. Compaction, cementation and dolomitisation may change their porosity and permeability considerably, but the main post-depositional process is called *karstification,* after the Kras region in the Slovenian Dinaric Alps, where the *'karst'* phenomenon was already being studied at the end of the nineteenth century (Cvijić, 1893)[2]. Depending on present and past climatic conditions, on the age of the outcropping rocks and on tectonic evolution, carbonate rocks show different degrees of dissolution, with effects ranging from slightly widened joints and fractures to caverns of large dimensions. These karst features have marked hydrological and hydrogeological implications. Karst areas have a very special surface morphology (see Figures 3.8 and 3.9) and their hydrography is characterised by low surface water drainage densities and abnormal flow features, such as sudden losses of surface water flow and closed surface water basins. *'Karst aquifers'* – as the aquifers in karstic rock are called – are discontinuous, very special because of their underground networks of caves and channels of varying complexity and limited hierarchy, and locally highly conductive, but heterogeneous in their capacity to store groundwater (Burger and Dubertret, 1975 and 1984; Waltham and Fookes, 2003; White, 2003; Ford and Williams, 2007).

A block diagram is shown in Figure 3.8. In most cases, the aquifers are phreatic and thick (from one hundred to several hundreds of metres). Caves in emerged zones (where groundwater is no longer present) can be entered by humans and are popular for speleological exploration. Some *'palaeokarst'* may also lie confined under non-karstic formations. Karst formations are usually compartmentalised and subdivided into elementary aquifer units of limited size, even in the largest complexes – whether their structure is tabular or more or less tectonically deformed and folded. The boundaries of karst aquifer systems depend mainly on geological and structural conditions. However, they may also be influenced by morphological and hydraulic

---

2 Karst can occur as well in non-carbonate rocks, such as rock salt and gypsum, but these are less important from a hydrogeological point of view.

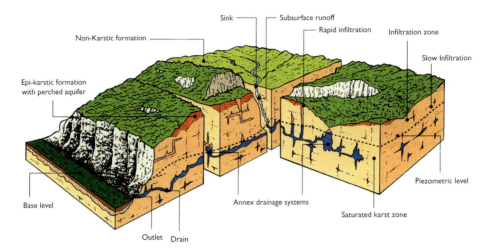

Figure 3.8 Block diagram of a karst aquifer system with tabular structure. [*Source*: M. Bakalowicz (2005)].

Figure 3.9 Karst landscape near Minerve, France. (*Photo*: Hugo Soria; GNU Free Documentation License version 1.2).

conditions – these are sometimes variable (temporary emergencies) and not always clearly identifiable.

Networks of karst conduits, sometimes very large and complex, can join and drain adjacent rock compartments. They replace the surface water drainage systems observed in other regions and have a similar function. In general, karst aquifers receive most of the local precipitation surplus (often 80 to 90%), which tends to be relatively high anyhow. Indeed, karst environments form mostly plateaus or mountain massifs well endowed with rain. For example, the karst areas of southern Europe enjoy up to 900 mm of average annual recharge. These aquifers thus intercept much of the total regional flux. In addition, they often even receive water that is lost from streams that rise in adjacent, non-karstic areas. Their discharge is generally concentrated in space, and consequently often large (Figure 3.10). Most of the time, there is only one

Figure 3.10 Karst spring of La Loue, Jura, France. (*Photo*: Philippe Chrochet).

main discharge outlet per aquifer system. These outlets form the largest springs on Earth, with mean discharge rates of tens of cubic metres per second (see the examples in Table 3.3). However, their irregular regimes with pronounced differences between peak flows and low flows are rather similar to those of streams in regions without aquifers.

Table 3.3 Large karst springs on Earth (*records refer to average flow rates above 10 m³/s*).

Name	Location	Yield (m³/s) Mean annual	Range
Matali	Papua New Guinea	90	20->240
Trebisnjica	Bosnia–Herzegovina	80	2–300
Bussento	Salerno, Italy	80	
Dumanli	Taurus, Turkey	~50	25->100
Ljubljanica	Slovenia	39	4–132
Khabour Ras-el-Ain	Syria	39	
Tisu	China	38	4–545
Stella	Friuli, Italy	37	
Ombla	Croatia	34	2.3–154
Chingsui	China	33	4–390
Frio	Mexico	28	6–515
Pivsko Oko	Serbia–Montenegro	25	
Coy	Mexico	24	13–200
Buna	Bosnia–Herzegovina	24	3–123
Silver Spring	Florida, USA	23	15–36
Kirkgözler	Turkey	22	10–35
Fontaine de Vaucluse	France	22	4.5–120
Sinjac	Piva, Serbia–Montenegro	21	1.4–154
Bunica	Bosnia–Herzegovina	20	0.7–207
Grab-Ruda	Croatia	20	2–105
Peschiera	Latium, Italy	19	
Bregava	Bosnia–Herzegovina	17.5	0.5–59
Timavo	Trieste, Italy	17.4	9–130
Waïkoropupu	New Zealand	15	5–21
Maligne	Canada	14	1–45
Sette Comuni	Italy	14	1–105
Zarga (Orontes)	Hermel, Lebanon	13	5–26
Big Spring	Missouri, USA	12.5	
Rijecina	Croatia	12.4	1.2–80
Sinn	Syria	10.5	

Source: Data collected in collaboration with Michel Bakalowicz.

Coastal and submarine springs are mainly produced by coastal karst aquifers, especially in the Mediterranean region, Florida (USA) and the Yucatan Peninsula (Mexico). In the Mediterranean area, coastal karst aquifer outflows were estimated to average 53 km³ per year (Zektser, 1988), which represents 17% of the total groundwater flux of the Mediterranean basin.

Because of their discontinuous nature and the often large depth to the saturated zone, the accessibility of karst aquifers is fairly random. They can nevertheless be very productive. Their development often concentrates on abstractions in their natural discharge zones and makes use of special types of collecting devices (galleries, adjustable valves, etc.).

The geography of karst – and thus of the Earth's karst aquifers – is not known everywhere to the same degree of detail. The most exploited karst aquifers belong to one of the following categories:

- Geologically old carbonate rocks, especially those of the Palaeozoic age, that have undergone a long period of karstification. Examples are the karst aquifers of South China (Guangxi), of several central regions in the United States (Ozark Plateaus), of the Russian Platform (Moscow Basin) and of southern Africa.
- Geologically younger limestone formations (Mesozoic or Tertiary age) affected by Alpine tectonics, especially in the Mediterranean region (Dinaric Alps). Karstification in this region has been enhanced to a large extent by recent tectonics – in some cases still ongoing –, resulting in dislocation, uplifts next to deep incisions by erosion, extension and decompression (Avias, 1977), as well as by wide variations in the level of the Mediterranean Sea, as during the Messinian regression (1 000 m).

A third category, although less prominent in karstic features than the previous two, also includes important aquifers:

- Geologically young carbonate formations that are not or only slightly tectonically disturbed, such as the limestones of Yucatán, Florida and the Bahamas, the chalk of England and France, and some Miocene carbonate deposits in the Mediterranean.

WHYMAP's map of Groundwater Resources of the World does not show karst aquifers explicitly and separately, but includes them generally in the category 'areas with complex hydrogeological structure' (presented in green).

The role of karst aquifers in the dynamics of terrestrial water, in particular of global groundwater flow, is more important than their relative size suggests. The same is true for the share they have in the renewal of groundwater resources. For example, in the four provinces of South-western China, the natural recharge of the karst aquifers is estimated at 127 km^3/year, or 53% of the total groundwater recharge, of which 48 km^3/year is exclusively within Guangxi (where it corresponds to 62% of the total groundwater recharge). The groundwater flux in karst aquifers in China has been estimated at 204 km^3/year, or 23% of the estimated total natural groundwater recharge rate of 874 km^3/year (Chen Mengxiong and Zektser, 1985; Zektser and Everett, 2004).

3.4.4 Volcanic aquifers

Most volcanic rock formations on Earth, even relatively young ones, are poorly permeable. However, productive but heterogeneous and discontinuous aquifers can be found in Quaternary volcanic rocks, or even in older ones (of Tertiary or Mesozoic age). They consist of an alternation of fissured lavas (especially basalts) and porous pyroclasts (such as ashes and tuffs), generally very permeable, but often anisotropic (horizontal permeability higher than vertical permeability). These aquifers are scattered over the many volcanic massifs of our planet, form more extensive aquifers on some lava plateaus and are sometimes interbedded in sedimentary basins (such as the Paraná Basin in South America).

Usually outcropping, volcanic rocks favour the infiltration of rainfall and snow-melt. That is the reason why, under humid climatic conditions, the more rainfall and snowmelt is triggered by their often pronounced relief, the more recharge the volcanic aquifers receive, even though rainfall enhances surface runoff as well. To illustrate this, the estimated mean annual recharge is 1 120 mm for the aquifer on Reunion, 700 mm for the aquifer on Tahiti and 800 mm for the aquifers on the Hawaiian Islands. The aquifer at Mount Etna, Sicily, receives an average recharge of 520 mm/year (95% of the massif's total flow) and its springs yield 2 m³/s. Recharge is less in the sub-arctic zone (230 mm/year in Iceland) and often low to very low in the arid zone, e.g. 0.5 to 1 mm/year in Djibouti. Nevertheless, topography and geographic position may cause large variations over rather short distances, such as shown by mean recharge figures for the Canary Islands, ranging from 600 mm/year in the highlands of Gran Canaria, Tenerife, La Palma and Gomera to less than 2 mm/year in the arid southern areas and on Lanzarote and Fuerteventura.

Volcanic aquifers combine water-table aquifers with groundwater movement under confined conditions, the latter especially in the deep layers of basins. This combination of conductive layers and zones of storage capacity shows some similarity to karst aquifers. Even on the large plateaus the aquifers are relatively fragmented and generally composed of local water systems. In humid climate zones, volcanic aquifers feed numerous springs discharging water often of excellent quality. They are frequently accompanied by thermal features (e.g. geysers) in regions where volcanism is active, such as in Iceland, New Zealand and Yellowstone National Park (USA).

The most extensive volcanic aquifer complexes of our planet are listed below (see also Figure 3.11):

- The Columbia Plateau in the USA (650 000 km²), with a very thick lava cover (over 1 000 m).
- The Deccan Traps in India, covering more than 400 000 km².
- The Rift Valley in East Africa (Djibouti, Ethiopia, Kenya, Tanzania), containing several endorheic aquifers as shallow basins.

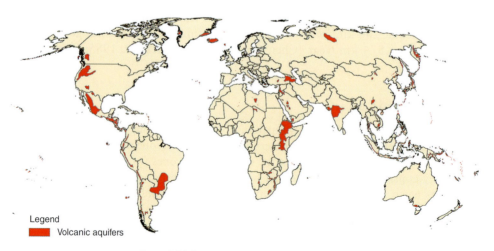

Legend
■ Volcanic aquifers

Figure 3.11 Principal volcanic aquifers on Earth.

Figure 3.12 Volcanic rocks in the Central Highlands of Yemen – poorly to moderately productive aquifers (*Photo*: Charles Dufour).

Figure 3.13 Basalt aquifer outcrop near Lake Baringo, Rift Valley, Kenya. (*Photo*: Slavek Vasak).

- The Sierra Madre Occidental in Mexico, with extensions along the axial zone of Central America, from Guatemala to Panama, and further South by the andesites almost all along the Andes chain.
- Basalts of the Serra Geral formation interbedded in the Paraná Basin (together with the Guaraní aquifer system) in Brazil and Argentina, one million km^2 in size and up to 1 600 m thick.
- All of Iceland (103 000 km^2).
- The Transcaucasian region and Mount Ararat.
- A series of basalt plateaus and massifs on the Arabian Peninsula, from Syria to Yemen.
- The many volcanic areas of the West Pacific Belt from Kamchatka and Japan to the Philippines, Papua New Guinea, the Solomon Islands and the islands of the Indonesian Archipelago.

The distribution of local volcanic aquifers corresponds to the locations of the volcanoes in the world. They are particularly numerous along the Pacific Ring of Fire and at different plate boundaries, for example in the Mesogean belt, along the plate edges in Indonesia and in the island arc of the Antilles. Many volcanic aquifers are found on islands scattered over all oceans. Small islands off the continental shelves are almost all volcanic:

- In the Atlantic: Macaronesia (Azores, Madeira, Savage Islands, Canary Islands, Cape Verde) and the Lesser Antilles.
- In the Pacific: Hawaiian Islands, Galapagos Islands and most of the islands of Oceania.
- In the Indian Ocean: Comoros, Reunion and Mauritius.

Consequently, many small island states and territories depend for their water resources on volcanic aquifers (see Table 3.4).

Because of their complexity or their small size (Figure 3.11 shows only the larger ones), volcanic aquifers are not shown individually on WHYMAP's map, but included within the category of 'areas with complex hydrogeological structure', shown in green.

Table 3.4 Groundwater resources of volcanic islands.

Islands	Natural renewable groundwater resources (mean annual values in km^3/year)	Share of groundwater in the total water resources (%)
Canary Islands (Spain)	0.7?	73
Cape Verde	0.12	40
Comoros Islands (without Mayotte)	1.0	83
Hawaii (USA)	13.2	72
Iceland	24	14
Mauritius	0.89	32
Mayotte (France)	0.075	25
Reunion (France)	2.8	56
Tahiti (France)	~ 0.73	29

References: AQUASTAT and national sources.

3.4.5 Basement aquifers

The so-called *basement* or *crystalline basement*, consisting of igneous and metamorphic rocks, underlies the sedimentary and volcanic rock sequences on all continents. In addition, basement rocks outcrop in over 30% of the continents, in all climate zones. Basement outcrops are of variable relief and receive on average more precipitation than the areas covered by sedimentary rocks. These outcrops store more groundwater than has been previously recognised, particularly in regions with a humid climate (in sub-arctic, temperate and tropical zones).

The basement outcrops contain very special aquifers, with conductive and storage functions which are to some extent spatially separated (see Figure 3.14):

- The deeper part, in fresh rock with fissures or fractures to a certain depth (about one hundred metres, sometimes more), forms discontinuous aquifer zones. This is especially so in areas where recent tectonic evolution has produced fracturing, – as is the case in the Alpine mountain chain areas or in the Mediterranean region where ophiolites are fractured to depths of 150 to 200 m. This part of the aquifer system is capable of transmitting groundwater locally, but the capacity to store groundwater is generally low. Properly planned drilled wells can have substantial productivity, with yields of at least a few and sometimes tens of cubic metres per hour.
- The weathered shallow horizons (*saprolite*) of variable thickness (often tens of metres, depending on present and past climatic conditions), is a less productive part of the aquifer system in terms of transmitting groundwater, but it has a higher capacity to store water. Without this storage reservoir, wells in the underlying fractured rock would quickly run dry. Recharge varies according to climate.

The water-bearing saprolite and the fissured bedrock below it (see Figure 3.14) together form exploitable aquifer systems. These systems are generally of limited size and their boundaries are poorly defined. However, they are scattered widely over all basement massifs, often in patterns defined by the natural drainage networks. These aquifers provide modest but rather widely distributed groundwater resources, used for rural domestic or animal water supply and even for small-scale irrigation. Finally, these aquifers contribute to perennial flow in the often very dense stream networks found in the humid zone.

The global distribution of basement aquifers is shown in Figure 3.15. Most of these aquifers belong to the third hydrogeological category distinguished on WHYMAP's map ('areas with local and shallow aquifers', shown in brown).

The largest complexes of basement aquifers on our planet are found in the following areas:

- *In the humid zones*: In the Northern hemisphere these complexes include the Canadian, Scandinavian, Finnish-Karelian and central Siberian (Angara) shields, where permafrost conditions severely restrict the possibilities of groundwater renewal, as well as the West African massifs and the Indian shield (that forms a large part of the Deccan Plateau). The main complexes in the Southern hemisphere are the Guyana shield, the Brazilian shield and the coastal mountain ranges of Central and Southern Africa, including Madagascar. Several of these

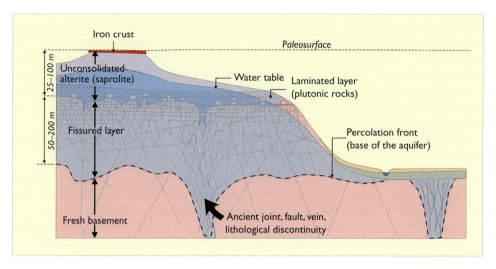

Figure 3.14 Basement aquifer consisting of residual regoliths and a fissured zone. [*Source:* After Lachassagne and Wyns (2005)].

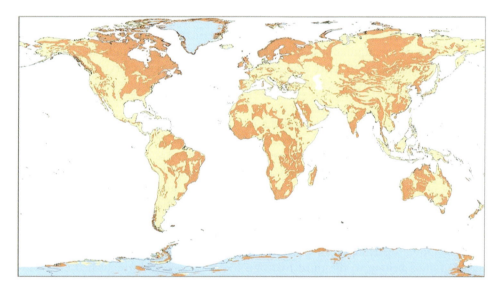

Figure 3.15 Global distribution of basement aquifers. [*Source*: After Lachassagne and Wyns (2005)].

basement aquifer complexes in the humid tropics enjoy very high recharge rates (see Chapter 2). Estimated mean annual recharge may be more than 300 mm, and locally up to 500 mm in the massifs of Guyana and North Brazil, West African Guinea and Madagascar.

• *In arid or semi-arid zones*: Here the massifs of the Sahara, Western Arabia and West and Central Australia are particularly noteworthy. Saprolite aquifer zones are generally absent, hence the storage capacity of the basement aquifers is very limited. Recharge is much less than that of basement aquifers in humid zones and is mainly produced by local ephemeral runoff.

Since the 1980s, increased attention has been paid to this type of aquifer in the search for low-cost drinking-water sources, especially in Peninsular India and Sub-Saharan Africa. This has contributed to significantly improved understanding of their characteristics (Foster, 2012).

3.5 A DIFFERENT ANGLE OF VIEW: GROUNDWATER PROVINCES AND GLOBAL GROUNDWATER REGIONS

The geographic variation in groundwater setting and groundwater state can be shown on maps of different scale and design. Classical hydrogeological maps as well as many other groundwater-related maps tend to depict – in the degree of detail allowed by their scale – selected characteristics (such as aquifer rock type, aquifer productivity, piezometric level, groundwater salinity, mean annual recharge rate, etc) spread across a geographic space that is in principle continuous and not subdivided into predefined spatial units. However, in some cases or for some purposes it is more convenient or useful to organise and present the information on the basis of predefined discrete spatial units, such as countries or aquifer systems. More convenient, if available data are already collected in this way (by country or aquifer). More useful, if different types of information have to be integrated and linked to spatial units that are well-defined and well-known by those who want to use the information. The corresponding maps belong to the category of *choropleth maps*: they assign one single value, class or colour to each of the predefined spatial units. Examples presented in previous sections are the figures 2.5, 2.8, 2.9 and 2.12.

Defining spatial units on the basis of administrative boundaries is common practice and does not need clarification. The advantages of using administrative units are evident, in particular because they are clearly identifiable and linked to government mandates and responsibilities. A disadvantage is that their boundaries are usually unrelated to groundwater-related features and processes. Hydrogeologists therefore prefer spatial units with boundaries based on hydrogeological criteria, in particular aquifer systems or – for more detail – their components, such as specific aquifer beds and interbedded aquitards, or recharge versus discharge zones. The aquifer concept has proven to be very successful, not only because an aquifer is a convenient and meaningful spatial unit for analysing relevant processes and presenting information, but also because it is a unit with a clear identity (aquifers have a name), facilitating efficient communication and exchange of information.

There are countless aquifers in the world. Besides very large ones, discussed before, there are numerous medium-sized aquifers and many more aquifers of very limited extent. For an overview of groundwater systems in a larger geographical area, such as the territory of a large country, upscaling may be needed to produce a clear picture. For this purpose, Meinzer introduced almost a century ago the concept of 'groundwater provinces' and delineated such provinces in the conterminous United States of America (see Figure 3.16). These provinces are relatively large areas, each

Figure 3.16 Meinzer's groundwater provinces in the conterminous USA. [*Source:* After Meinzer (1923)].

having a broad uniformity of hydrogeological and geological conditions. They allow a macroscopic picture of groundwater conditions in large territories where many aquifers are present; hence they are suitable spatial units for analysis and communication on groundwater at that scale level. Meinzer's groundwater provinces for the USA have been modified afterwards (Heath, 1982) and groundwater provinces have been delineated in other parts of the world as well, e.g. in Australia and in South America.

For analysing, exchanging, presenting, disseminating and discussing groundwater information at a global or continental scale, even groundwater provinces may be too small a spatial unit to be used. Therefore, IGRAC has made an attempt at a next step in spatial aggregating, by defining so-called '*global groundwater regions*' (Figure 3.17). Criteria for aggregating are similar to those used for groundwater provinces, but at a different scale level. With a large degree of generalisation, the thirty-six resulting global groundwater regions may be subdivided into four categories, marked in Figure 3.17 by different colours:

- *Basement regions* (red): These are characterised by the predominance of geologically very old basement rocks at or near the surface. In large parts of these regions, groundwater is only present at shallow depths, often in fractured or weathered zones in consolidated rocks or in rather thin layers of unconsolidated sediments overlying basement. Stored volumes of groundwater are limited.
- *Sedimentary basin regions* (yellow): These regions contain extensive and very thick accumulations of sediments that may be permeable down to considerable

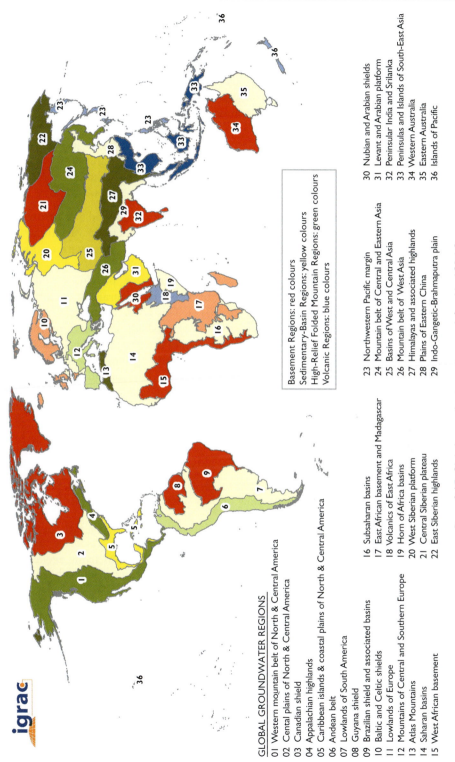

GLOBAL GROUNDWATER REGIONS

01 Western mountain belt of North & Central America
02 Cental plains of North & Central America
03 Canadian shield
04 Appalachian highlands
05 Caribbean islands & coastal plains of North & Central America
06 Andean belt
07 Lowlands of South America
08 Guyana shield
09 Brazilian shield and associated basins
10 Baltic and Celtic shields
11 Lowlands of Europe
12 Mountains of Central and Southern Europe
13 Atlas Mountains
14 Saharan basins
15 West African basement

16 Subsaharan basins
17 East African basement and Madagascar
18 Volcanics of East Africa
19 Horn of Africa basins
20 West Siberian platform
21 Central Siberian plateau
22 East Siberian highlands

Basement Regions: red colours
Sedimentary-Basin Regions: yellow colours
High-Relief Folded Mountain Regions: green colours
Volcanic Regions: blue colours

23 Northwestern Pacific margin
24 Mountain belt of Central and Eastern Asia
25 Basins of West and Central Asia
26 Mountain belt of West Asia
27 Himalayas and associated highlands
28 Plains of Eastern China
29 Indo-Gangetic-Brahmaputra plain

30 Nubian and Arabian shields
31 Levant and Arabian platform
32 Peninsular India and Srilanka
33 Peninsulas and Islands of South-East Asia
34 Western Australia
35 Eastern Australia
36 Islands of Pacific

Figure 3.17 Global groundwater regions as defined by IGRAC.

depths. Consequently, they may form huge groundwater reservoirs. By far the largest part of all groundwater reserves on earth are found in these regions.

- *High-relief folded mountain regions* (green): Folded mountains with irregular topography are the predominant geological feature of these regions. Various types of rocks are arranged in complex structures, leading to a rather fragmented occurrence of groundwater. Productive aquifers are often found in alluvial deposits in valleys and plains, even more so if these are located in downfaulted depressions.
- *Volcanic regions* (blue): Groundwater occurrence and features in these regions are markedly affected by relatively recent volcanism (Quaternary or Tertiary age). Aquifers are limited to porous or fractured lavas and sediments interbedded between lava flows. Zones of high permeability are not exceptional.

It goes without saying that this classification is highly simplified and focused only on macroscopically dominant features. There is still considerable variation within regions of such large spatial dimensions: each region includes zones with characteristics that are very different from the region's dominant features.

In IGRAC's approach, aquifer systems, groundwater provinces and global groundwater systems are three linked levels of a coherent hierarchical approach to scale-dependent zoning (Van der Gun *et al.*, 2011). This approach has some similarity to differentiating between river basins, secondary watersheds (e.g. Mediterranean Sea watershed) and primary watersheds (e.g. watershed of the Atlantic and Arctic Oceans) as different levels of spatial aggregation of surface water systems. A brief description of each of the 36 global groundwater regions is presented in Appendix 4. The description includes – *inter alia* – a listing of the provisionally defined groundwater provinces within each of these regions.

REFERENCES

Abdurrahman, W., 2002. *Development and management of groundwater in Saudi Arabia.* GW-MATE/Unesco Expert Group Meeting, Socially sustainable management of groundwater mining from aquifer storage. Paris, 6 p.

Anonymous, 2003. *Water resources and use in Australia.* Unpublished, accessed at www.farmweb.au.com by IGRAC [July 2005].

Araújo, L.M., A.B. França & P.E. Potter, 1999. Hydrogeology of the Mercosul aquifer system in the Paraná and Chaco-Paraná basins, South America, and comparison with the Navajo-Nugget aquifer system, USA. *Hydrogeology Journal*, Vol. 7, No. 3, pp. 317–336.

Arnold, G.E. & W.J. Willems, 1996. Aperçu général des aquifères européens. In: *Eur. Water Pollut. Control*, Vol. 6, No. 5, pp. 11–18.

Avias, J.V., 1977. Globotectonic control of perimediterranean karstic terranes main aquifers. In: Tolson J.S. & F.L. Doyle (eds), *Karst hydrogeology*. University of Alabama, Huntsville, USA, pp. 57–72.

Bakalowicz, M., 2005. Ressources en eau du kars, in *Géosciences*, 2, BRGM, Sept. 2005.

Bakhbakhi, M., 2002. *Hydrogeological framework of the Nubian Sandstone Aquifer System.* GW-MATE/UNESCO Expert Group Meeting, Socially sustainable management of groundwater mining from aquifer storage. Paris, 6 p.

Bear, J., 1979. *Hydraulics of groundwater.* New York, McGraw-Hill International Book Company, 567 p.

Bocanegra, E., M. Hernandez & E. Usunoff (eds), 2005. *Groundwater and Human Development.* IAH Selected papers on Hydrogeology No. 6, Balkema Publishers, 278 p.

Burger, A. & L. Dubertret (eds), 1975. *Hydrogeology of karstic terrains.* IAH, Paris, 190 p.

Burger, A. & L. Dubertret (eds), 1984. *Hydrogeology of karstic terrains. Case Histories.* International Contributions to Hydrogeology, Vol. 1, IAH, UNESCO, IUGS, Heise, Hannover, 264 p.

CBLT, 2002. Information supplied by Commission du Bassin du Lac Tchad (Lake Chad Basin Commission).

CEDARE/IFAD, 2002. *Regional Strategy for the Utilisation of the Nubian Sandstone Aquifer System,* Vol. II. CEDARE, Heliopolis Barhya Cairo, Egypt.

Cvijić, J., 1893. *Das Karstphänomen.* PhD thesis Vienna University, Vienna, Austria.

Diagana, B., 1997. *Gestion des eaux internationales en Afrique Sub-Saharieenne. Bilon diagnostic de la gestion integrée des eaux et des contraintes environnementales dans la vallée du fleuve Sénégal.* A/PNUD/DASDG/RAF/94/01C-SAT1.

Dodo, A., 1992. *Etude des circulations profondes dans le grand bassin sédimentaire du Niger: identification des aquifères et comprehension de leurs fonctionnements.* PhD thesis, Univ. of Neuchâtel, Switzerland.

European Commission, 2008. *Groundwater Protection in Europe. The new groundwater directive – consolidating the EU regulatory framework.*

FAO, 2003. *Review of world water resources by country.* Water Report 23. Rome, FAO, 112 p.

Ford, D. & P. Williams, 2007. *Karst Hydrogeology and Geomorphology.* Wiley, 576 p.

Foster, S., 2012. Hard-rock aquifers in tropical regions: using science to inform development and management policy. *Hydrogeology Journal,* Vol. 20, pp. 659–672.

Foster, S. & H. Garduño, 2004. *Towards sustainable groundwater resource use for irrigated agriculture on the North China Plain.* GW-MATE Case Profile Collection Number 8, The World Bank, Washington, 16 p.

Freeze, R.A. & J.A. Cherry, 1979. *Groundwater.* Englewood Cliffs, N.J., Prentice-Hall, Inc., 604 p.

Habermehl, M., 2002. *Groundwater development in the Great Artesian Basin, Australia.* GW-MATE UNESCO Expert Group meeting, Socially Sustainable Management of Groundwater Mining from Aquifer Storage, 35 p., Paris.

ICID, 1983. *Irrigation and Drainage in the World.* Vol. III, Sudan., New Delhi, Third ed. 1993, pp. 1262–1280.

ICID, 2000. *Irrigation and drainage in the world.* Chapter Pakistan, New Delhi, 12 p.

IGRAC, 2009. *Map of Global Groundwater Regions* (revised 29 April 2009, after the original version of 2004).

IME, 2008. *Les aquifères fossiles au sud de a Méditerranée.* Institut Méditerranéen de l'Eau, Marseille, France, 30 p.

Lachassagne, P. & R. Wyns, 2005. Aquifères de socle: nouveaux concepts. In: *Géosciences,* No. 2, sept. 2005, Orleans, pp. 32–37.

Lamarck, J-B., 1802. *Hydrogéologie.* Paris, Agasse, Impr-Lib., 268 p.

Lau, J.E., D.P. Commander & G. Jacobson, 1987. *Hydrogeology of Australia.* Bureau of Mineral Resources, Geology and Geophysics, Bull. 227, Canberra, Australian Gov. Publ. Service.

Margat, J., K. Frenken & J-M. Faurès, 2005. *Key water resources statistics in AQUASTAT.* IWG-Env., Int. Work Session on Water Statistics, Vienna, June 20–22, 2005.

Meinzer, O.E., 1923: *The Occurrence of Groundwater in the United States.* USGS Water Supply Paper 489, Washington DC.

Ndengu, S., 2004. International shared aquifers in Namibia. In: Appelgren, B. (ed), 2004. *Managing Shared Aquifer Resources in Africa,* Proceedings of the International ISARM-Africa Workshop at Tripoli, Libya, 2–4 June 2002, UNESCO-IHP-VI, Series on Groundwater No. 8, pp. 117–122.

OSS, 2003. *Système aquifère du Sahara septentrional.* Rapport de synthèse. Observatoire du Sahara et du Sahel, Tunis, 129 pp.

OSS, 2005. *See:* UNESCO/OSS, 2005.

Rebouças, A., 1976. Le grand basin hydrogéologique du Maranhao, Brésil. Perspectives sur l'exploitation. *Memoires XIth IAH Congress,* Budapest, pp. 448–458.

Rebouças, A., 1988. Groundwater in Brazil. *Episodes,* Vol. 11, No. 3, pp. 209–214.

Safar-Zitoun, M., 1993a. *Notice explicative de la carte hydrogéologique international de l'Afrique, Feuille 3.* Association Africaine de Cartographie, Alger.

Safar-Zitoun, M., 1993b. *Notice explicative de la carte hydrogéologique international de l'Afrique, Feuille 4.* Association Africaine de Cartographie, Alger.

Schneider, J.L., 2001. *Carte de valorisation des eaux souterraines de la République du Tchad (à 1/1500000) et géology-archéologie-hydrologie de la République du Tchad.* Two volumes, 1100 p.

Shiklomanov, I.A., 1998. *World Water Resources at the Beginning of the Twenty-First Century.* Draft of a book planned to be published in 2001, UNESCO, Int. Hydrological Series.

Southwest Florida Water Management District, 2011. *Sinkholes.* WaterMatters. Available from: *http://www.swfwmd.state.fl.us/hydrology/sinkholes/brochure.pdf.*

Struckmeier, W.F. & J. Margat, 1995. Hydrogeological Maps: A Guide and Standard Legend. *IAH Intern. Contr. to Hydrogeology,* Vol. 17, Heise, Hannover, 177 p.

Tóth, J., 1963. A theoretical analysis of flow in small drainage basins. *Journal of Geophysical Research,* Vol. 68, pp. 4795–4812.

UNESCO, CGMW, IAH, IAEA & BGR, 2008. *Groundwater Resources of the World, 1: 25 M (WHYMAP).* UNESCO & BGR, Paris, Hannover.

UNESCO/OSS, 2005. *Ressources en eau et gestion des aquifères transfrontaliers de l'Afrique du Nord et du Sahel.* ISARM-Africa, UNESCO-IHP IV, Series on Groundwater No 11, Paris.

UN, 1976. *Groundwater in the Western hemisphere.* UN-DTCD, New York, Natural Resources Water Series no. 4.

UN, 1982. *Groundwater in the Eastern Mediterranean and Western Asia.* UN-DTCD, New York, Natural Resources Water Series no. 9.

UN, 1983. *Groundwater in the Pacific Region.* UN-DTCD, New York, Natural Resources Water Series no. 12.

UN, 1986. *Groundwater in Continental Asia.* UN-DTCD, New York, Natural Resources Water Series no. 15.

UN, 1987. *Groundwater in Africa.* UN-DTCD, New York, Natural Resources Water Series no. 18.

UN, 1990. *Groundwater in Europe.* UN-DTCD, New York, Natural Resources Water Series no. 19.

USGS, 2003. Principal Aquifers. In: *National Atlas of the United States of America, scale 1:5 000 000.* USGS, revised 2003 version.

USGS, 2011. *Ground Water Atlas of the United States.* Available from: http://pubs.usgs.gov/ha/ha730/index.html [accessed in 2011].

Van der Gun, Jac, Slavek Vasak & Josef Reckman, 2011. Geography of the world's groundwater: a hierarchical approach to scale-dependent zoning. In: J.A.A. Jones (ed.): *Sustaining Groundwater Resources.* Initiative of the International Year of Planet Earth. Dordrecht, Springer, pp. 131–158.

Waltham, A.C. & P.G. Fookes, 2003.Engineering classification of karst ground conditions. *Quarterly Journal of Engineering Geology and Hydrogeology,* Vol. 36, pp 101–118.

Wang, R., Ren, H. & Z. Ouyang, 2000. *China Water Vision: the ecosphere of water, life, environment and development.* Edited by Rusong Wang, Beijing, China Meteorological Press, 178 pp.

WHYMAP, 2008. *Groundwater Resources of the World.* Map at scale 1: 25 M. BGR & UNESCO, Hannover, Paris.

White, W.B., 2003. Conceptual models for karstic aquifers. In: *Speleogenesis and Evolution of Karst Aquifers,* Vol. 1 No 1, January 2003, pp. 1–6.

Zektser, I.S. & L.G. Everett (eds), 2004. *Groundwater resources of the world and their use.* UNESCO, IHP-VI, Series on Groundwater No 6. Paris, UNESCO, 346 p.

Chapter 4

Groundwater resources

- *Should all groundwater be considered as a resource?*
- *Interdependence between groundwater resources and surface water resources*
- *Exploitation strategies defining which groundwater to be exploited and used*
- *Natural groundwater quality*
- *Groundwater dependent ecosystems*
- *Global distribution of renewable and non-renewable groundwater resources*
- *Access to groundwater and how to withdraw it*
- *How fragile and vulnerable are groundwater resources?*
- *Augmentation of groundwater resources*

4.1 GROUNDWATER: A NATURAL RESOURCE, BUT ONLY PARTIALLY EXPLOITABLE

A natural resource is commonly defined as something found in nature (usually a substance or 'raw material') and necessary or useful to humans. Typical examples are solar energy (a perpetual resource), forests and water (renewable resources), or fossil fuels and minerals (non-renewable resources). Obviously, a natural resource does not exist without someone to use it, so the concept is anthropocentric. To complicate the picture, different people may value resources differently, depending on their cultural background, views of nature, social change, resource scarcity and technological and economic factors (Cutter *et al.*, 1991).

In a generic sense, groundwater is a natural resource. This does not mean, however, that all groundwater on Earth has the real potential to satisfy human needs or desires. In the first place, some of it has to be discarded because its quality does not meet the water quality standards for its proposed uses, even after conventional treatment. Secondly, the exploitation of some groundwater bodies may be technically or economically unfeasible. Thirdly, the exploitation of groundwater may be subject to significant environmental constraints resulting from aspirations to conserve surface water, protect ecosystems or maintain the stability of the land surface. Therefore, only part of the '*theoretical*' groundwater resources, represented by total flows and stocks, can be considered as

Box 4.1: Are surface water and groundwater independent resources?

Running or stagnant surface water in rivers, streams and lakes is very different from groundwater in aquifers, in terms of where and how it moves at a given moment. In addition, unlike surface water, groundwater is invisible. The knowledge we have about each of the two is under the competence of different specialists: hydrologists versus hydrogeologists. The techniques to exploit them are also quite different: diversions and dams versus wells and boreholes. The differentiation according to exploitation technique is helpful for keeping separate records of their abstraction. However, it is important not to confuse source of supply with resource.

Surface water and groundwater interact and exchange a lot of water during their way down through the continents (see Section 2.3). In humid areas, where only a minor share of groundwater flows directly into the sea, groundwater and the baseflow of streams (fed by groundwater) do indeed form one single resource. In arid regions we observe the opposite: short-lived surface runoff events or inflows of surface water originating in more humid regions elsewhere (like the Nile in Egypt) are the main sources of recharge to the aquifers and part of their fluxes will evaporate in closed depressions. Here again, surface water and groundwater are the same resource, although it is sometimes used twice in succession, as is the case in the Nile Valley.

The larger the territory in question is, the greater usually the part of mobile water that alternates between surface water and groundwater. According to estimates based on country statistics in FAO's AQUASTAT, cited already in Chapter 2, the following may be concluded for the globally aggregated flows:

•   approximately 25% of the natural surface water flux comes from groundwater;
•   almost 90% of the natural groundwater recharge finally joins rivers and forms the stable part of the surface water flows (baseflows).

These connections between aquifers and streams, in particular the transformation of groundwater flow into the baseflow of streams, are well understood by hydrologists, but not yet sufficiently known by others. They are often absent in diagrams of the water cycle and sometimes ignored in water resources assessments.

These connections explain why separate statistics on surface water fluxes, based on measured or calculated hydrological data, and those on groundwater fluxes, based on the estimated recharge of the aquifers, ultimately result in double-counting and therefore are not fully additive. As has been pointed out in Chapter 2 (Box 2.4), shared fluxes ('overlap' in the FAO's AQUASTAT statistics) have to be subtracted from their sum, except for special cases of reuse where abstracted surface water after being used recharges groundwater and then is used again.

Thus, for a defined area, in particular a river basin, the renewable surface water and groundwater resources should not be estimated on the basis of the assumed origin of the mobile water (by runoff or by groundwater recharge). Rather the total water resources should be the point of departure for assessing these two water resources components. How to subdivide the total water resources into shares of surface water and groundwater depends on whether it is preferable to withdraw the required water by surface water diversions or by pumping from aquifers. Such preferences are based on different practical and economic criteria, as well as on the care needed for controlling competition and conflicts between users.

Finally, only the *exploitable* surface water and groundwater resources should be considered in integrated water resources management.

'exploitable' groundwater resources. The concept of exploitable groundwater resources provides a realistic basis for planning groundwater exploitation and management.

Understanding the role of groundwater in the global water cycle and estimating the fluxes at global, regional and national levels is necessary but not sufficient to correctly assess the exploitable groundwater resources. As previously pointed out, the assessment should also take into account the relevant technical, economic, water quality and environmental constraints. This introduces a certain degree of subjectivity and may also reflect the competitiveness of groundwater compared to surface water resources and even the political preferences of the moment. In addition, assessed quantities of groundwater and surface water resources are generally not independent of each other, due to the physical continuity between groundwater and surface water. Hence, exploitable resources of both types of water should be assessed in a coordinated way (see Box 4.1).

In practice, groundwater provides primarily *renewable* resources (fluxes that can be captured) and, in more exceptional cases, *non-renewable* resources (groundwater reserves that can be exploited and depleted, see Section 4.6). The groundwater reserves also act as an important buffer, by making the exploitation regime of the renewable resources less dependent on the natural rhythm of the recharge.

In most countries, the groundwater resources originate from within. More rarely than in the case of surface water resources, a proportion of them may nevertheless come from a neighbouring country, if transboundary aquifers are present. In some countries (e.g. Egypt, Bahrain, Qatar, Israel, Jordan, Syria) groundwater resources of external origin may be considerable. Conflicts about the exploitation of transboundary aquifers may arise and their joint management may require agreements between neighbouring states.

The diversity in size, structure, fluxes and reserves of the aquifers, in their links with atmospheric and surface water, and in their water quality is reflected in an enormous variation in groundwater resources. It also leads to large differences in accessibility, in usability and in competitiveness compared to surface water resources. This diversity is also reflected in the sensitivity of the groundwater systems to factors influencing their regime and, finally, in the constraints to be considered in their management.

Nevertheless, assessing the *exploitable* groundwater resources of an aquifer – i.e. the resources that really matter in practice – depends very much on the chosen exploitation strategies, in other words: on *groundwater resources management*.

4.2 WHICH EXPLOITATION STRATEGIES?

Groundwater exploitation strategies may vary according to their relative reliance on flux or on stock. In particular the role assigned to the groundwater reserves (stock) introduces a marked differentiation between management strategies. An adequate groundwater governance infrastructure is a prerequisite for the development, adoption and implementation of groundwater exploitation strategies. Which strategy to adopt depends on aquifer setting, local conditions and preferences of the parties involved. Three main strategy types can be distinguished (see Fig. 4.1):

a   *Intercepting fluxes only: sustainable yield strategy*
This strategy aims to keep the groundwater flow regime in a state of *dynamic equilibrium*, in the short as well as in the longer term. After an unavoidable limited depletion of storage upon starting groundwater withdrawal, the stored volume remains stable – with minor fluctuations over time – while the abstracted groundwater is balanced by changes in the aquifer's in- and outflows. The aggregate withdrawal may correspond to a larger or smaller fraction of the mean natural groundwater flux – depending on the buffer capacity of the aquifer and on the constraints imposed by any possible aspirations to control water levels or maintain a certain natural groundwater discharge. What this strategy aims for is *harvesting inflows*. The management of the reserves under this strategy is similar to managing a revolving fund deposited in a bank.

b   *Planned storage depletion during a limited period, followed by abstracting at a sustainable rate in the longer term: a 'mixed' strategy*
Interception of the flux under equilibrium flow conditions is pursued in the longer term, but a controlled imbalance is allowed in the short or medium term. During this period of imbalance, intensive groundwater abstraction in response to water demands takes place, leading to depletion of some of the stored groundwater. This depletion may be a *loan* from the reserve, followed by a *reimbursement* meant to let the stock recover, at least partly (e.g. by artificial recharge). Alternatively, the depletion may be

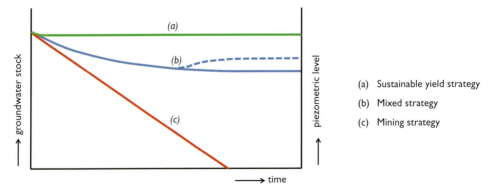

Figure 4.1 Schematic picture of the evolution of groundwater reserves and groundwater levels under the three main strategic approaches to groundwater exploitation.

followed merely by *stabilisation of the reserves* after adjusting the withdrawals to a level compatible with a dynamic equilibrium. In both cases, a drastic reduction in the rate of abstraction is needed to restore equilibrium in the latter phase.

This 'mixed' strategy is similar to a 'mining' strategy (described below) during its first phase and then returns to a 'sustainable yield' strategy at a later stage. It may be a deliberately chosen strategy, but in many cases it is a *'de facto'* strategy, attempting to stop or reverse groundwater level declines that have spontaneously developed. In the past and too often still today, these declines may result from pumping unintentionally at aggregate rates beyond those that the aquifer can sustainably provide (*'hydrological overabstraction'*, see Section 6.3 and Box 6.5). This applies at least to the first phase, while restoring the equilibrium during the second phase aims to correct this.

c *Progressive depletion of the reserves: mining strategy*
In this strategy groundwater is exploited under conditions of a permanently unbalanced groundwater regime, even in the long term. The stock is progressively reduced, which makes the exploitation of groundwater in this case very similar to that of a mineral resource such as rock salt or gold. The depletion of stored groundwater causes the natural groundwater discharge to cease relatively rapidly because of capture. This strategy entails *mining* of non-renewable resources, which is equivalent to consumption of an asset.

As mentioned before, the selection of a certain type of strategy depends on several factors and is not only dependent on aquifer conditions. Nevertheless, aquifer conditions are an important indicator of which type of strategy is likely to be appropriate and which not. First of all, non-renewable groundwater resources call for a mining strategy (type *c*), whereas for aquifers with renewable groundwater a strategy that includes a sustainable yield component (types *a* and *b*) is usually more appropriate. In humid zones with plentiful water resources and with significant *in situ* environmental functions of groundwater it is generally convenient and feasible to opt for a sustainable yield strategy (type *a*). In arid regions under the pressure of high water demands, without powerful water resources management institutions and in the absence of alternative sources of water, a mixed strategy is usually more realistic (type *b*). It requires, however, sufficient groundwater storage in the aquifer to bridge a transitional mining period without exhausting the aquifer in significant parts of its territory.

In some areas, groundwater is the dominant water resource or even the only one present; elsewhere it may be a minor or complementary resource. Nevertheless, it is important to exploit and manage the groundwater resources in conjunction with surface water resources.

Assessing the exploitable groundwater resources of an aquifer or region is tuned to the applicable exploitation strategy. Under a sustainable yield strategy and in the later phase of a mixed strategy, the assessment focuses on the renewable groundwater resources. Under a mining strategy and in the early phase of a mixed strategy, however, it focuses on the groundwater reserves. In both cases, the outcomes of the assessment depend not only on available groundwater, but also on the relevant technical, economic, water quality and environmental constraints.

4.3 NATURAL GROUNDWATER QUALITY

Water quality is an essential aspect of all water resources, also of groundwater resources. It indicates how suitable the water is for different types of use, each with their own criteria expressed as water quality standards.

The natural composition of groundwater (i.e. the substances dissolved in it) reflects the original composition of the recharge water, the mineral composition of the subsurface medium and the climate. Modifications of the composition start during infiltration into the soil. Subsequent changes during the groundwater's passage through the rock units depend very much on the soluble constituents present in these rocks, the groundwater flow paths and the velocity of groundwater movement (residence time). Climatic conditions may result in higher concentrations due to evaporation, while the exchange of water between aquifers and streams may also leave its mark.

Poor water quality, in particular restricting the water's suitability for drinking purposes, is in general due to excessive concentrations of dissolved solids, but may also be related to naturally occurring humic and fulvic acids, tannins and bacteria. The most common natural deficiencies in groundwater quality include:

*   Often high salinity (in particular NaCl) in arid regions, in deep aquifers, near salt-bearing rock formations or in coastal aquifers fed by rainfall with considerable chloride content.
*   Sometimes excessive hardness in carbonate formations or high sulphate content in gypsiferous formations.
*   Often high iron and manganese concentrations in basement aquifers and in other formations rich in organic matter or other sources of carbon dioxide.
*   Often significant acidity in crystalline rocks.
*   Significant fluoride levels in some sedimentary rocks (phosphates) or crystalline rocks (e.g. in India).
*   Significant arsenic content, in particular in some Quaternary sedimentary formations (e.g. in Bangladesh).
*   Naturally occurring bacteria.

However, water quality is not simply defined by analysis results that specify the chemical composition of water samples or their physical or even biotic parameters. An important step is to link the analysis data to water quality classes, by comparing the observed values of the parameters to standards for different types of use. The many types of water use (drinking water, irrigation, different types of industry, etc.) all have their own water quality requirements. This means that water quality has to be interpreted in the context of various potential uses, for each of which a full range of water suitability classes has been defined. Different groundwater quality maps thus are needed if different types of water use are considered. Furthermore, hydrochemical or groundwater quality maps at best represent only one aquifer at a time – usually the upper one, – suggesting that the vertical hydrochemical stratification may be ignored, which is often not justified. The vertical stratification or variation of groundwater quality can be visualised in cross sections or by producing sets of maps for different depth intervals or for different superimposed aquifer units.

Groundwater quality has been assessed and is being monitored in numerous aquifers around the world. This has resulted in considerable knowledge on the natural quality of groundwater in many aquifers, as a function of the original quality of the water that has entered the subsurface and of the subsequent processes in the soil, unsaturated zone, aquifers and aquitards. It is difficult to produce groundwater quality maps at a global scale, because global inventories on groundwater quality are scarce and excessive concentrations of most parameters are usually rather local. At that scale, it has nevertheless been proven feasible to map relatively large areas where all groundwater is saline (mostly in arid regions), as shown on the world map by WHYMAP (2008). Other global maps, where the water quality parameters are more detailed, are briefly presented and discussed below. They deal with the global occurrence of brackish and saline groundwater, excessive arsenic and excessive fluoride in groundwater, respectively. Only shallow and intermediate depths are considered, corresponding to the depths relevant for the active hydrological cycle and for conventional groundwater withdrawal (on average, down to some 500 m).

Finally, it should not be overlooked that groundwater is vulnerable with respect to its quality and is exposed to water quality degradation risks related to anthropogenic pollution (see Sections 4.8 and 6.5).

4.3.1 Saline and brackish groundwater

Water is called saline if the total concentration of dissolved solids (TDS) exceeds 10 000 milligram per litre and brackish if its TDS is between 1 000 and 10 000 milligram per litre. While groundwater at a great depth is mostly saline, it is usually fresh in the upper part of the subsurface, the domain where the subsurface flux of the water cycle is mainly concentrated. The thickness of this fresh groundwater zone varies, but in most cases it is a few hundred metres, usually less than 500 metres. However, in some zones groundwater is not fresh at these shallow and intermediate depths, but brackish or saline. Results of a global inventory carried out by IGRAC (Van Weert *et al.*, 2009) show the provisional location of such zones and the origin of the relatively high mineral content of their groundwater (Figure 4.2). It is a provisional picture, because this picture was derived only from information that was easily accessible during the inventory, and hence there is considerable scope for improvement. In addition, some of the zones were delineated on the basis of area-wide observed brackish or saline groundwater, whereas other zones should instead be interpreted as zones where brackish or saline groundwater is likely to occur, as inferred from prevailing geological or water use conditions.

Regarding the origin of salinity, a distinction has been made between three genetic categories: marine, natural terrestrial and anthropogenic terrestrial. Saline groundwater of marine origin has entered the subsurface either simultaneously with marine sediments (connate saline groundwater) or later – during marine transgression periods in geologic history, by short-duration incidental inundations by the sea, by seawater sprays joining recharge water or by seawater intrusion (often triggered by intensive groundwater withdrawal). Groundwater salinity of terrestrial origin is subdivided into natural and anthropogenic. Natural mechanisms that may cause a greatly increased content of dissolved solids in groundwater include evaporation at or near land surface (important in hot arid climates), dissolution from evaporates (halites) and carbonate

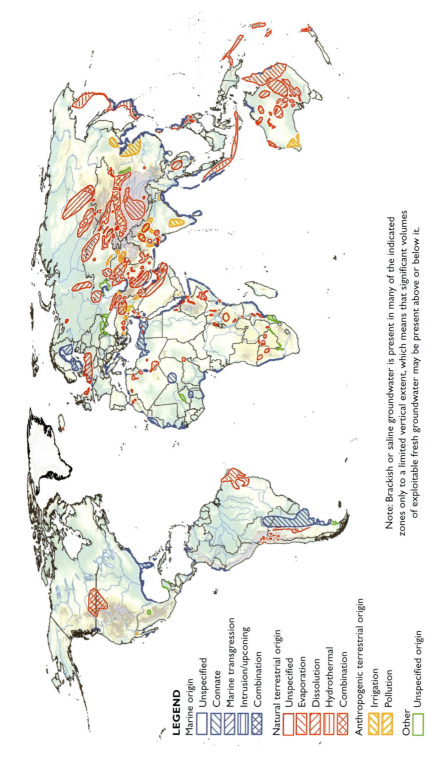

Figure 4.2 Inventoried zones where brackish or saline groundwater occurs or is likely to occur at shallow to moderate depths (< 500 m). [*Source:* IGRAC (Van Weert *et al.*, 2009)].

Note: Brackish or saline groundwater is present in many of the indicated zones only to a limited vertical extent, which means that significant volumes of exploitable fresh groundwater may be present above or below it.

LEGEND

Marine origin
- Unspecified
- Connate
- Marine transgression
- Intrusion/upconing
- Combination

Natural terrestrial origin
- Unspecified
- Evaporation
- Dissolution
- Hydrothermal
- Combination

Anthropogenic terrestrial origin
- Irrigation
- Pollution

Other
- Unspecified origin

formations, membrane effects (salt filtering, ultra-filtration or hyper-filtration), and hydrothermal igneous activities (juvenile water). Anthropogenic causes of salinization are irrigation and pollution by numerous types of waste. The position and vertical extent of the saline or brackish body of water inside an aquifer depends on its genesis. For example, relict saline water from marine transgressions and inundations are likely to be found in the shallow part of the subsurface, and the same is true for salinization by irrigation and by most types of anthropogenic pollution. Seawater intrusion wedges and saline groundwater produced by membrane effects, on the other hand, are usually found in deeper zones, below fresh groundwater.

The content of dissolved minerals limits the suitability of slightly brackish water for drinking purposes and many other types of human water use. Saline groundwater and brackish groundwater with a mineral content above 5 000 milligram per litre are even unfit for almost all uses. Most of the saline and brackish groundwater bodies are rather stagnant. Seawater intrusion and anthropogenic terrestrial salinization are accompanied by movement of saline water, but even in these cases the fluxes are relatively small.

4.3.2 Arsenic in groundwater

Arsenic (As) is usually present in very low concentrations in groundwater. However, the WHO provisional guideline for drinking water of 1993 recommends that the water to be consumed should contain no more than 10 microgram arsenic per litre, a value that has replaced the previous standard of 50 microgram per litre. Excessive arsenic in drinking water poses a severe threat to human health, among others due to its carcinogenic effects, especially in cases of long-term exposure. High arsenic contents in irrigation water are likely to lead to higher levels of arsenic in the edible parts of crops, but the associated risks have not yet been studied in detail.

Knowledge on the occurrence and behaviour of arsenic in groundwater has expanded rapidly over the last few decades. Elements of a summary paper by Smedley (2008) are presented below. Arsenic can be mobilised naturally in water and soils through weathering reactions and microbiological activity. This is generally considered to be the dominant mechanism. Arsenic mobilisation can also be initiated or acerbated by anthropogenic activities such as metal mining, groundwater abstraction and the use of pesticides.

The worldwide distribution of documented cases of excessive arsenic in groundwater (here taken as exceeding 50 microgram per litre) is shown in Figure 4.3. Such excessive concentrations tend to occur mainly in three particular settings:

- *Areas of metalliferous mineralisation and related mining* (particularly in connection with gold): Examples are the Rhon Phibun District in Thailand (arsenopyrite and pyrite-rich waste piles from mining activities) and the Yatenga and Zondoma provinces in northern Burkina Faso (gold mining).
- *Areas of geothermal activity*: located at active plate margins (e.g. the Pacific Rim), continental hotspots (e.g. Yellowstone), oceanic hotspots (e.g. Hawaii) and within plate rift zones (e.g. the East African Rift).
- *Major alluvial/deltaic plains and inland basins* composed of young (Quaternary) sediments, under either considerably reducing or oxic and high-pH conditions. Well-known examples are the Bengal Basin (Bangladesh) and the Huhhoth Basin in Inner Mongolia (China), both under reducing conditions, and the Chaco-Pampanean Plain (Argentina) where alkaline oxidising conditions trigger the mobilisation of arsenic.

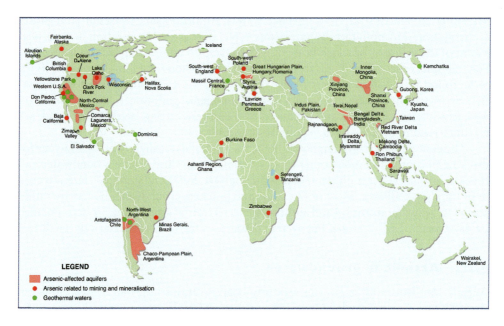

Figure 4.3 Summarised world distribution of documented problems of excessive arsenic in ground-water and the environment (concentrations higher than 50 microgram per litre). [*Source*: Smedley (2008). Reproduced with the permission of the British Geological Survey © NERC. All rights reserved].

Except for mining, it is not yet fully understood to what extent human activities (such as groundwater pumping) may contribute to arsenic problems in aquifers.

4.3.3   Fluoride in groundwater

Fluoride is another potentially hazardous natural constituent of groundwater. In spite of a significant mitigation effect of fluoride against dental caries if its concentration in drinking water is 1 milligram per litre, the WHO guideline value of 2004 for maximum fluoride content in drinking water is 1.5 milligram per litre. Continued consumption of drinking water with higher fluoride concentrations may cause dental fluorosis and in extreme cases even skeletal fluorosis (Feenstra *et al.*, 2007). Freeze and Lehr (2009) describe in detail the related persistent debate about fluoridation of drinking water in the USA.

High fluoride concentrations in groundwater are closely correlated to geology and climate, which motivated IGRAC to prepare a tentative world map showing the probability of occurrence of elevated fluoride in groundwater (Brunt *et al.*, 2004; Vasak *et al.*, 2008). According to these authors, high fluoride concentrations are generally associated with sodium-bicarbonate water found in weathered alkaline igneous or metamorphous rocks, in coastal aquifers where cation-exchange is observed or in aquifers affected by evaporation. In dry climates, groundwater fluxes are usually

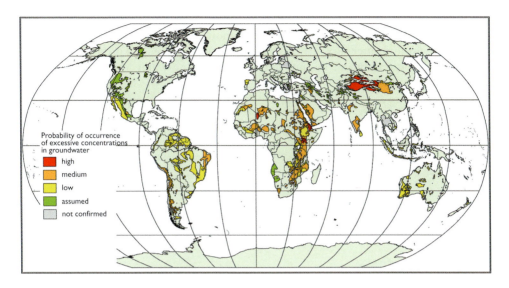

Figure 4.4 Global picture of likely occurrence of excessive fluoride in groundwater. [*Source*: IGRAC (Vasak, Brunt and Griffioen, 2006)].

too small to produce much dilution of the concentrations and evaporation may even aggravate the situation.

This map is shown in Figure 4.4. It was based on a classification of probability of fluoride occurrence on the basis of geological and climatological criteria, as outlined above. Furthermore, reported cases from 44 countries on excessive fluoride in groundwater were used, the geological map of the world (CGMW and UNESCO, 2000) and climate maps showing the degree of aridity. Regions of medium to high probability of excessive fluoride are abundant in South America (the Andes region and Eastern Brazil), Asia (North-western and central China, Northwestern and Southern India, Arabian Peninsula) and Africa (Rift Valley, Western Africa).

High fluoride in groundwater need not have an exclusively natural origin. Use of phosphatic fertilizers in agriculture and industrial activities such as the ceramic industry and burning coal may also contribute to the fluoride content.

4.4 GROUNDWATER-RELATED ECOSYSTEMS

Groundwater plays an important role in sustaining certain types of terrestrial, aquatic and coastal ecosystems – which produces *in situ* benefits but at the same time poses restrictions on exploitable groundwater resources. These ecosystems include wetlands, terrestrial vegetation (depending on shallow groundwater), terrestrial fauna (using groundwater as a source of drinking water), springs and rivers (baseflows), as well as estuarine and near-shore marine systems (mangroves, salt marshes, sea grass beds, marine fauna). Figure 4.5 shows typical examples of groundwater-dependent ecosystems, while an illustrative picture is presented in Figure 4.6. The dependency of ecosystems on groundwater is based on one or more of the following basic groundwater features: groundwater flux, groundwater level or pressure, and groundwater quality. The role of groundwater lies not only in maintaining water levels and supplying

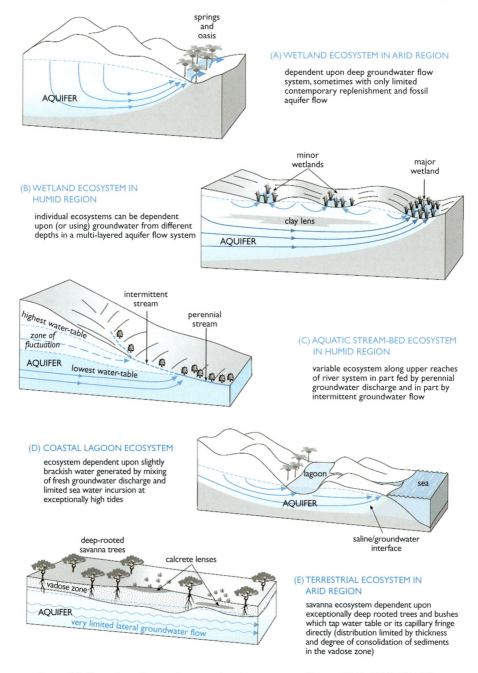

Figure 4.5 Examples of groundwater-related ecosystems. [*Source*: GW-MATE (2006)].

quantities of water, but also often in regulating the thermal and chemical regimes of the ecosystems.

Some groundwater-related ecosystems are fully groundwater-dependent, while other ones only use groundwater but would not disappear without it (Gibert *et al.*, 1999). Although groundwater may be crucial for the sustainability of certain ecosystems, it is usually only one of the relevant factors. The response of ecosystems to change in the conditions of the above-mentioned groundwater features is variable. In some cases there is a threshold response, whereby an ecosystem collapses completely if conditions change beyond a critical level. In other cases there is a more gradual change in the health, composition or ecological function of the ecosystem considered. It is the general experience in many countries that groundwater-dependent ecosystems are quickly affected by the changes in groundwater regime that have been produced by intensive groundwater withdrawal since the middle of the twentieth century. Regeneration of ecosystems that have degraded or disappeared due to changes in groundwater regimes is often very difficult. Desiccation of peat may even trigger irreversible damage by peat-moor fires.

There are numerous groundwater-related ecosystems around the world, in humid and dry climates alike, the large majority of them too small to be mapped on a global scale. The Ramsar list of 'Wetlands of International Importance' (Ramsar, 2009) includes many groundwater-dependent ecosystems, although they are not explicitly classified according to their dependency on groundwater.

Finally, more frequently than we may generally think, groundwater itself can be considered as an underground aquatic ecosystem. Box 4.2 explains that this includes much more than only caves in karst aquifer systems.

Figure 4.6 Wetland ecosystem (springs and oasis) in semi-arid North Horr, Kenya. [*Photo:* Slavek Vasak].

Box 4.2: Aquifers as underground aquatic ecosystems?

Although less obvious and less rich than in running or stagnant surface water, life is not absent in groundwater, in spite of severe ecological constraints. It is not limited to what lives in karstic caves, such as the famous Proteus discovered in the seventeenth century in Carniola (Slovenia).

Geo-microbiological and hereto-biological investigations during recent decades have revealed that many aquifers, especially alluvial aquifers connected to streams, contain a diverse microfauna and even macrofauna, such as bacteria, worms, crustaceans, insects and molluscs (Gibert et al., 1994). According to Griebler and Lueders (2009), the enormous volume of the saturated terrestrial underground forms the largest habitat for micro-organisms on earth. It harbours a degree of microbiological diversity only marginally explored to date, although first observations of groundwater microbiodata date back to Antonie van Leeuwenhoek in 1667.

These subsurface organisms – living below land surface, in the ground, in caves or in groundwater – may reveal connections with surface water and like physical and chemical characteristics they serve as biotic indicators of groundwater quality. Moreover, they play a role in groundwater self-purification (by degradation of contaminants).

4.5 GEOGRAPHY OF GROUNDWATER RESOURCES

4.5.1 Global distribution

The global distribution of exploitable and usable groundwater resources depends in the first place on two natural factors: climate and hydrogeological conditions (see chapters 2 and 3). However, it also depends – as outlined above – on the chosen exploitation strategies, on water quality and on the degree to which groundwater and surface water are connected and interdependent.

From this perspective, a preliminary global picture can be based on the distribution by climatic zones:

- *Groundwater resources in the humid temperate zone* are mainly renewable, very widespread, often abundant, accessible and generally of adequate quality for most uses. Despite some local water quality problems (hardness, iron and manganese, fluoride, salinity of some deep aquifer zones), they are competitive with surface water for a diversity of uses. They are strongly linked to permanent surface water resources and for that reason the management of all water resources should be at least coordinated but preferably integrated. Aquifers of sufficiently large storage capacity are effective multi-annual buffers and also offer significant opportunities to mitigate periodic shortages of rainfall (droughts). In that respect, they can compete with surface water storage reservoirs, but without requiring comparable invest- ments and provisions. Deep groundwater in sedimentary basins offers resources of secondary importance that are less actively recharged. However, these are more independent from surface water and in some areas they may be attractive because of their high productivity, their lack of sensitivity to pollution generated at the land surface and sometimes even their artesian conditions that make exploitation financially and technically favourable during the initial phase of development.

- Renewable groundwater resources *in the arid and semi-arid zone* are less widespread and of variable but generally limited magnitude. However, they are often the only local permanent water resources, except when rivers carry water from elsewhere (external surface water resources, such as supplied by the Euphrates, Indus, Nile, etc.). They are thus an exclusive source for a wide range of uses, in spite of frequent quality limitations (salinity). The groundwater resources are also dependent on surface water, but in these climates they are its 'by-product' rather than its buffer. This is also true for alluvial aquifers along rivers that come from more humid areas, if groundwater and surface water have their natural discharge in separate ways. Unlike in humid climates, the share of surface water recharging the aquifers largely defines the groundwater resources and thus this share should be subtracted in the assessment of surface water resources, except in cases where this surface water is used first before it contributes to aquifer recharge.

These are also the areas, more than elsewhere, where aquifers, through appropriate arrangements such as artificial recharge, may contribute to better control of irregular and scarce surface water resources – which, again, underlines the importance of integrated management of water resources. It is also in the arid and semi-arid zones that non-renewable resources of large aquifers are of crucial importance. These resources are huge in some countries and they pose special management problems, touched upon already in Section 4.2 (under 'mining strategy') and discussed in more detail in Section 4.6. The geographical distribution of the various types of groundwater resources predominant in the arid zone is depicted in Figure 4.7.

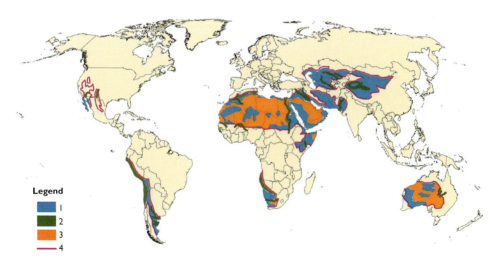

Figure 4.7 Distribution of the main types of groundwater resources in the arid zone *1. Very poor local internal renewable resources, generally dependent on periodic local runoff. 2. Local renewable resources dependent on streams of external origin (including 'secondary' resources). Alluvial or piedmont aquifers. 3. Non-renewable resources – deep aquifers in sedimentary basins (fossil water). 4. Limits of the arid zone.* [*Source:* UNESCO, see Figure 2.14].

- *In the humid tropical and equatorial zone,* groundwater resources are often abundant, but are more irregularly distributed because of the occurrence of geological formations that are ancient and typically constitute aquifers with limited storage and transmissive properties. They can only compete with surface water for meeting spatially scattered demand (rural water supply) – especially because of their generally better quality. Sometimes they are also used to meet more concentrated demand, because they are rarely subject to a constraint related to conserving surface water since the latter tends to be more than abundant.
- Finally, groundwater resources *in the arctic and subarctic zones* of the Northern Hemisphere are highly restricted, even if aquifers are present, due to widespread continuous permafrost conditions. These permafrost conditions may reach down to several hundred meters of depth, impede recharge and restrict access to groundwater. This is the case in Siberia in particular, where thick and extensive sedimentary aquifers exist (see Figure 3.4), but containing mostly saline water (Zektser and Everett, 2004).

Table 4.1 Groundwater resources of the Mediterranean countries.

Country	(1) Total groundwater flux (internal plus external) in km³/year [a]	(2) Estimated exploitable renewable groundwater resources in km³/year [b]	(3) Ratio (1)/(2) in %
Spain	29.9	4.5	15
France	100	30	30
Italy	43	13	30
Malta	0.05	0.015	30
Greece	10.3	2.45	24
Cyprus	0.41	0.2	49
Turkey	69	12	17
Syria	5.4	3.8	70
Lebanon	3.2	0.685	21
Israel	1.075	1.0	93
Palestine:			
West Bank	0.68	0.54	79
Gaza	[c] 0.056	0.05	89
Egypt	[c] 2.3	[c] 0.7	30
Libya	0.6	0.5	83
Tunisia	1.21	1.15	95
Algeria	1.5	~1	67
Morocco	10	4	40

Source: Plan Bleu (2004), by compilation of national sources.
Explanatory remarks:
[a] *'Groundwater produced internally'* + *'External Groundwater inflow'* as in FAO's AQUASTAT.
Note: External groundwater resources, produced by transboundary groundwater flow, are only significant in Egypt, Israel and Syria.
[b] Based on a non-homogeneous set of criteria, tailor-made for each of the countries.
[c] Additional recharge by irrigation return flows or by leakage from distribution networks (secondary resources) not included. In Egypt, the exploitable and effectively exploited resources are close to 7 km³/year (Attia, 2005).

4.5.2   Renewable groundwater resources by country: from abundance to scarcity

The internal groundwater flux of a country gives an indication of its renewable groundwater resources, but usually this flux overestimates its exploitable groundwater resources.

Estimates of exploitable groundwater resources are made in too few countries and using too different criteria to allow a consistent and fairly comprehensive global picture to be drawn, comparable to that of the natural fluxes (Figures 2.6 and 2.8). The estimated exploitable resources are only a fraction of these 'theoretical' natural resources or internal groundwater flux (referred to as *'Groundwater Produced Internally'* in FAO's AQUASTAT). As Table 4.1 shows, the lower these theoretical natural resources, the higher this fraction tends to be.

The concept of exploitable groundwater resources has some similarity with that of *safe yield* (Meinzer, 1920). The latter is defined as the amount of water that can be withdrawn from an aquifer annually without producing an undesired result. Originally, this potential undesired result was interpreted as synonymous with storage depletion, and hence sustainability of pumping was the only criterion. For a long time, most hydrogeologists have persisted in the opinion that safe yield according to this basic interpretation equals natural recharge, but such a view ignores the dynamics of a groundwater system. After all, the groundwater balance equation dictates that the formal condition for zero storage depletion is that groundwater abstraction is balanced by the sum of increase in recharge and decrease in discharge as induced by pumping (Bredehoeft *et al.*, 1982). At the level of an aquifer, the maximum abstraction rate that can be sustained without causing progressive storage depletion can be conveniently defined by modelling. Groundwater recharge is only a proxy for this form of safe yield and often overestimates its value. Depending on country and on investigators, other criteria have often been added, while over time the term safe yield has gradually made way for the term *sustainable yield*. These added criteria include economic and environmental criteria, in particular constraints related to the conservation of surface water and groundwater-dependent wetlands. These wider interpretations also underly the concept of overexploitation (see Custodio, 1993 and 2000; and Section 6.3).

The methods and criteria used for assessing exploitable renewable groundwater resources obviously vary by country, which together with the diversity of natural conditions results in a broad spectrum of estimates. These correspond to fractions of the natural resources (internal or total groundwater fluxes) ranging from a few percent to one hundred percent[1]. This is illustrated by the selected data presented in the Tables 4.1 and 4.2, with the caveat that this does not appear to have been produced by comparable or consistent approaches. A more homogeneous global overview of the exploitable renewable groundwater resources remains to be developed on the basis of criteria that are more explicitly specified, but it is unlikely that they will ever form a uniform data set.

---

1 In some cases (e.g. if groundwater abstraction causes large induced recharge from surface water bodies) it is even possible that they are equivalent to more than 100% of the natural recharge.

Table 4.2 Exploitable renewable groundwater resources of selected countries. Selected national data, based on a non-homogeneous set of criteria, tailor-made for each of the countries (complementary to the Mediterranean countries covered already in Table 4.1).

Country	Estimated mean annual exploitable water resources in km³/year	Part of the natural groundwater flux (see AQUASTAT) in %	References
AFRICA			
Botswana	1	59	FAO 1995
Ethiopia	2.6	13	FAO 1995
Kenya	0.6	17	FAO 1995
Mauritania	0.88	44	Shahin 1989
Sudan	0.7	10	FAO 1995
Zimbabwe	1 to 2	17 to 33	FAO 1995
AMERICA			
Dominican Republic	0.5	33	ICID 2000
Uruguay	2	9	ICID 2000
USA (contiguous states)	106	8	UN-ECE 1981
ASIA			
China	290	33	Zektser & Everett 2004; Li 2001
India	396	92	ICID 2000
Japan	~ 10	36	UN/ESCAP 1993
Pakistan	47	76	Kahlown & Majeed 2002
Russia (Asian part)	8.94	3	Zektser & Everett 2004
EUROPE			
Denmark	2	51	Eurostat 1998
Iceland	6	25	Eurostat 1998
Ireland	3	28	Eurostat 1998
Netherlands	1.9	42	UN-ECE 1981
Romania	4.5	54	UN-ECE 1981
United Kingdom	7	71	Eurostat 1998
Russia (European part)	317	40	Caponenco, 2000
Sweden	3	15	Eurostat 1998
Ukraine	5.7	29	ICID 2000
United Kingdom	7	71	Eurostat 1998
Australia	26	36	Commonwe. of Australia 2003

4.6   THE PARTICULAR CASE OF NON-RENEWABLE GROUNDWATER RESOURCES

4.6.1   What does renewability or non-renewability of water resources mean?

In nature, water both flows and accumulates. Accumulation is especially evident if we focus on groundwater. Aquifers convey water but are at the same time water reservoirs, and hence the amounts of their water can be defined in terms of flow and stock (see Chapter 2). The renewal of water in a natural system, in particular an aquifer and the resources it presents, may be conceptually understood by comparing flux and stock of the system with the help of two indicators that are more or less the inverse of each other:

a *Mean annual renewal ratio*, defined as mean annual flux volume divided by the stock (fraction or in%)
b *Mean renewal time*, defined as stock divided by mean annual flux (in years)[2]

Given the extreme range in magnitude of flows and stocks of aquifers in the world, there is very large variation in renewal ratios and renewal times. The annual renewal ratio varies from more than 100% to less than 0.001%, which corresponds with mean renewal times (mean residence times) ranging from less than one year to more than 100 000 years. A low renewal ratio may have different causes: very limited recharge (in the arid zone or in the case of completely confined aquifers), a very large volume of stored groundwater, or a combination of both.

A low mean renewal ratio – and thus a long mean renewal time – implies that a substantial part or even most of the water in storage is ancient, which usually means that it entered the aquifer before the modern water cycle was in place. Such water is commonly referred to as 'fossil' water.

The groundwater resources of an aquifer may be called non-renewable[3] in the particular case that the renewal of groundwater is very limited, while its stock is large. It does not mean that such an aquifer does not receive any recharge at all, or that it is entirely isolated from systems that do contain renewable water resources. 'Non-renewability' of groundwater in an absolute sense – which corresponds to zero recharge and zero mean renewal ratio – is rare. Groundwater specialists use the concept of 'non-renewable groundwater resources' in a relative sense. It only implies that the mean renewal time is very long compared to that in aquifers with renewable resources, hence that – in contrast to renewable groundwater resources – it does make sense to base their exploitation on the stock rather than on the flux. The mean annual renewal ratio may be used as a criterion whether to call groundwater non-renewable or not. Given, however, the many factors at play, there is no single value that will be satisfactory in all cases. Some hydrogeologists qualify groundwater resources as non-renewable if the mean annual renewal ratio is less than 0.1% (mean renewal time in the order of at least one thousand years); others put the upper limit at even much smaller values. The annual renewal ratio criterion should preferably be combined with a criterion on the mean annual recharge (e.g. mean recharge less than 5 mm per year), in order to prevent fairly to well recharged aquifers with large stored volume being classified as 'non-renewable'.

To what extent non-renewable groundwater is an exploitable resource depends on the technical and economic feasibility of its exploitation, which in this case means mining. It should be possible to extract from the reserve for a sufficiently long period of time – on a human scale – a quantity of water much higher than what would be

---

2 The term 'renewal time' (or: 'mean residence time', 'turnover time') assumes first-order stochastic stationarity, i.e. the long-term mean values of flux and stock are supposed to remain constant over time.

3 Renewable or renewed? The commonly used qualification 'renewable' should refer to the potential for water to be abstracted from the aquifer, rather than related to the resource. It is more correct to use 'non-renewed' instead of 'non-renewable'.

obtained if only the average flux of the aquifer (or part of it) were intercepted. The non-renewable groundwater resources are thus defined, like other non-renewable mineral raw materials, in terms of extractable stock, even though this stock can be converted to a mean productive flux during a defined finite period. The term 'extractable' refers to technical and economic feasibility criteria and implies the absence of significant external impacts.

This extractable reserve, converted to an exploitable resource for a defined limited period of time, is usually much less than the total reserve of an aquifer. In particular, the huge reserves of very large aquifers in the arid or semiarid zone, including 100 000 km³ present below a number of deserts, as cited by Issar and Nativ (1998), are impressive compared to what surface water reservoirs store. However, these reserves can only be partially recovered: it is impossible to capture all the water stored in an aquifer in much the same way that it is impossible to extract all the oil stored in an oil reservoir. Only their extractable fraction is an exploitable non-renewable resource, and the latter may even be subject to additional constraints.

The main constraint is the decline of water levels caused by withdrawal. Such declines may reach several hundreds of metres after having withdrawn only a small fraction of total water volume. This happens especially in the case of confined groundwater, where exploitation tends to produce a marked drop in pressure. Extractable volumes of groundwater are often in the order of a hundredth of the volumes stored.

Developing non-renewable groundwater resources requires aquifers with considerable exploitable reserves that can be withdrawn without causing a significant impact on surface water or on renewable groundwater resources. Particularly in the arid zone – where internal renewable water resources are very low or zero – such options do exist and do produce their benefits.

The evaluation of these resources is inseparably linked to the choice of an exploitation plan for the medium or long term. This includes the choice of the desired duration of the exploitation, in other words the selected management strategy, and is similar to plans for exploiting a deposit of mineral raw materials (oil, minerals).

4.6.2   Geographical distribution

The distribution of non-renewable groundwater resources on Earth is controlled by two major factors:

* *Structural hydrogeological conditions*: existence of aquifers of high storage capacity and low exposure to factors that favour recharge. These conditions correspond mainly to deep and extensive confined aquifers, usually hydraulically isolated from more active parts of the water cycle by aquitards.
* *Climatic conditions*: aridity, which generally causes low inputs to all water systems (surface water and groundwater) and therefore tends to reduce recharge of aquifers in all categories.

The distribution of non-renewable groundwater resources does not simply coincide with the major aquifers located mainly in large sedimentary basins (see Chapter 3). Their distribution corresponds essentially to that of deep aquifers inside the sedimentary basins in the arid and semi-arid zones. This is also where they are more important than elsewhere, because renewable water resources are scarce and water demands high in climatologically dry regions. In addition, their exploitation is likely to interfere less with renewable water resources than groundwater withdrawal of similar intensity elsewhere. Non-renewable groundwater resources have already been most intensely exploited in these arid and semi-arid regions for the last few decades, (e.g. in the Arabian region) and in several countries they are the main source of water supply.

Incidentally, other conditions may prevent aquifers from being recharged. Widespread occurrence of permafrost in arctic regions causes groundwater in several sedimentary basins of Siberia to be non-renewable.

Non-renewable groundwater resources are not absent in the humid zones. However, their relative importance is less, except that in some basins deep water can temporarily be withdrawn by flowing wells, due to artesian pressure conditions. The risk that their exploitation has impacts on renewable resources is greater than in dry regions.

Well-known aquifers containing non-renewable resources, where usually groundwater mining is already ongoing, are found mainly in Africa (Sahara and, to a less extent, Southern Africa), on the Arabian Peninsula, in Australia (Great Artesian Basin) and in Siberia (Russia). In the Great Artesian Basin the discharge of all flowing artesian wells combined has declined markedly since 1918, in spite of the fact that the number of wells has more than doubled since then. The decline has been caused by steady reduction in artesian pressure; recent management activities are now attempting to recover pressure and reduce losses of water discharged by uncontrolled artesian wells. Less explored and less studied non-renewable groundwater is found in the southern Sahara and the Sahel (from Senegal to Chad), as well as in East Africa (see Table 4.3, Table 4.4 and the sketch map of Figure 4.8).

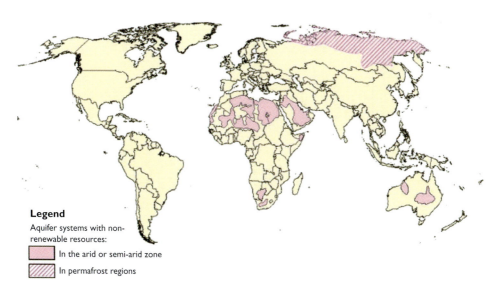

Legend

Aquifer systems with non-renewable resources:

- In the arid or semi-arid zone
- In permafrost regions

Figure 4.8 Non-renewable groundwater resources: main occurrences on Earth. [*Sources:* Zektser (1998); WHYMAP (2008)].

Table 4.3 Selection of large aquifer systems in the world with non-renewable groundwater.

Code (See Fig 3.5 and Table 3.2)	Aquifer	Countries	Size in thousands of km²	Volume of the reserves		Estimated mean annual renewal ratio	Exploitation	References
				Theoretical total (km³)	Estimated exploitable (km³)			
1	Nubian Sandstone Aquifer system	Egypt Libya Sudan Chad	2 176	542 000 (freshwater)	Egypt: 5 367 Libya: 4 850 Sudan: 2 610 Chad: 1 630	$1.3.10^{-5}$	1.6 km³/yr in 2000; ~ 40 km³ abstracted during 1960–2000	OSS 1998; Khouri 1990; Bakhbakhi 2002; CEDARE/IFAD 2002; Foster & Loucks 2006
2	North Western Sahara Aquifer System (NWSAS), multilayer	Algeria Libya Tunisia	1 019	60 000	1 280 (CI: 780; CT: 500)	$1.4.10^{-5}$	2.5 km³/yr in 2000; of which 0.46 km³/yr in Tunisia and 1.68 km³/yr in Algeria	UNESCO 1972; OSS 2003; Mamou 1999; Pallas & Salem 1999;
3	Murzuk Basin, multilayer	Algeria Libya Niger	450	4 800 in Libya	Libya: 60–80		1.75 km³/yr in 2000 in Libya	El Gheriani 2002; Pallas & Salem 1999
5	SenegaloMauritanian (Maestrichtien aquifer)	Mauritania Senegal Guinea Bissau	300	300 to 400 in Senegal	Mauritania: 3 Senegal: 10–20	9.10^{-5}	~ 10 million m³/yr in the 1970s	Diagana 2005
6	Iullemeden Aquifer System (IAS) (Continental Intercalaire and Continental Terminal)	Mali Niger Nigeria	635	10 000 to 15 000	Niger: 250–550 (CI; 40–100 m of depth) Mali: ~ 100	5 to 8.10^{-5}	~ 0.1 km³/yr in 2000 (Niger and Nigeria)	BRGM 1976; Dodo 1992; UNESCO/Project IAS 2003
7	Lake Chad Basin, multilayer	Niger Nigeria Chad Cameroun	1 917		Niger: 45 Chad: 220			BRGM 1976; Terap & Kabaina 1992; CBLT/OSS 2005
12,13 (partly)	Karoo/Ntane Sandstone, Central Kalahari	Botswana	90	275	86	4.10^{-5}	Potential: 2.89 km³/yr 'extractable mining' for 30 years	Foster & Loucks 2006 Carlsson et al. 1993

No.	Aquifer	Country	Area (10³ km²)	Volume / depth		Rate	Abstraction	References
22	Saq Aquifer and Arabian platform multilayer aquifer system	Saudi Arabia, Bahrain, Qatar, UAE	1 485 (in Saudi Arabia)	~ 500 to 2 185 until 300 m in depth; Qatar: 2.5 UAE: 5		3.10^{-5}	13.5 km^3/yr ~ 1995 in Saudi Arabia, 0.14 km^3/yr in Qatar, ~ 1.5 km^3/yr in UAE, 0.09 in Bahrain	Min. of Agriculture and Water of Saudi Arabia 1985; Neuland 1988; Ghurm Alghamdi 2002; FAO 1997; Foster & Loucks 2006
22 (partly)	Disi Aquifer	Jordan	~ 3	6.25			0.17 km^3/yr in 1994, predicted 0.125 km^3 for 50 years	Garber & Salameh 1992
—	Basin and Range Aquifers (Lowlands of Arizona)	USA	Σ ~ 60	1 500 (until 366 m in depth)		$2.5.10^{-4}$	7 km^3/yr in 1975; 225 km^3 abstracted between 1920 and 1980, including 202 km^3 of storage depletion	MacGuinness 1963; Foster 1977; USWRC 1983
36	Great Artesian Basin, multilayer aquifer system	Australia	1 700	~ 20 000	~ 170	5.10^{-5}	35 km^3 abstracted between 1880 and 1973, including 2.5 km^3 of storage depletion. Peak years: 0.75 km^3/yr in 1917 0.6 km^3/yr in 2000	Habermehl 1980, 2001, 2002
25	West Siberian Artesian Basin (partly under permafrost)	Russia (Siberia)	3 200	1 000 000		$5.5.10^{-5}$		UN 1986

Table 4.4 Non-renewable groundwater resources assessed by country. *Source:* Foster and Loucks (2006).

Country	Aquifer	Estimated volume of exploitable reserves in km³*	Current rate of exploitation in km³/year** (reference year)
Algeria	North Western Sahara Aquifer System (NWSAS)	900	1.68 (2000)
Australia	Great Artesian Basin	170	0.63
Botswana	Karoo Sandstone Central Kalahari	86	–
Chad	Nubian Sandstone Aquifer System	1 630 ⎫ 1 850	- ε
	Chad Basin Aquifer System		
	(CT Hamadien + CI)	220 ⎭	- 0.11 (2000)
Egypt	Nubian Sandstone Aquifer System (including post-Nubian Aquifer System)	5 367	0.9
Israel	Nubian Sandstone Aquifer / Negev	(total: 20)	0.05 (2000)
Jordan	Disi Aquifer (Saq)	6.25	0.35 (1998)
Libya	Nubian Sandstone Aquifer System (including post-Nubian Aquifer System)	4 850 ⎫	0.9 ⎫
	North Western Sahara Aquifer System (NWSAS)	250 ⎬ 5 170	0.55 ⎬ 3.2 (2000)
	Murzuk Basin	60–80 ⎭	1.75 ⎭
Mali	Iullemeden Multilayer Aquifer System (CI)	~ 100	0.0013 (2000)
	Taoudeni – Tanezrouft Basin	–	0.2 (2000)
Mauritania	Maestrichtien Aquifer	~ 3	0.09 (2003)
	Taoudeni – Tanezrouft Basin	–	ε
Niger	Iullemeden Multilayered Aquifer System (CI)	250–550	0.008 (2000)
	Chad Basin Aquifer System (CT Hamadien) and Manga Pliocene	45	ε
Qatar	Arabian Platform Multilayer Aquifer System	2.5	0.14 (1995)
Saudi Arabia	Arabian Platform Multilayer Aquifer System (including Saq Aquifer)	500–2 000	3.5 (1995) 20.47 (2000)
Senegal	Maestrichtien Aquifer	10–20	0.18 (2003)
Sudan	Nubian Sandstone Aquifer System	2 610	0.41 (1998)
Tunisia	North Western Sahara Aquifer System (NWSAS)	130	0.46 (2000)
United Arab Emirates	Arabian Multilayer Aquifer System	5	1.57 (1995–1996)

* Each country evidently uses its own assessment criteria; hence, these criteria vary.
** The symbol ε means an insignificantly small rate.

Aquifers with both valuable renewable resources and large reserves are sometimes subject to prolonged intensive exploitation causing significant groundwater depletion, which in practice may be difficult to reverse. Examples are the Ogallala aquifer of the High Plains region (Mid-West USA), the Basins and Range Lowland aquifers of Utah, Nevada and Arizona (USA) and the multi-layer aquifer of the Heilonggan basin (Huang Huai Hai plain, China). However, these aquifers do not meet the criteria for classifying them as aquifers with exploitable non-renewable groundwater resources. That is why they are not included in this overview.

Finally, it should be emphasised that large aquifers with non-renewable groundwater resources are often crossed by international boundaries and thus are transboundary aquifers, especially in Africa (Sahara) and the Middle East. For assessing the exploitable non-renewable resources of such aquifers, the national fractions of these resources (as shown in Table 4.4) have to be summed. If an aquifer is transboundary, then this imposes special challenges to its management, to be addressed by international consultations, co-ordination and co-operation (see also Chapter 7).

4.7   HOW TO GET ACCESS TO GROUNDWATER AND WITHDRAW IT

4.7.1   A diversity of techniques

The exploitable groundwater resources cannot be assessed properly without taking into account the practicalities of getting access to groundwater and withdrawing it. Both are highly dependent on local aquifer characteristics – decisive for the productivity of the abstraction works – and on the depth where groundwater can be tapped. These conditions are as important as the existing groundwater volume and its recharge. Together with the socio-economic context, they largely dictate how exploitation can best be undertaken. The following is an overview of the related options.

In terms of their nature and function, two main categories of groundwater abstraction works currently in use can be distinguished:

- *Gravity-based abstraction works*. This category includes spring capturing works, drains, infiltration galleries, flowing wells (artesian wells) and subsurface dams. They offer the advantage that no energy supply is required for their operation, but they are subject to site constraints and they are less easy to control (e.g. by valves). This is why they often flow permanently, thus producing discharge beyond demand, which means that water is wasted. In fact, artesian wells are an intermediate category, in the sense that their output decreases in response to the pressure declines they cause. Artesian wells are therefore not sustainable and their usual fate is that sooner or later they will be transformed into pumped wells.
- *Abstraction works that require an external energy source to lift groundwater to the surface*. This category includes dug and drilled wells, usually pumped. Pumping leads to an energy cost, thus to higher production costs than in the case of gravity-based works, but pumped wells offer more flexibility in implementation and operation. They tend to be more efficiently managed because of the need to supply energy and to bear the related energy costs[4].

In hydraulic terms, the works in the first category impose a fixed water level, usually on a permanent basis, and collect water at a variable rate. They can be considered as artificial springs. Those in the latter category impose a flow, usually discontinuous in time and variable according to demands. They produce a relatively pronounced drawdown of the water level inside the well and in the aquifer zone in its immediate surroundings.

---

4 Wind – and solar-powered pumps are exceptions to this.

All abstraction works deplete a fraction of the aquifer's reserves during the initial phase of abstraction, until a new dynamic hydrological equilibrium is established. But those in the second category contribute more to depletion, especially when they are present in large numbers and abstract considerable quantities of groundwater.

The main difference is energy consumption: it is zero for the first category, but necessary and variable for the second one, whatever source of energy is used (human or animal traction, wind, fossil fuel or electricity). Another difference is the operational degree of freedom in controlling flow, which is obviously greater for the second category. Indeed, the yield depends on the local regime of the exploited groundwater system and is like a 'harvest' for the first category, while for the second category it is a decision variable depending on the withdrawal capacity of the device and on how it is operated. Their construction allows the abstraction works of the first category (except for the artesian wells) to withdraw renewable groundwater resources only, without the risk of causing a permanent state of non-equilibrium in the aquifer ('hydrological over-abstraction', see Section 6.3 and Box 6.5). However, those in the second category can make better use of the buffer capacity of the aquifer's reserves and control the abstraction accordingly. They do not offer protection against the risk of hydrological overabstraction and are also capable of exploiting non-renewable groundwater resources.

4.7.2   Factors influencing the type of abstraction work to be chosen

The different types of groundwater abstraction work are not adapted to the same degree to the wide variety of natural hydrogeological conditions (outlined in Chapter 3). This aspect is briefly reviewed in the following sections for wells, for infiltration galleries and for subsurface dams:

- *Wells*

Dug and drilled wells (see Figures 4.9 and 4.10) are the most widespread groundwater abstraction works. They are more suitable than any other type of abstraction works in cases of thick and/or deep aquifers, deep static water levels in unconfined aquifers or – in contrast – artesian water levels in confined aquifers. They tend to be less suitable for abstracting groundwater from shallow aquifers of very small thickness, from thin fresh groundwater layers overlying saline water or from poorly productive and/or discontinuous aquifer formations.

What are the criteria for choosing between a dug well or a drilled one? These are the available technical expertise in the area concerned, the desired depth (dug wells rarely exceed 50 to 100 m, drilled wells may go to 1 000 m in depth or deeper), the properties of the rock to be penetrated (how hard and how cohesive is it?), the anticipated well yield[5], the required contact surface with the aquifer[6], the usefulness

---

5 The hydraulic effect of a larger well diameter is less than often believed. All other parameters remaining equal, well yield is not proportional to its diameter: for a given drawdown in the well, a ten times larger diameter may increase well yield by several tens of percents, but doubling the yield is unusual.

6 This contact surface needs to be larger in poorly permeable or discontinuous aquifers. In such a case, dug wells – and in particular collector wells - are more appropriate.

Figure 4.9 A traditional dug well. [*Photo*: WWC, 1998].

Figure 4.10 Drilling a deep borehole. [*Photo*: Ton Negenman].

of stored water inside the well[7], the construction cost[8], the ease of operation and the protection of the well head (easier for drilled wells), as well as access to the site during well construction[9].

In numerous cases, both types may be competitive. However, when an aquifer is very thick (more than 100 metres) then only a drilled well could penetrate it more or less completely and achieve a maximum productivity, often more than 100 cubic metres per hour, or up to 400–500 cubic metres per hour in the most productive aquifers. In addition, only drilled wells are appropriate in the case of confined aquifers where hydrostatic pressure enables artesian flow.

A special case is formed by *collector wells*. Their design includes a number of horizontal perforated collector pipes, in a radial pattern connected to the main body of the well. This results in an increased contact surface between the well and the aquifer (up to 1 000 m²) which greatly increases the well's productivity, commonly to 500–1 000 cubic metres per hour, but sometimes more (e.g. a collector well in Milan is known to yield 4 500 cubic metres per hour). Such a collector well thus cancels out the need for a larger piece of land to accommodate different wells, wellhouses and pumping facilities. A collector well is particularly appropriate in alluvial aquifers, where it often compares favourably with a group of wells. It may also serve to abstract groundwater at a given rate against minimal drawdown (which is needed in the case of fresh groundwater overlying saline water) or to increase well yield if drawdown is physically limited, such as in the case of aquifers of very limited thickness.

• *Infiltration galleries*

Infiltration galleries are appropriate for groundwater withdrawal from poorly permeable or discontinuous aquifers (limestone, crystalline rocks or fractured volcanics), from various types of aquifers of limited thickness or from those containing a fresh-saline groundwater interface. A modest uniform slope and the absence of significant relief are preferred topographic conditions. Infiltration galleries are suitable in particular situations where it is attractive to avoid energy consumption and where water demands are fairly regular and continuous. In addition, their yield adapts to hydrological conditions, which means that infiltration galleries do not produce a progressive decline in groundwater levels, but favour a dynamic equilibrium between the aquifer's inflows and outflows.

• *Subsurface dams*

Subsurface dams are used mainly in local alluvial aquifers that are of a small width and thickness. Their role is similar to surface water dams and they function by intercepting groundwater flow and/or by storing surplus water underground.

Their passive role is twofold. First of all, they force groundwater to emerge at the surface, where it can be captured like water from a spring. Secondly, they improve the yield of abstraction works in the neighbourhood (wells or infiltration galleries) by

---

7 Stored water inside the well makes the exploitation more flexible and reduces pumping duration. In a poorly permeable aquifer a dug well is therefore preferred.
8 Dug wells are often cheaper due to smaller depth.
9 Obstacles may hamper the transport of heavy drilling equipment.

increasing the saturated thickness of the aquifer. This permits these abstraction works to be perennial, thanks to the accumulated reserves – which is particularly an advantage in the arid zone where even the underflow in river beds may be intermittent.

Subsurface dams offer the same advantages (flow under gravity) and disadvantages (continuous production) as galleries, but there are more restrictions in finding suitable sites.

How groundwater is exploited and the types of abstraction works can also be classified according to the categories of economic actors – individuals or groups – who have the means to implement them, and based on the need for additional equipment and the cost of it. Such a classification corresponds fairly well with the distinction between extensive exploitation (mainly by individuals) versus intensive and concentrated (mainly collective). If one combines the two most important criteria, energy and socio-economic factors, then the modalities of groundwater exploitation can be divided into four groups, as presented in Table 4.5.

4.7.3   Bringing groundwater to the surface: a vital art

The art of capturing groundwater is very old, as testified by wells already present at the dawn of history (6000 BC in Mesopotamia). Due to many inventions it has evolved over time and diversified around the world, while it has benefited at the same time from advances in techniques for prospecting and exploring the subsurface for other purposes. Its evolution is a central theme in the history of technology. From ancient times up until the present industrial era, each of the main techniques for the exploitation of groundwater has been predominant by turn for a period of time.

For millennia, only the techniques that function by gravity (group 'a' in Table 4.5) or by lifting water at a low rate (1b) have been implemented, withdrawing water from shallow aquifers only (phreatic groundwater). Undoubtedly the efforts to capture groundwater were oldest and most noteworthy in arid and semi-arid regions. This is where human ingenuity excelled in developing energy-friendly techniques for lifting water from wells to the surface. In addition to the simple and ubiquitous *rope-and-bag* lifting technique, these include the *shaduf*, the water wheel (*noria* or *saqiyah*[10]), the *chain-and-buckets pump* and the *arhor*[11], each with its own geographical distribution (Figure 4.12).

These are also the regions where large *infiltration galleries* have been dug and implemented, with a total length of hundreds of thousands of kilometres. This technique, inspired by mining activities, was probably invented in Armenia (Kingdom of Urartu) where in the eighth century BC these galleries were called *kankan*. A typical infiltration gallery is constructed as a slightly sloping tunnel, tapping groundwater at its upflow end (often by means of a 'mother well') and conveying the water under gravity – often over large distances – into an open section, from which it can be diverted

---

10 Strictly speaking, noria seems to be the more correct name for the abstraction device, while saqiyah originally refered to the outflow canal (acequia in Spanish).
11 Funnel-shaped water bag, lifted, closed, opened and emptied with support of a double pulley; widespread in North-West Africa.

Table 4.5 Classification of modalities and techniques of groundwater abstraction.

		Economic setting for implementation	
		1 *initiative of individual economic actors (users)*	*2* *predominantly initiative of economic collectives, in particular public agencies*
Hydraulic conditions (energy)	*a* *gravity*	*1a* small infiltration gallery, shallow to medium-deep flowing drilled well, horizontal drain	*2a* large infiltration gallery, network of drains, very deep flowing well (several hundreds of m), subsurface dam
	b *lifting, pumping*	*1b* dug well, collector well, tube well of small to moderate depth + lifting by human or animal traction, or by pumping	*2b* collector wells, group of high-production dug wells (well field), very deep drilled wells + pump

Note: The productivity of each type of abstraction work depends in practice much more on aquifer characteristics than on the properties of these works. This is why the productivity of each type shows substantial variation, as shown in Figure 4.11.

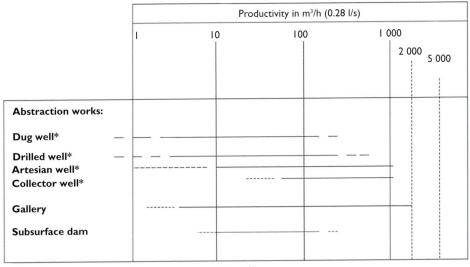

*Assuming that pumping equipment does not restrict yield.

Figure 4.11 Productivity range of different types of groundwater abstraction works.

for its use (see Figures 4.13 and 4.14). Under the name *qanats* they were implemented by the thousands in Iran and throughout the Persian Empire, from which they spread westward as far as the Maghreb and Spain and eastward up to China (Xinjiang) (see Table 4.6 and Figure 4.16). The majority of them are in unconsolidated formations and they have become known under different names, such as foggara and rhettara in Northern Africa, karez in Afghanistan and surrounding countries, ghayl in Yemen and falaj in Oman. Infiltration galleries are also present in the Western hemisphere, notably

(a) Dug well with pulley, rope and bag (Yemen).

(b) Saqiyah or noria (Morocco).

(c) Shaduf (Morocco).

(d) Arhor with double pulley (Morocco).

Figure 4.12 Ancient water-lifting devices for wells. [Sources: Jac van der Gun (a); Jean Margat (b, c and d)].

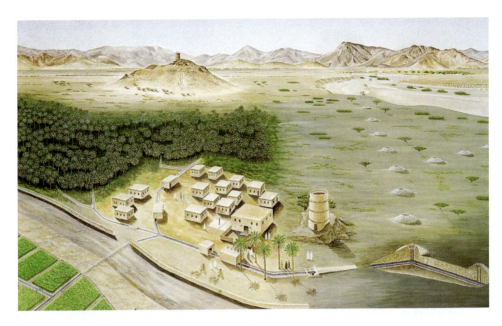

Figure 4.13 Schematic drawing of an infiltration gallery system (falaj) and its use. [*Source*: Oman Geologic Heritage – MRMWR (2008)]. *The main canal is excavated from the source of the falaj in mountains and wadis to cross wide areas as buried canals with holes for airing and maintenance. It discharges into open canals near the village from where it supplies irrigation ditches.*

Figure 4.14 Inside a qanat in Bam, Iran. [*Photo*: Majid Labbaf Khaneiki].

Figure 4.15 Artesian well on the Bolivian Altiplano. [*Photo*: Jac van der Gun]. *The casing extends several metres above ground surface to ensure that no groundwater is spilt after the horizontal discharge pipe is closed.*

Table 4.6. Main infiltration gallery networks around the world.

Country	Local name	Total length* (km)	Number* (and date)	Total yield** (m³/s)	Reference
Iran	Qanat		21 000 oper.		Bahrami 1955
			25 000	380	A. Massoumi 1966
				500	UN/ECAFE
			38 000 (1966)		Ahmadi 2006
			20 000 (1998)		Ahmadi 2006
			18 000		Ghayour 2006
		350 000		500	ICID 1975
		80 000 oper.	18 400	239	Balland 1992
			30 796	280	Hassanzadeh 1992
			34 355 oper.	260	Semsar Yazdi 2009
Afghanistan	Karez	12 000	~ 6 000 (1967)	~ 100	Balland 1992
			6 741		Bank & Soldal 2002
		16 374	8 370		Semsar Yazdi 2009
			(5 984 oper.)		
Kazakhstan	Karez	124	261 abandoned	0	Deom & Sala 2006
Pakistan Baluchistan	Karez		1000 (730 oper.)		Mian & Abdullah 2000
Oman	Falaj (pl.aflaj)		4 112 (3 017 oper.)		MRMWR 2008
			4 112 (3 008 oper.)	14.6	Semsar Yazdi 2009
Syria	Qanat		Several hundreds	20	Goblot 1979
China	Kariz, Kanerjing	~ 5 000	984	11.4	Nasier & Wang 1990
Xinjiang, Tourfan			736	10.8	World Wat. Eng. 1991
			900 (350 oper.)	9.5	Turpan Museum 2006
Algeria Gouara, Touat	Foggara	1 378 (oper.)	777 (oper.)	2.9	SOFRETEN 1963
		1 377	572 (oper.)	3.7	El Hebil 1988
		1 417		3 (1970)	ERESS 1997
Tidikelt		525	175 (1900) 125 (currently)		Goblot 1979
All Sahara				3.6 (1950) 3.1 (2000)	OSS 2003
Morocco Haouz	Rhettara	~ 2 000	500 (~ 1970)	5	Pascon 1977
		700		5	El Hebil 1988
Ziz-Rheris Basin			275	0.86	ORMVAT 2008
Spain	Galería	1 621	1 001 (1980)	5.6	Gonzalez Perez 1987
Tenerife		1 621	1 047	5.16	Plan hidrológico 1989

* oper. = operational
** 1 m³/s = 0.0315 km³/year

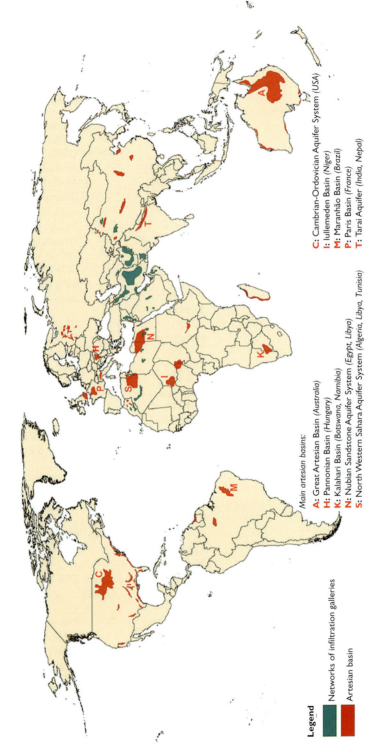

Legend

Networks of infiltration galleries

Artesian basin

Main artesian basins:

A: Great Artesian Basin (Australia)
H: Pannonian Basin (Hungary)
K: Kalahari Basin (Botswana, Namibia)
N: Nubian Sandstone Aquifer System (Egypt, Libya)
S: North Western Sahara Aquifer System (Algeria, Libya, Tunisia)

C: Cambrian-Ordovician Aquifer System (USA)
I: Iullemeden Basin (Niger)
M: Maranhão Basin (Brazil)
P: Paris Basin (France)
T: Tarai Aquifer (India, Nepal)

Figure 4.16 Groundwater withdrawal without energy cost: main infiltration gallery networks (qanats, etc.) and explored artesian basins (with conditions initially allowing flowing wells).

in the USA, Mexico, Ecuador, Peru and Chile (ICQHS, 2009). These 'horizontal wells' have also proved to be suitable for capturing groundwater on volcanic islands (Bermuda, Canary Islands, Madeira, Cape Verde, Hawaii). Many of these works are still in use, despite their gradual disappearance resulting from deteriorating socio-economic conditions on which their feasibility depends, high labour costs for their maintenance, and competition by pumped abstraction. Their current global discharge is difficult to quantify, but it is probably around 500 m³/s, or approximately 16 km³/year.

At the onset of the industrial era, engineers developed unprecedented techniques for the exploration and exploitation of groundwater. New drilling techniques (rotary, 1890; rapid percussion) increased the number of drilled and cased wells, while new pump technology (centrifugal and turbine pumps, 1910–1930) intensified abstractions due to the disseminated use of mechanical power. Groundwater exploitation from greater depths and at higher rates of productivity started in the nineteenth century and expanded enormously during the twentieth century. In several countries, these advances since the mid-nineteenth century have produced as a first spectacular result the discovery and the exploitation of deep and large confined aquifers, allowing artesian wells to be drilled (see Figures 4.15 and 4.16). This relatively modern method of groundwater withdrawal by gravity (category *2a* in Table 4.5) is facilitated by often very high initial water pressures, causing piezometric levels to be far above ground surface. For example, in the Continental Intercalaire aquifer in the Lower Sahara in Algeria the pressure reaches 35 bar at the well head, equivalent to a static groundwater level of 350 m above ground level, which is a world record (see also Box 4.3). However, these artesian conditions do not last for ever: their duration depends on the initial pressure and on the increase in the number of productive wells and their yields. The possibility to exploit an artesian basin by flowing drilled wells rarely lasts for more than a century.

Artesian wells had their heyday during the twentieth century. The proliferation of indiscriminate drilling in most of the exploited basins has accelerated the normally inevitable pressure drops and thus led to decreasing well yields, followed by decreasing total groundwater abstraction, sometimes very quickly. In the artesian basin of Denver (Colorado, USA), within a few years of its discovery in 1884 some 400 wells had been drilled in an area of 65 km², but only six of them were still flowing in 1890. More generally, average well yields have been decreasing steadily in artesian basins. An example of this is in the Great Artesian Basin in Australia where the total output of the flowing wells peaked in 1917, with 24 m³/s discharged by 1 600 wells. But despite the increasing number of wells, this discharge has steadily decreased ever since: the 3 400 artesian wells present in the year 2000 had a total discharge of no more than 11 m³/s.

How these large confined aquifers, exposed to decompression, function hydraulically has remained not properly understood for a long period of time. Therefore, in groundwater development planning, which only became common practice in many countries during the second half of the twentieth century, there was no guidance on how to deal with such aquifers.

4.7.4 From drawing groundwater to pumping

Through the centuries, a rich diversity of ingenious techniques has been developed and used to bring groundwater to the surface. Nowadays, however, shallow and deep

Box 4.3: Artesian groundwater

Groundwater may rise to the surface and gush into the air if wells tap a confined aquifer under sufficiently high pressure. This was known and used for practical purposes long before this phenomenon was named after the French region Artois. Cassini described already in 1671 the 'naturally emerging fountains' of Modena in Italy, while the Arabian scholar Bīrūni explained the phenomenon of artesian wells correctly even more than six centuries earlier.

The development of artesian wells has been facilitated since the nineteenth century by technology for drilling deep wells:

- The first mechanically drilled borehole, conducted in the United States (West Virginia) in 1823, produced a flowing well.
- In Europe, the first boreholes were drilled in France in the Paris Basin (Tours, 1830; Grenelle wells in Paris at 548 m, from 1833 to 1841). They were followed around 1840 by boreholes in Russia (in the Moscow artesian basin and the Ukrainian Dnieper-Donets basin) and in 1856 in Venice, Italy.
- In Africa, the first drilled artesian wells were made in 1856 in the oasis of Wadi Rhir in the Algerian Sahara, in succession to traditional flowing dug wells (already described in the eighteenth century).
- In Australia, the first drilled well in the Great Artesian Basin dates from 1878.

During the twentieth century, advances in geological and geophysical exploration methods and in drilling techniques for oil exploration contributed to the exploitation of deep groundwater (deeper than 1 000 m) and permitted various sedimentary basins that were at that point unknown to be discovered and information on the geometry of other basins to be improved (Figure 4.16). An example is the discovery of the deep Continental Intercalaire aquifer in the Northern Sahara basin by means of the Zelfana borehole in 1949, at 1 167 m deep. Artesian production wells have mushroomed in several countries:

- In the United States, 4 811 artesian wells with a total yield of 38 m^3/s were reported in 1930.
- In Australia, 8 500 wells were drilled before 1945 – including 3 000 flowing wells with a total yield of 17 m^3/s.
- In the Algerian and Tunisian Sahara, 2 295 artesian wells were operational in 1970.

groundwater is mainly withdrawn by pumping from dug and drilled wells. A distinction can be made between hand pumps – using human muscular energy – and mechanically powered pumps. Hand pumps have already been in use for centuries for the domestic water supply of communities or individual households and they continue to be very important for similar purposes in various developing countries, especially in rural areas (see Figure 4.17). In this context, the VLOM concept (Village Level Operation

(a) Old hand pump for community water supply in Vianen, The Netherlands. [*Photo*: Jac van der Gun].

(b) Modern hand pump for community water supply in Gambia. (FairWater Blue Pump). [*Photo*: FairWater].

Figure 4.17 Examples of old and modern hand pumps for community water supply.

and Maintenance) was introduced during the 1980s as guidance for making the use of hand pumps for community water supply successful and sustainable (Arlosoroff *et al.*, 1987). This concept was initially applied to the hardware only (easy maintenance, in-country manufactured pumps, robustness, reliability and cost-effectiveness), but was later expanded to also include the management aspects of maintenance (community decision making on when to service and by whom, and direct payment of repairers by the community).

Today, the vast majority of all groundwater withdrawn in the world is lifted to the surface by mechanically powered pumps (Figure 4.18). Equipped with electrical or diesel pumps, shallow dug wells and drilled tube wells have proliferated in many countries during the twentieth century, in particular for irrigation by private farmers. For example, in California there were 47 000 wells in 1930 and 100 000 in 1965, pumping 650 m^3/s; in South Africa the number of wells increased from 250 000 in 1950 to one million in 1990 and since then 20 000 to 30 000 more wells have been drilled each year; and in Pakistan, more than 175 000 tube wells existed in 1981 (UN, 1986). Keeping count is nowadays becoming problematic. There are at present millions of pumped wells in China, especially on the Northern Plains (Huang Huai Hai and Song Liao) and in the deltas of the Yangtze River and the Pearl River (3 036 000 'electric wells'), with a total installed capacity of 38.2 million kW (Wang *et al.*, 2000). Inventoried wells in India rose from 4 million in 1951 to 17 million in 1997 (Zektser and Everett, 2004) and 19 million around 2000, according to IWMI, and their number probably exceeds 23 million now – undoubtedly a world record.

(a) Public water supply well with electrically powered line shaft turbine pump, USA [*Photo*: USGS].

(b) Diesel-powered pumped irrigation well in Yemen. [*Photo*: Jac van der Gun].

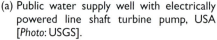

Figure 4.18 Examples of mechanically powered pumps on wells.

The number of diesel or electric pump sets in India has increased even faster, from 87 000 in 1950 and 12 580 000 in 1990 (Central Groundwater Board) to 20 million in 2003 (FAO).

In most of the deep aquifers with predominantly non-renewable groundwater resources, pumped deep wells have replaced artesian wells after the latter had lost the pressure required to lift water to the ground surface. For instance, in 1950 only 4% of the total groundwater abstraction from the deep Jeffara aquifer in Tunisia – initially artesian – was pumped, but this share rose to 43% in 1990 (Mamou, 1990). In the Algerian and Tunesian parts of the North Western Sahara Aquifer System, 71% of the total withdrawal of 60 m³/s was pumped in the year 2000. The large deep Nubian Sandstone Aquifer, shared by Egypt and Libya, was initially exploited by flowing wells, during the 1960s. In 1998, however, 97% of the abstracted 44 m³/s – thus almost all of it – was pumped.

4.7.5   Energy consumption

Energy consumption by pumping (fuel, electricity) is increasing and appears to be a significant part of the total energy consumption in the countries where groundwater is exploited intensively (India, China, USA). However, in the absence of adequate statistics, this share can rarely be quantified. Nevertheless, two examples from the Mediterranean area can be presented. In Spain, according to the National Hydrological Plan of 1993, groundwater pumping (5.5 km³/year) consumed about 3 billion kWh/year. In Israel, 8% of the national electricity production in 1995 was used for pumping groundwater.

4.8   FRAGILITY AND VULNERABILITY
OF THE RESOURCES

Although usually renewable, groundwater resources are not invulnerable. Various human activities can affect their quantity or quality, as well as their usability, but the impacts of these activities vary according to local conditions. As they are not equally exploitable, so the resources are also different from the point of view of fragility and vulnerability. The specific geological conditions on which the resources in the different categories of aquifers depend create differences in natural defence against anthropogenic threats (see also Chapter 6).

Phreatic aquifers are more sensitive than any other aquifers to disturbances and reductions in the rates of recharge. These are due to changes in land use, especially by urbanisation that makes large areas impermeable or by afforestation through which evapotranspiration increases. This can result in decreases in the flow rates of springs or perennial streams and in the exploitable renewable water resources. These aquifers are also sensitive to hydraulic interventions at the surface (stream bed deepening, dredging, embankment works, bank clogging, digging canals) that modify conditions at their boundaries. In addition, they are threatened by certain underground constructions (tunnels functioning as a drain, screens) that change the aquifer dynamics. Alluvial aquifers in arid or semiarid regions that are mainly recharged by surface water floods are particularly sensitive to changes in the regime of the corresponding streams by hydraulic works, such as diversion structures and regulating reservoirs upstream. In contrast, shallow phreatic groundwater in plains irrigated by surface water is susceptible to water table rise, which may induce salinity. The valleys of the Euphrates and the Indus are well known examples.

Groundwater resources are commonly vulnerable to pollution, which may degrade their quality[12]. The *intrinsic vulnerability*, depending exclusively on natural conditions at and below ground surface, is independent of the actual or potential presence of pollutants, but it is sensitive to different processes of potential pollution, local or diffuse, temporary or chronic, and to the various pollutants (see Chapter 6, Box 6.6). This vulnerability indicates how directly and quickly pollutants may reach groundwater. Varying degrees of vulnerability can be distinguished according to the depth to the water table, soil permeability and conditions at the land surface. A number of factors that are assumed to be correlated with relative protection – be it by retardation, dilution or even absorption – defines the general intrinsic vulnerability, but in a simplistic way. Unlike flowing surface water, groundwater has low resilience to pollution, which increases its vulnerability to pollution and makes the effects of pollution persistent.

---

12 The concept 'groundwater vulnerability to pollution' was formulated for the first time in France in 1968. Since that time the concept has spread widely and has been applied in different methods to analyse and map this property as a function of hydrogeological factors (Vrba and Zaporozec, 1994; Zaporozec, 2002; Morris *et al.*, 2003).

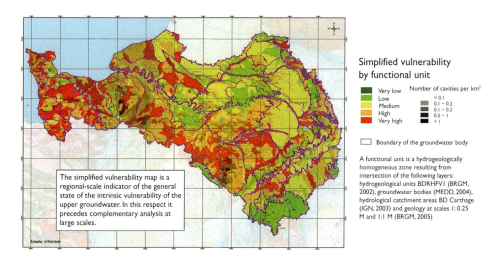

Figure 4.19 Map of the intrinsic vulnerability to pollution of groundwater in the Seine-Normandy Basin (France). [*Source:* BRGM (2005)].

Maps showing levels of intrinsic vulnerability (usually of the first aquifer below the ground surface) have been prepared at various scales and particularly in Europe (Germany, Spain, France, Italy, United Kingdom, Czech Republic). The main goal of the maps is raising awareness among those involved in land use planning, in order to guide them and to adjust precautionary measures. An example of such maps is shown in Figure 4.19. At the country level, these maps make it possible to identify the most vulnerable areas and to compare their relative importance, based on which the level of corresponding constraints to land use is defined. For example, in Spain, aquifer zoning according to three potential risk classes showed that the areas classified as most vulnerable to groundwater pollution cover 141 068 km², or about 28% of the national surface area (*Libro blanco de las aguas subterráneas*, 1994). This type of assessment is not feasible on a global scale.

Groundwater in shallow phreatic aquifers and in karst aquifers is most vulnerable to all pollution processes, no matter whether these are local and chronic (industrial waste, effluent waste, losses from sewers), accidental (transport) or diffuse over large areas (excess fertilizers and pesticides in agriculture). In addition, these aquifers may receive surface water of degraded quality. Caves or natural sinkholes in karst formations often become uncontrolled waste dumps.

In principle, deep confined aquifers are protected against pollution generated at the surface. However, they can be penetrated by boreholes that bring water of different qualities into contact. Moreover, they may not be protected against the effects of anything stored or injected underground.

Groundwater resources are not only vulnerable to various external human threats, they are also sensitive to how and at what rate they are abstracted. As a general rule, exploiting an aquifer changes its dynamic regime. Among others, an increase in abstraction produces feedback to the aquifer's exploitability. Intensive exploitation of unconfined aquifers causes significant declines in the water table, with the result that the unsaturated zone becomes thicker and that eventually the hydraulic connection with surface water bodies that contribute to groundwater recharge is lost. In any case, water level declines lead to higher energy costs and thus make groundwater exploitation more expensive. This is most clearly shown by the transition from flowing wells to pumped wells in aquifers that were exploited initially under artesian conditions. These declines may also cause compaction and irreversible degradation of the hydraulic characteristics of poorly consolidated aquifers and therefore reduce the productivity of groundwater abstraction facilities. In addition, they may cause environmental impacts that limit *de facto* the groundwater resources' exploitability. More generally, no aquifer has a natural protection against the risk of overexploitation (see Chapter 6), except those that are difficult to exploit for reasons such as very deep groundwater levels, low productivity, access problems or unsuitable water quality. Coastal aquifers are particularly vulnerable, because the interface between freshwater and salt water is very sensitive to even minor disturbances of the dynamic equilibrium. Such disturbances may lead to seawater intrusion.

It may be concluded that evaluating the fragility and vulnerability of groundwater systems, in whatever circumstances, is an integral part of assessing the exploitable groundwater resources from a human perspective.

4.9 CAN THE GROUNDWATER RESOURCES BE AUGMENTED?

4.9.1 Approaches and techniques

There are two main ways to augment the exploitable groundwater resources, sometimes even to make their flux larger than the rate of natural groundwater renewal. Both consist of transforming surface water into groundwater, but they differ as to whether human activity produces recharge in a direct way (*artificial recharge*) or indirectly by triggering the recharge process (*induced recharge*):

- *Induced recharge* is produced as a side-effect of groundwater abstraction. This phenomenon is fairly widespread in exploited zones of alluvial aquifers directly along perennial streams. Declines in the groundwater level in the aquifer may reverse the exchange between groundwater and surface water: from groundwater under natural conditions usually flowing towards the river to surface water inflow from the river into the aquifer. The side-effects of abstraction thus induce an additional resource. Many groundwater abstractions for urban water supply in the world benefit from induced recharge. In fact, it is an indirect withdrawal of surface water and is often recognised as such in water withdrawal statistics. However, clogging of the stream bed and river banks may become a limiting factor in the longer term.
- *Artificial recharge* (see e.g. Bouwer, 2002) is a technique to control surface water (the source to be used) and at the same time to augment the exploitable groundwater resources artificially. Depending on the particular case, it allows:
    - An increase in the flux withdrawn which improves the performance of abstraction facilities;

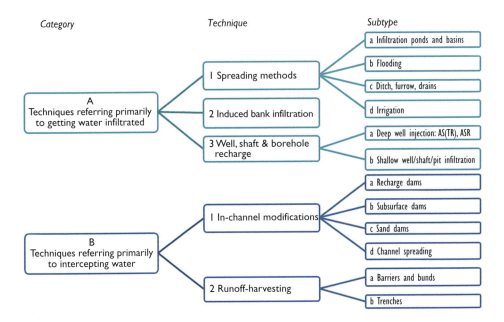

Figure 4.20 Overview of MAR techniques and subtypes (after IGRAC, 2007).

– Storage of water, in competition with surface storage, subject to other site-specific constraints. Storage is particularly important in arid areas where the control of very irregular surface runoff (erratic floods) is more difficult than elsewhere and where protecting stored water against evaporation is most needed.

In recent years, the term 'artificial recharge' has gradually been making way for *managed aquifer recharge*' (MAR). MAR tends to be interpreted more comprehensively than artificial recharge, by also including induced recharge and sometimes also the recovery of the water stored. A classification of managed aquifer recharge techniques, prepared in the framework of a dedicated inventory project organised by IGRAC[13], is shown in Figure 4.20.

Artificial recharge can help restore the balance of an excessively exploited aquifer, or create conditions of surplus, which could allow more groundwater to be withdrawn than is warranted by the natural recharge. It can also be a useful component in a wastewater reuse chain, by facilitating the storage of wastewater and by contributing to its purification. Furthermore, artificial recharge is in some cases primarily intended to control or improve water quality.

Aquifer characteristics define to a significant extent the options for artificial recharge and the effects to be expected. *Unconfined aquifers* with relatively deep water levels offer the largest storage capacity. They are very favourable because they

---

13 This project was carried out with the assistance of the Acacia Institute (The Netherlands) and in close cooperation with the IAH-MAR Commission and UNESCO-IHP.

combine the capacity to store rather large volumes of water in a short time interval (because recharge water may be available for a short while only) with the capacity to keep these volumes long enough in storage (attenuation at the boundaries): water should enter quickly and leave slowly. *Confined aquifers* offer above all a rapid long-distance transmission of the pressure restoration that artificial recharge is aimed for in order to recover the productivity of abstraction facilities – and sometimes to stop induced land subsidence. However, their storage capacity is low and their depth requires injection as an exclusive implementation technique for artificial recharge.

The different artificial recharge techniques shown in Figure 4.20 can be subdivided into extensive and intensive techniques. So-called *extensive techniques* (ponding, infiltration basins) are cheapest and better adapted to arid conditions, but require more space. *Intensive techniques* (shallow well, shaft, pit or deep well injection) are more expensive, less durable (due to clogging) and more localised. Which technique to chose depends on the source of water to be used (flow regime, physical quality), the type of aquifer, available means and the desired result.

4.9.2   Implementation in different parts of the world

Artificial recharge facilities (see Figure 4.21) are operational in many countries in Europe, in the United States, India, Australia, the Maghreb, the Middle East and Africa (see e.g. Detay, 1998; Petry *et al.*, 1998; ISMAR5, 2005; ISMAR6, 2007; Fernández Escalante, 2008; Van Steenbergen and Tuinhof, 2010). Many of them are located near a well field, in order to ensure or increase its production, but many others are primarily intended to increase the resources of an entire regional or local aquifer system. Examples of the latter are recharge facilities in the Santa Clara Valley in California, the large recharge dams in Oman and numerous recharge schemes scattered over rural India.

In arid or semiarid regions, the main purpose of measures to control the very irregular and ephemeral surface water flows by dams or dikes may be to increase the recharge of aquifers that already depend on this type of flows under natural conditions. This is particularly the case in Saudi Arabia, the United Arab Emirates (35 dams with a total capacity of 80 million m^3 in 1995), Oman (15 dams, total capacity of 58 million m^3). These dams are capable of retaining the volume of the flood peaks and transferring it to the groundwater reservoirs.

It also happens that artificial recharge is less a goal than a factual result. This is observed in all aquifers underlying plains where intensive surface water irrigation takes place. Infiltration of excess irrigation water there may contribute much more to recharge than rainfall, especially in arid or semiarid zones. Consequently, the exploitable groundwater resources of many alluvial plain aquifers in the arid zone are largely *secondary resources*, by-products of irrigation. Examples are the Indus Valley in Pakistan, the Nile Delta and Nile Valley in Egypt, several plains of Northern China and even Mediterranean alluvial plains in Europe, especially the plain of Crau near the Rhone delta in France and many coastal plains along the Spanish Mediterranean coast and on Gran Canaria. The same is true for many oases in the foothill areas of the northern Sahara or in Central Asia, irrigated by flood waters.

For example, in Egypt, the secondary groundwater resources of the Nile Valley and Nile Delta are estimated at about 8 km^3/year (the total fresh groundwater poten-

(a) Small sand dam in Kitui district, Kenya. [*Photo*: Sander de Haas]. *The dam has created a small sand reservoir in the stream bed where flood water is stored. This can be abstracted to meet modest local water demands.*

(b) Large recharge dam in the Batinah Plain, Oman [*Photo*: Jac van der Gun]. *Flood water is retained by the dam and is subsequently released in a controlled way to recharge the aquifer zone downstream.*

Figure 4.21 Examples of recharge dams.

tial is 8.4 km³/year, according to Khater, 2005). In the Indus Valley in Pakistan, where annually 130 km³ of surface water goes to irrigation, the estimated 35 to 40% losses of water from canals and irrigated lands account for more than 80% of the potential of the alluvial aquifer, estimated at about 60 km³/year.

In northern China, the secondary resources of several aquifers constitute a major part of their recharge: 88% in the Yinchuan plain and 51% in the Hetao plain.

However, such a massive additional recharge does not only have beneficial impacts, by facilitating intensive exploitation of aquifers. The significant groundwater level rise produced often also has negative impacts: waterlogging and soil salinization in arid zones, which require active drainage, especially in nearly horizontal plains. 'Vertical drainage' by pumped drilled wells is practiced in such areas, for example in Pakistan and Uzbekistan.

REFERENCES

Arlosoroff, S., G. Tschannerl, D. Grey, W. Journey, A. Karp, O. Langeneffer & R. Roche, 1987. *Community water supply: The handpump option.* The World Bank, Washington, 202 p.

Attia, F.A.R., 2005. Non-Renewable Groundwater and Sustainable Agriculture: Current Status and Future Perspectives. *Internat. Conf. Water, Land and Food Security in Arid and Semi-Arid Regions.* CIHEAM, Valenzano-Bari, Italy, 6–11 Sept., Key-Note Paper, pp. 127–141.

Bakhbakhi, M., 2002. *Hydrogeological framework of the Nubian Sandstone Aquifer System.* GW-MATE/UNESCO Expert Group Meeting, Socially sustainable management of groundwater mining from aquifer storage. Paris, 6 p.

Balland, D. (ed.), 1992. *Les eaux cachées. Etudes géographiques sur les galeries drainantes souterraines.* Paris, Université de Paris, Département de Géographie, 144 p.

Bank, D., & O. Soldal, 2002. Towards a policy for sustainable use of groundwater by non-governmental organisations in Afghanistan. *Hydrogeology Journal,* Vol. 10, No. 3, pp. 377–392.

Bouwer H., 2002. Artificial recharge of groundwater: hydrogeology and engineering. *Hydrogeology Journal,* Vol. 10, No. 1, pp. 121–142.

Bredehoeft, J.D., S.S.Papadopoulos & H.H. Cooper Jr., 1982. The water budget myth. In: *Scientific Basis of Water Resources Management Studies in Geophysics,* National Academy Press, Washington DC, USA, pp. 51–57.

BRGM, 1976. Carte de planification des ressources en eau souterraine des états members du CIEH de l 'Afrique soudano-sahélienne. Note (118 p.) and three maps at scale 1: 1 500 000.

BRGM, 2005. *Cartographie de la vulnérabilité intrinsèque des eaux souterraines.* Agence de l'Eau Seine-Normandie/Seine-Normandy Water Agency. Available at: http://sigessn.brgm.fr/spip.php?rubrique10.

Brunt, R., L. Vasak & J. Griffioen, 2004. *Arsenic in groundwater: Probability of occurrence of excessive concentration on global scale.* Report SP 2004–1, IGRAC, Utrecht, The Netherlands.

Carlsson L., E. Selaolo & M. Von Hoyer, 1993. Assessment of groundwater resources in Botswana – Experience from the National Water Master Plan Study. *Proc. Intl. Conf. Africa Needs Groundwater, Sept. 1993, Witwatersrand,* Geological Society of South Africa.

Caponenco, N., 2000. *The Russian Water Vision.* World Water Council.

CBLT/OSS, 2005. Système aquifère du basin du lac Tchad. In: ISARM-Africa: *Ressources en eau et gestion des aquifères transfrontaliers de l'Afrique du Nord et du Sahel,* IHP-IV, Series on groundwater No. 11, Paris.

CEDARE/IFAD, 2002. Regional Strategy for the Utilisation of the Nubian Sandstone Aquifer System, Volume II. CEDARE, Heliopolis Bahry, Cairo, Egypt.

CGMW/UNESCO, 2000. *Geological Map of the World at 1:25 million*. Commission for the Geological Map ofthe World, UNESCO, September 2000 (second edition).

Commonwealth of Australia, 2003. *Water resources and use in Australia*. Unpublished document, accessed at www.farmweb.au.com by IGRAC [July 2005].

Custodio E., 1993. Aquifer Intensive Exploitation and Over-Exploitation with respect to Sustainable Management, Environmental Pollution. ICEP2, European Centre for Pollution Research, pp. 509–516.

Custodio E., 2000. *The complex concept of overexploited aquifer*. Papeles, Proyecto Aguas Subterráneas, Series A2, Fundación Marcelino Botín, Santander, Spain.

Custodio E., 2002. Aquifer overexploitation – What does it mean? In: *Hydrogeology. J.*, Vol. 10, No. 2, pp. 254–277.

Cutter, S.L., H.L. Renwick & W.H. Renwick, 1991. *Exploitation, conservation, preservation: a geographic perspective on natural resources use*. Second edition, New York, John Wiley & Sons.

Deom, J.M., & R. Sala, 2006. *Contribution to Urumqi Conference*, 2006.

Detay, M., 1998. La gestion active des aquifères. IAHR, Proceedings of the Conference on Coping with Water Scarcity, Hurghada, Egypt, 1998.

Diagana, B., 2005. Système aquifère sénégalo-mauritanien. In: ISARM-Africa: *Ressources en eau et gestion des aquifères transfrontaliers de l'Afrique du Nord et du Sahel*, OSS/UNESCO, IHP-IV, Series on groundwater No. 11, Paris.

Dodo, A., 1992. Etude des circulations profondes dans le grand bassin sédimentaire du Niger: identification des aquifères et comprehension de leurs fonctionnements. PhD thesis, Univ. of Neuchâtel, Switzerland.

El Gheriani, A., 2002. *The Great Man Made River project*. Colloque Eau et Économie, Soc. Hydrot. France, Paris, Sept. 2002.

El-Hebil, A., 1988. Foggara et khettara, un système millénaire de captage des eaux. In: *L'eau et le Maghreb. Un aperçu sur le present, l'héritage et l'avenir*, PNUD, New York.

Enrich, E.C., 2001. Groundwater: A renewable resource? – Focus on Sahara and Sahel. In: *British Geological Survey*. Wallingford, UK, for EC ENV4-CT97–0591.

EUROSTAT, 1998. *Water in Europe. Part 1. Renewable water resources*. Office for Official Publication of the European Communities, Luxembourg.

FAO, 1995. *Irrigation in Africa in figures*. FAO Water Reports 7, Rome.

FAO, 1997. *Irrigation in the Near East region in figures*. FAO Water Reports 9, Saudi Arabia 205212, Rome.

Feenstra, L., L. Vasak and J. Griffioen, 2007. *Fluoride in groundwater: Overview and evaluation of removal methods*. IGRAC, Utrecht, The Netherlands, Report SP 2007–1, 25 p.

Fernández Escalante, E., 2008. Posters of the ISMAR5 Conference, Berlin, Germany, 2005.

Foster, K.E., 1977. *Managing a finite groundwater supply in an arid area: the Santa Cruz basin example in Arizona*. Proceedings Conf on Alternative Strategies for Desert Development, May 1977, Sacramento, California, 21 p.

Foster, S.S.D., & P. Loucks (eds), 2006. Non-renewable Groundwater Resources. A guidebook on socially-sustainable management for water policy-makers. Paris, UNESCO-IHP, 103 p.

Freeze, R.A., & J.H. Lehr, 2009. The fluoride wars: How a modest public health measure became America's longest-running poltical drama. Hoboken, N.J, Wiley, ix+383 p.

Garber, A., & E. Salameh (eds), 1992. Jordan's Water Resources and their Future Potential. In: *Proc. Symp. Water Resources*, Oct. 1992, Amman, University of Jordan, 120 p.

Ghurm Alghamdi, A.A., 2002. *Sustainable Management for Deep Aquifer Storage in the Kingdom of Saudi Arabia*. GW-MATE UNESCO Expert Group meeting, Socially Sustainable Management of Groundwater Mining from Aquifer Storage, 9 p. Paris.

Gibert, J., D. Danielopol & J. Stanford (eds), 1994. *Groundwater Ecology.* London, Academic Press, 571 p.

Gibert, J., J. Mathieu & F. Fournier (eds), 1999. Groundwater/Surface Water Ecotones: Biological and Hydrological Interactions and Management Options. Cambrige, Cambridge University Press, 246 p.

Goblot, H., 1979. Les qanats, une technique d'acquisition de l'eau. Paris, Mouton, 336 p.

Gonzalez Perez, 1987. *La sobre-explotación de acuíferos: Canarias.* Canarias Agua 2000, April 1987, Tenerife, Spain.

Griebler, C., & T. Lueders, 2009. Microbial diversity in groundwater ecosystems. In: *Freshwater Biology* (2009), Vol. 54, pp. 649–677. Available from: doi: 10.1111/j.1365–2427.2008.02013.x

GW-MATE, 2006. Groundwater dependent ecosystems – the challenge of balanced assessment and adequate conservation. Briefing Note Series, Note 15, GW-MATE/World Bank, Washington DC.

Habermehl, M.A., 1980. *The Great Artesian Basin, Australia.* BMR Journal of Australian Geology & Geophysics, 5, pp. 9–38.

Habermehl, M.A., 2001. Hydrogeology and environmental geology of the Great Artesian Basin, Australia. Chapter 11 in: Gostin, V.A. (editor), 2001, *Gondwana to Greenhouse – Australian Environmental Science.* Geological Society of Australia Inc., Special Publication 21, pp. 127–143, 344–346.

Habermehl, M., 2002. *Groundwater development in the Great Artesian Basin, Australia.* GW-MATE UNESCO Expert Group meeting, Socially Sustainable Management of Groundwater Mining from Aquifer Storage, Paris, 35 p.

Hairy, I., 2011. *Du Nil à Alexandrie, histoires d'eau.* 2nd edition, revised and enlarged. Centre d'Études Alexandrines. Alexandria, Egypt, Sahara.

Hassanzadeh, Y., 1992. Optimisation des ressources hydrauliques. In: Soc. Hydrot de France: *L'Avenir de l'eau*, 22e journées de l'Hydraulique, 15–17 sept 1992, Paris.

ICID, 1975. *Iran.* ICID Jubilee Commemoration Volume, July 1975, pp. 240–245.

ICID, 1983. *Irrigation and Drainage in the World. Vol. III, Sudan.* Third edition, New Delhi, pp. 1262–1280.

ICID, 2000. Irrigation and Drainage in the World. New Delhi.

ICQHS, 2009. An Introduction to International Center on Qanats and Historic Hydraulic Structures. ICQHS, UNESCO, March 2009, 24 pages.

IGRAC, 2007. *Global inventory of artificial recharge.* Available from: http://www.un-igrac.org/publications/155.

ISMAR5, 2005. Fifth International Symposium on Managed Aquifer Recharge. *Proceedings of ISMAR5 Conference, Berlin, Germany,* 10–16 June 2005.

ISMAR6, 2007. Sixth International Symposium on Managed Aquifer Recharge. Proceedings of ISMAR6 Conference, Phoenix, Arizona, 28 October – 2 November 2007.

Issar, A.S., & R. Nativ, 1998. Water Beneath Deserts: Key to the Past, A resource for the present. In: *Episodes,* Vol. 11, No. 4, pp. 256–262.

Kahlown, M.A., & A. Majeed, 2002. Water resources situation in Pakistan: challenges and future strategies. In: *Science Vision,* Vol. 7, No. 3, Islamabad, Pakistan, pp. 33–45.

Khouri, J. 1990. *Arab Water Security – A Regional Strategy for Horizon 2030.* International Seminar 'Stratégies de gestion des eaux dans les pays mediterranéens', Algiers, May 1990. CCE/Gov.Algeria/CEFIGRE, 68 p.

Li, Lierong, 2001. *Groundwater Exploitation in China.* Los Alamos National Laboratory, University of California.

MacGuinnes, 1963. *The role of ground-water in the national water situation.* USGS Water Supply Paper 1800, Arizona, pp. 138–163.

Mamou, A., 1990. Charactéristiques et évaluation des ressources en eau du Sud-Tunesien. Thesis, University of Paris XI.

Mamou, A., 1999. *Gestion des ressources en eau du système aquifère du Sahara Septentrional.* Proceedings International Conference, Tripoli, Libya, 20–24 Nov. 1999, UNESCO-IHP V, Technical Documents in Hydrology, No. 42, Paris.

Margat, J., 1998. Les eaux souterraines dans le bassin Méditerranéen. Ressources et utilisations. Plan Bleu and BRGM, Orleans, BRGM Documents No. 282, 110 p.

Massoumi, A., 1966. *Ground-water production in Iran.* Second Ground-water Seminar, Teheran. UN/ECAFE, UNESCO, Water Resources Series 33, 1967.

Meinzer, O.E., 1920. Quantitative methods for estimating ground-water supplies. In: *Bull. Geol. Soc. Amer.*, Vol. 31, pp. 329–338.

Mian, B.A. and M. Abdullah, 2000. Impact of groundwater development by tubewell technology on karez in Baluchistan, Pakistan. In: *Qanat, The First International Symposium*, 8–11 May 2000, Yazd, Iran.

Ministry of Agriculture and Water of Saudi Arabia, 1985. *Water, agriculture and soil studies of Saq and overlying aquifers, Final Report.* Report of BRGM and CNABRL.

Morris, B., A. Lawrence, J. Chilton, B. Adams, R. Calow & B. Klinck, 2003. *Groundwater and its susceptibility to degradation.* UNEP, Early Warning and Assessment Report Series, RS 03–03, Nairobi, Kenya, 140 p.

MRMWR, 2008. *Aflaj Oman in the World Heritage List.* Ministry of Regional Municipalities and Water Resources, Muscat, Sultanate of Oman.

Nasier & Wang Xin, 1990. *The present situation and future perspective of Xinjiang Karez.* Communication presented at the international colloquium on irrigation by karez, Urumqi.

Neuland, H., 1988. Foodstuff production target in arid zones: issues and prospects from the Arabian Peninsula. In: *Proceedings Sixth Congress IWRA/AIRE, Ottawa, May–June 1988,* Vol III, pp. 1150–1206.

OSS, 2003. *Système aquifère du Shara septentrional.* Synthesis report, Observatoire du Sahara et du Sahel, Tunis, 129 p.

OSS/Himida, I.H., 1992. *Overview of the development and utilization of Nubian artesian basin, North East Africa.* Observatoire du Sahara et du Sahel, Project launch workshop 'Aquifères des grands bassins', Cairo, 22–25 Nov. 1992.

OSS/UNESCO, 2005. Ressources en eau et gestion des aquifères transfrontaliers de l'Afrique du Nord et du Sahel. OSS/UNESCO, ISARM-Africa, IHP-IV, Series on groundwater No. 11, Paris.

Pallas, Ph., & O. Salem, 1999. Water Resources Utilisation and Management of the Socialist People Arab Jamahiriya. In: *Proc. Int. Conf. Regional aquifer Systems – Managing Non-renewable Resources, Tripoli, Nov. 1999,* UNESCO IHP-V Technical Documents in Hydrology No. 42.

Pascon, P., 1977. *Le Haouz de Marrakech.* PhD thesis., Tanger, 2 volumes.

Peters, J. (ed.), 1998. Artificial recharge of Groundwater. In: Proceedings of the Third International Symposium on Artificial Recharge of Groundwater, Amsterdam. Rotterdam, Balkema, 474 p.

Petry, B, J. van der Gun & P. Boeriu, 1998. Coping with water scarcety: a case study from Oman. In: IAHR, African Division: *Proceedings of the Conference on Coping with Water Scarcity at Hurghada, Egypt.*

Plan Bleu, 2004. *L'eau des méditerranéens. Situation et perspectives.* PNUE/PAM, MAP Technical Reports Series No. 158, Athens, 366 p.

Plan Hidrológico Insular de Tenerife, 1989. *Avance:bases para el planeamento hidrogéologico.* Gobierno de Canarias, Dirección General de Aguas.

Ramsar, 2009. List of Wetlands of International Importance. Available from: http://www.ramsar. org/cda/en/ramsar-about-sites/main/ramsar/1-36-55_4000_0__ [accessed February 2012].

Semsar Yazdi, A., 2009. An introduction to qanat and an analysis of its situation in the world. In: *Watarid2 Conference 2009, Yazd*. Teheran, Hermann, 2011.

Semsar Yazdi, A., & M. Labbaf Khaneiki (eds), 2012. Qanat in its cradle. Situation of qanat (kariz, karez, falaj) in the world. Volume 1. UNESCO, ICQHS, Yazd, Iran.

Shahin, M., 1989. Review and Assessment of Water Resources in the Arab Region. In: *Water International, Vol. 14*, IWRA, pp. 206–219.

Smedley, P.L., 2008. Sources and distribution of arsenic in groundwater. In: T. Appelo and J.P. Heederik, *Arsenic in Groundwater – a world problem*, Utrecht Seminar NNC-IAH, November 2006, Utrecht, The Netherlands.

Terap, M.M., & B.W. Kaibana, 1992. *Ressources en eau souterraine du Tchad*. Observatoire du Sahara et du Sahel, Project launch workshop 'Aquifères des grands bassins', Cairo, 22–25 Nov. 1992.

UN, 1986. Groundwater in Continental Asia. UN-DTCD, New York, Natural Resources Water Series no. 15.

UN-ECE, 1981. Long-term perspectives for water resources use and water supply in the ECE region. ECE/Water/26, New York.

UN-ESCAP, 1993. Study on assessment of water resources of member countries and demand by user sector. Japan: Water resources and their use, p. 37.

UNESCO, 2001. Proceedings of the International Conference on Regional aquifer systems in arid zones – Managing non-renewable resources, Tripoli, Libya, 20–24 November 1999. Paris, UNESCO, Technical Documents in Hydrology, No. 42.

USWRC, 1983. *See:* USGS, 1985

USGS, 1985. Arizona Ground Water Resources. In: National Water Summary 1984. USGS Water Supply Paper 2275, Washington, pp. 135–140.

Van Weert, F., J. van der Gun & J. Reckman, 2009. *Global overview of saline groundwater occurrence and genesis*. IGRAC, Utrecht, Report GP-2009–1.

Van Steenbergen, F., & A. Tuinhof, 2010. Managing the water buffer for development and climate change adaptation: Groundwater recharge, retention, reuse and rainwater storage. BGR, CPWC and Netherlands Committee IHP-HWRP.

Vasak, L., R. Brunt & J. Griffioen, 2008. Mapping of hazardous substances in groundwater on a global scale. In: T. Appelo and J.P. Heederik, *Arsenic in Groundwater – a world problem*, Utrecht NNC-IAH Seminar, November 2006, Utrecht, The Netherlands.

Vrba, J., & A. Zaporozec, 1994. *Guidebook on Mapping Groundwater Vulnerability*. IAH, International Contributions to Hydrogeology, Vol 16, Heise, 131 p.

Wang, R., Ren, H., & Ouyang, Z. (eds), 2000. *China Water Vision. The ecosphere of water, life, environment and development*. Beijing, China Meteorological Press, 178 p.

WHYMAP, 2008. *Groundwater Resources of the World*. Map at scale 1: 25 M. BGR & UNESCO, Hannover, Paris.

Winter, T.C., J.W. Harvey, O.L. Franke & W.M. Alley, 1998. *Groundwater and surface water: A single resource*. US Geological Survey Circular 1139, Washington DC.

Zaporozec, A., 2003. Groundwater and its susceptibility to degradation. UNEP.

Zaporozec, A. (ed.), 2002. *Groundwater contamination inventory. A methodological guide*. UNESCO, IHP-VI, Series on Groundwater No. 2, 161 p.

Zektser, I.S., 1998. Groundwater resources of the world and their use. In: Water: a looming crisis? Proceedings of the International Conference at the Beginning of the Twenty-first Century, Paris.

Zektser, I.S., & L.G. Everett (eds), 2004. *Groundwater resources of the world and their use*. UNESCO, IHP-VI, Series on Groundwater No. 6. Paris, UNESCO, 346 p.

Chapter 5

Groundwater withdrawal and use

- *Groundwater as a special source of water*
- *Groundwater withdrawal and use: how much, for what purposes, where and by whom?*
- *Trends over time in groundwater withdrawal*
- *Depletion of groundwater reserves*
- *Proportion of groundwater in water supplies, by country and by water use sector*

5.1 GROUNDWATER: A SPECIAL AND OFTEN PREFERRED SOURCE OF WATER

Since time immemorial and in many regions around the world, people have known how to abstract a substantial part – sometimes even the majority – of the water they need from underground. Wherever exploitable groundwater has been found, it has often appeared to be more voluminous and more reliable than surface water – especially in regions where surface water is scarce and has an irregular regime – and more easily accessible to larger numbers of people. These factors contribute to the special role of groundwater in the water economy.

Groundwater is exploited in different ways, it contributes in different degrees to a better quality of life and its share in the water economy varies from one area to another. In spite of these variations, from the point of view of abstraction and use there are a number of generally shared characteristics that distinguish groundwater from surface water. These largely explain why in many specific cases groundwater is considered to be a preferential source for water supply, but in other cases not. Box 5.1 presents an overview of these characteristics, in terms of general advantages of groundwater as a source of water supply and of disadvantages or constraints to its withdrawal and use.

The following aspects underline the special nature of groundwater as a source of water:

a   *Groundwater is a widely accessible source of water*
Groundwater is accessible to many actors in the local economy. In principle, it is accessible to anybody living above an aquifer, thus to many more people than only those living close to rivers and therefore with direct access to surface water. Households,

Box 5.1: Groundwater as a source of water supply: advantages and disadvantages

Spatial distribution	Widespread resource, facilitating abstraction close to the place of use, thus minimising cost of conveyance. Depending on productivity, however, a number of abstraction works rather than a single one may be needed to meet high local demands.
Availability over time	Permanent resource usually with a large volume of water in storage (buffer) and only limited variation in flow. Hence, more resilient to climatic variation and climate hazards than surface water, thus ensuring a safer water supply. This natural resource does not usually need any regulating facilities for meeting demands that vary considerably over time.
Resources assessment	The assessment of this invisible resource may require the use of methods that are rather complex and more expensive than those for surface water resources assessment, but usually they require less time.
Natural quality	Constant or slightly variable quality, which is convenient for the treatment of the pumped 'raw' water meant to adjust certain parameters that are excessive for some types of use (potable water, process water): hardness, iron content, manganese content. Salinity often increases at greater depth in the aquifer and with a higher degree of climatic aridity.
Vulnerability to pollution	Deep groundwater is generally relatively invulnerable, except in cases of improper drilling or deliberate injection of polluting substances. Groundwater in phreatic aquifers is more vulnerable to diffuse and point pollution than surface water: although pollution arrives less directly, it stays much longer (less resilience).
Production cost	On average, investment and operational costs are lower than those related to surface water, but they vary considerably depending on the local aquifer conditions (depth, aquifer productivity). The operational costs are sensitive to variations in energy prices, except in cases of withdrawal by gravity. The yield decreases according to the overall rate of aquifer exploitation.
Flexibility in implementation	Compared to the hydraulic works required for surface water withdrawal, groundwater allows a higher flexibility to adapt the technical infrastructure progressively to changing water demands. Return on investment is quicker and the latter can be spread over time more easily.

Risk of overexploitation	Groundwater is exploited much more on an uncoordinated individual basis than surface water. In general, there is little awareness of the cumulative implications of intensive groundwater withdrawal with respect to sustainability. Feedback from excessively exploited groundwater systems is delayed and so there is no immediate urge to reduce or stop abstractions. As a result, intensely exploited aquifers easily become overexploited.

farmers, industries, local communities and water supply companies – all of them may have the means and the right to exploit groundwater. If both surface water and groundwater is available, then people tend to prefer groundwater because this source is often the cheapest, the most practical and best suited to individual exploitation. Bringing groundwater to where it is needed requires little or no collective facilities. This is unlike surface water, usually poorly accessible to the individual, where diversion works are needed and often also provisions for regulation and transport. Groundwater is a local resource *par excellence* and – as Deb Roy and Shah (2001) have called it – a *'democratic'* resource.

Users of groundwater, therefore, are the direct exploiters of their resource much more often than surface water users. Consequently, the role of water abstracting agencies and intermediary distributors is relatively small, except in the case of public water supply. Most of the time groundwater is a 'self-service' resource. This is why investments and operational costs for groundwater development are in general directly covered by the users, or by the intermediaries who distribute the water and pass these costs on to the end users. Mainly the private sector is involved, which is unlike the situation with surface water exploitation, where the public sector (national or local authorities) usually takes care of constructing and operating the hydraulic works meant to control and convey water.

b   *Most groundwater systems have a very significant buffer capacity*
Groundwater systems have relatively large volumes of water in storage – usually several hundred to many thousand times the volume of their mean annual recharge. This produces a unique buffer capacity, which is a major strength of groundwater systems. It allows periodic, seasonal or multi-annual dry periods to be conveniently bridged, thus creating conditions for survival in semi-arid and arid regions and reducing in general the risk of temporary water shortages (which enhances the value of groundwater). Furthermore, it smooths out water quality variations and causes part of the stored water (medium-deep to deep groundwater) to be relatively invulnerable to sudden disasters and thus suitable to serve as a source of emergency water supply.

c   *Groundwater is susceptible to human influences and is vulnerable to pollution in particular*
Those who abstract groundwater are not the only ones that pollute or otherwise influence it. In principle, all persons living or working in a certain area can affect the regime

or quality of the area's groundwater, especially phreatic groundwater. The following list includes the obvious actors and the ways they often influence groundwater:

- *Farmers:* reduction of groundwater recharge by tilling their land and groundwater pollution due to a surplus of manure, fertilizers and pesticides.
- *Industries:* storage of dangerous primary materials and products, as well as dumping waste; the industrial fabrication processes may harbour the risk of accidents harmful to the environment.
- *Exploiters of quarries and mining companies:* draining the subsurface to enable extraction of materials changes the groundwater regime and their waste management (e.g. tailing heaps) may affect groundwater quality.
- *Local authorities:* their sewerage networks (not free from leaks) and urban waste disposal practices may produce pollution.
- *Transporters of dangerous materials:* they expose groundwater to the risk of accidental pollution (on their own sites or on public roads).
- *Supervisors of public or private works related to urbanisation or hydraulic engineering:* they produce changes in terrain conditions, flow of water and relations between surface water and groundwater.
- *Users of the underground (underground urban zones, underground traffic, underground storage):* they may change the aquifer structure to some extent and modify the groundwater regime.

In all cases, these harmful consequences are unintended *external* side effects of human activities. Reducing or preventing these side effects requires extra efforts in addition to those strictly needed to achieve the direct economic goals of each activity. Such efforts are not spontaneously forthcoming, mainly because the side effects are externalities which means that they affect third-parties.

d   *Groundwater is a poorly known and poorly understood common property resource*
Whether as groundwater exploiters or as individuals influencing groundwater in a different way, most people living or working above an aquifer are unaware that groundwater is a common good, part of our natural heritage. Most people have at best an incomplete notion of the groundwater resource they use and influence, sometimes to the point of causing water use conflicts. That is the reason why they usually have no idea of the side effects of their abstractions or other activities until they experience negative feedback themselves. This lack of knowledge has more consequences when such side effects appear at a distance or are related to a system other than groundwater (e.g. a water course, aquatic ecosystem, wetland), or when they are strongly delayed – e.g. in a coastal zone where in the long run triggered seawater intrusion finally becomes evident. As a general rule, people do not adopt groundwater resources management as their individual objective, because that is beyond the scale of their own activities and the motives behind their activities are primarily micro-economic, not addressing the community's objectives.

In short, groundwater is a *de facto* common property, not recognised as such by the general public and without spontaneously emerging rules on how to coordinate withdrawals. That explains why, when the total withdrawal from an aquifer causes side effects that are harmful to third parties (e.g. to users of streams fed by groundwater, or to ecosystems), those who abstract groundwater do not feel collective responsibility. These aspects evidently have consequences for the objectives and approaches to be adopted for managing groundwater, both as an extractable resource and as a component of the natural environment.

5.2   HOW MUCH GROUNDWATER IS WITHDRAWN AND WHERE IS WITHDRAWAL MOST INTENSIVE?

Analysing the global patterns of groundwater withdrawal and the purposes for which groundwater is used requires data on the current groundwater abstraction in each country. Comparing this data with the exploitable natural groundwater resources gives an impression of the pressures these abstractions put on the groundwater systems. Demographic data and alternative sources of water supply should be taken into account if the socio-economic implications of these pressures have to be assessed. The database established for these purposes has been carefully based on different available national or international statistical sources[1], but due to data deficiencies it is not complete or up-to-date, nor completely homogeneous (see Box 5.2).

Even UN agencies themselves lack reliable data for several countries. The map presented in Figure 5.1 gives an overall impression of the status of the available information. Aggregated abstraction values have been found in the consulted databases for 149 countries, together representing 92% of the world population. Two-thirds of these values date from the year 2000 or from more recent years, while 17 of them date from before 1990. Regions where relatively recent data (reference year 2000 or later) on groundwater withdrawals are scarce or lacking are tropical Africa, South America and Central America. Most of the corresponding countries are richly endowed with

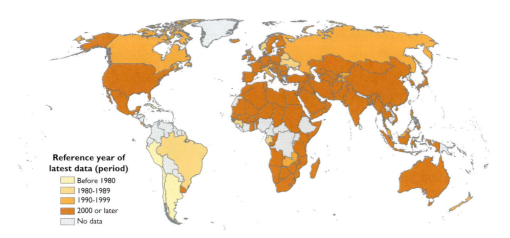

Figure 5.1 Status of data on groundwater abstraction by country, based on compiled data. [*Data source*: National and international databases (see also Appendix 5)].

---

1 In particular BRGM, EUROSTAT, FAO/AQUASTAT, IWMI, OECD, PLAN BLEU, UNESCO-IGRAC, WWAP and WRI.

Box 5.2: Is sufficient information available on groundwater abstractions?

In general, a distinction between groundwater and surface water can be made more clearly and more easily at the level of sources of water supply and abstractions than at the level of defining groundwater and surface water resources. Nevertheless, inventorying groundwater abstractions and assessing the amount of groundwater abstracted is by no means easy, because of the large number of exploiters and the variability of their operations, especially related to wells used for supplementary irrigation. The efforts to conduct a periodic and necessary inventory of groundwater abstraction vary among countries. The available statistics are thus not flawless, and even more so because of several other reasons:

- Groundwater abstraction is not defined consistently by all countries. Water diverted from springs is included in the statistics of some countries and excluded in those of others. The same is true for groundwater abstractions near rivers (capturing induced recharge from surface water), groundwater withdrawn by qanats or for the often significant quantities of drainage water from mines and quarries.
- The abstracted quantities are not estimated with the same precision. In particular, estimates of the abstraction by farmers are the most uncertain ones. Some statistics are derived from rather comprehensive counts and inventories, and others are based on indirect estimates.
- Withdrawals from renewable resources are not always clearly recorded separately from those from non-renewable resources, in countries where both exist. In the absence of rigorous uniform criteria, there are also differences in interpretation as to which resources to consider as non-renewable.
- Data collection is not synchronous and sufficiently long time series of groundwater abstraction are rare.

The distribution of abstractions by use sector is not always comparable. It is often ambiguous whether some types of abstractions have been ranked under the industrial water or public water supply sector.

rainfall, and hence surface water is abundant and irrigation absent or insignificant, thus groundwater withdrawal is likely to be modest. Time series of groundwater abstraction available for several countries show that these abstractions tend to change rather slowly over time and at predictable rates (see Section 5.3), which eases the drawback of having only relatively old data for a number of countries.

Table 5.1 lists abstraction data as collected for selected countries in the world with relatively large groundwater abstraction, for the most recent year available. It gives a provisional impression of the magnitude of groundwater withdrawals around the world. Next, the groundwater abstraction data found in national and international databases, including recent updates (IGRAC, 2010; AQUASTAT, 2011 and 2012;

Table 5.1 Groundwater abstraction and use in selected countries with significant groundwater exploitation *(according to national and international data sources).*

Country	Reference year	Total groundwater abstraction (km³/year)	Share (%) in total water abstraction (including reservoir evaporation losses)	Mean abstraction per capita (m³/year)	Use (% of total groundwater abstraction) Public water supply	Agriculture (Irrigation)	Industry (incl. mines and energy)
AFRICA							
Algeria	2000	2.60	39	85	32	66	2
Egypt	2000	2.21	3	33	59	41	0
Libya	2000	4.27	95	806	14	83	3
Morocco	2005	2.91	31	98	14	83	3
South Africa	2003	2.84	16	61	10	84	6
Tunisia	2005	1.88	66	192	19	74	7
NORTH AMERICA & THE CARIBBEAN							
Canada	1990	1.00	2	36	43	43	14
	1995	1.87	4	64	–	–	–
Cuba	1975	3.80	–	420	–	–	–
Mexico	2002	27.4	31	260	22	72	6
USA (conterm.)	2005	104.9	23	348	23	71	6
SOUTH AMERICA							
Argentina	1975	4.20	22	162	12	79	9
Brazil	1987	8.00	15	56	38	38	24
Peru	1973	2.00	–	140	25	60	15
Venezuela	–	1.40	17	–	60	10	30
ASIA							
Afghanistan	2000	5.3	23	233	6	94	0
Bangladesh	1990	10.70	73	100	13	86	1
China	2005	97.7	18	73	20	54	26
India	1990	190.0	29	223	9	89	2
Indonesia	2004	12.5	15	59	93	2	5
Iran	2004	53.1	55	764	11	87	2
Japan	1998	13.2	14	104	29	23	48
	2002	10.94	13	86	–	–	–
Kazakhstan	2000	2.40	7	140	21	71	8
Korea (South)	1995	2.50	10	56	83	17	0
	2002	3.40	13	72	–	–	–
Myanmar	2000	2.99	–	67	–	–	–
Pakistan	2000	55.0	32	389	6	94	0
	2008	61.1	–	366	–	–	–
Philippines	1994	5.86	10	87	36	0	64
Saudi Arabia	2006	21.37	92	870	5	92	3
Syria	2003	9.18	57	510	5	90	5
Tajikistan	1994	2.30	19	377	31	39	30
Thailand	2007	9.80	–	144	–	–	–
Turkey	2004	6.30	14	88	32	60	8
	2007	12.1	–	173	–	–	–

(Continued)

Table 5.1 (Continued)

Country	Reference year	Total groundwater abstraction (km³/year)	Share (%) in total water abstraction (including reservoir evaporation losses)	Mean abstraction per capita (m³/year)	Use (% of total groundwater abstraction)		
					Public water supply	Agriculture (Irrigation)	Industry (incl. mines and energy)
Un.Arab Emirates	1995	1.60	76	840	0	80	20
	2005	3.05	–	750	–	–	–
Uzbekistan	1994	7.40	12	325	32	57	11
Yemen	2000	2.40	71	131	11	86	3
EUROPE							
Belarus	1989	1.20	–	117	–	–	–
France	2009	6.02	18	99	60	19	21
Germany	2001	6.20	16	75	65	2	33
	2007	5.82	–	69			
Greece	2000	3.65	38	332	14	86	0
	2007	3.65	–	323			
Italy	1998	10.40	24	181	23	67	10
Poland	2002	2.66	23	69	54	39	7
Portugal	2000	4.75	54	474	7	89	4
Romania	2002	0.86	12	38	76	6	18
Russian Fed.	1996	14.6	19	100	79	3	18
	1999	11.6	–	79			
Spain	2008	5.7	16	131	23	72	5
Ukraine	1989	4.22	14	82	30	52	18
United Kingdom	2000	2.35	15	40	77	9	14
	2006	2.16	–	36	–	–	–
OCEANIA							
Australia	2000	4.96	14	260	47	52	1

EUROSTAT, 2011; Siebert *et al.*, 2010), have been extrapolated to the year 2010, taking into account the trends in the available time series of national groundwater abstraction. The results for individual countries are listed in Appendix 5. Estimates of the groundwater abstraction aggregated by continent and differentiated according to water use sector are presented in Table 5.2.

Figure 5.2 is a global picture based on these estimates for 2010, expressing the abstraction intensity as an equivalent layer of water, averaged over each country. It highlights the differences in abstraction between countries even better than Table 5.1. Like Table 5.2, it shows that the abstractions are unevenly distributed over the continents. Within each continent, however, the abstraction intensity does not only vary considerably from country to country (see Figure 5.2), but also inside each country.

In spite of flaws in the data, the world's total annual groundwater abstraction as of the year 2010 is estimated to be around 1 000 km³/year. Worldwide, this is more than the annual volume of any other raw material extracted from the subsurface. The current share of groundwater in the total water withdrawal on Earth is approximately 26%.

Table 5.2 Estimates of total groundwater abstraction by continent, based on extrapolation (reference year: 2010).

| Continent | Groundwater abstraction[1] | | | | | Compared to total water abstraction | |
| | Irrigation | Domestic | Industrial | Total | | Total water abstraction[2] | Share of groundwater |
	(km³/yr)	(km³/yr)	(km³/yr)	(km³/yr)	(%)	(km³/yr)	(%)
North America	102	33	8	143	14.6	524	27
Central America and the Caribbean	4	8	2	14	1.4	149	9
South America	13	8	5	26	2.6	182	14
Europe (including Russian Federation)	26	33	13	72	7.3	497	14
Africa	26	13	2	41	4.1	196	21
Asia	514	111	55	680	69.3	2 257	30
Oceania	3	3	0	7	0.7	26	25
WORLD	688	209	85	982	100.0	3 831	26

Data sources:

[1] Estimated on the basis of IGRAC (2010), AQUASTAT (2011), EUROSTAT (2011), Margat (2008) and Siebert *et al.* (2010)

[2] Average of the 1995 and 2025 'business as usual scenario' estimates presented by Alcamo *et al.* (2003)

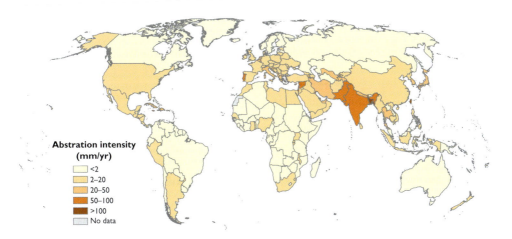

Figure 5.2 Intensity of groundwater abstraction by country for the year 2010, in mm/year (*1 mm/year = 1 000 m³/year per km²*). [*Data sources*: IGRAC (2010), AQUASTAT (2011), EUROSTAT (2011), Margat (2008) and Siebert *et al.* (2010)].

The ten countries that abstracted the largest quantities of groundwater throughout the year 2010 – according to the compiled and processed statistics – are listed in Table 5.3, in order of decreasing importance and with a specification of their estimated total withdrawals. Around 72% of the assumed total global groundwater withdrawal takes place in these ten countries. In fact, the exploitation of groundwater is

Table 5.3 Top 10 of groundwater abstracting countries (as per 2010).

	Country	Abstraction (km³/year)
1	India	251
2	China	112
3	USA	112
4	Pakistan	65
5	Iran	64
6	Bangladesh	30
7	Mexico	29
8	Saudi Arabia	24
9	Indonesia	15
10	Turkey	13

huge and highly concentrated in a limited number of regions, especially in Asia, while it is relatively modest in Africa (where only Egypt abstracts more than 5 km³ per year) and in South America (except Brazil).

Several factors may explain the uneven distribution of groundwater withdrawal around the world. In the first place, it goes without saying that total groundwater abstraction by country is positively correlated to the country's size. But after removing this effect – e.g. by conversion to groundwater abstraction intensity in mm/year (equivalent to 1 000 m³/year per km²) – there are still huge differences. These are related to factors such as population density, characteristics of the local economy, climate, water demands (e.g. is irrigation needed or not?), the presence or absence of surface water resources and the relative attractiveness of groundwater compared to surface water (cost, reliability, risk). Differences in local knowledge on the groundwater potential, in appropriate technical means, in traditions and in water governance also contribute to explaining the variation in groundwater abstraction intensities.

There are very large differences between countries in average groundwater withdrawal per capita, as shown in Table 5.1 and Figure 5.3. In general, countries abstracting the highest quantities of groundwater per capita are located in the arid zones, where surface water is scarce and unreliable and where irrigation is well developed. Values vary from 500 to 1 000 m³/year for several arid countries to only a few tens of m³/year or less for most countries in tropical Africa.

5.3 EVOLUTION OF GROUNDWATER WITHDRAWALS DURING THE TWENTIETH CENTURY

5.3.1 Evolution and trends

The available information only partially reveals how groundwater withdrawals in the world have evolved over time. The records are usually limited to the last few decades of the twentieth century. It is therefore often difficult to identify significant trends, even more so because most available time series consist of a limited number of values,

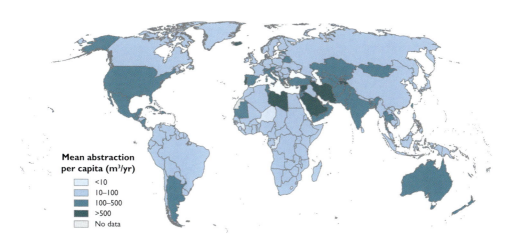

Figure 5.3 Countries classified according to mean groundwater abstraction per inhabitant. [*Data sources*: IGRAC (2010), AQUASTAT (2011), EUROSTAT (2011), Margat (2008) and Siebert *et al*. (2010)].

while the records may be heterogeneous and there are sometimes alternative data sets. In addition, differences between values for different years not only reflect trends but also short-term variations, e.g. in response to variable weather conditions.

Nevertheless, relatively well-defined time series of aggregated groundwater abstraction exist for a number of countries. They are presented graphically in Figure 5.4. The most plausible of identified alternative data sets have been selected for India, China, Pakistan and Mexico. The graph shows differences in information density between the countries, as well as some uncertainty on how to define the trends.

At first glance, groundwater abstraction during the second half of the twentieth century follows a pattern quite similar to that of total water withdrawals. The most pronounced increases have been observed in countries where current groundwater withdrawals are highest. Periods of maximum growth are not simultaneous, but vary from country to country, e.g. 1950–1970 in the United States, 1960–1990 in India and 1975–2000 in China. The observed boom in groundwater abstraction, driven by population growth and the associated increasing demands for water, food and income, has no precedent in history and it has proceeded in waves. It started during the first half of the 20th century in a limited number of countries such as Italy, Mexico, Spain and the USA, followed by a second wave since the 1960s in parts of South and East Asia, the Middle East and Northern Africa. A perceived third wave since the 1990s includes some South- and South-east Asian countries such as Sri Lanka and Viet Nam, as well as much of Sub-Saharan Africa (Shah *et al*., 2007).

In developed countries, a strong but variable increase in groundwater abstraction in the earlier stages has been followed by stabilising or declining trends. The USA provides a typical and most reliable example, with a growth of 144% in 30 years (from 1950 to 1980), followed by a temporary decline but a stable average between 1980 and 2005 (Figure 5.5). A similar case is offered by Japan, where abstraction increased by 60% in 30 years (from 1965 to 1995), before declining by 13% between 1995 and 1999.

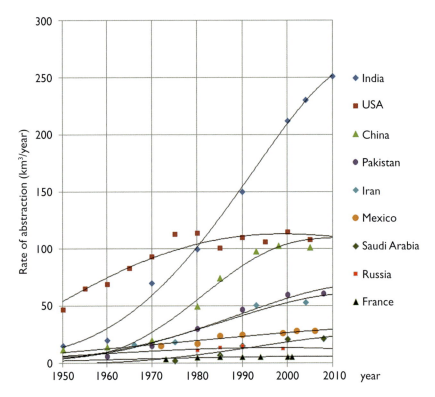

Figure 5.4 Evolution of aggregated groundwater abstraction (in km³/year) from 1950 onwards for a number of countries with intensive groundwater exploitation.

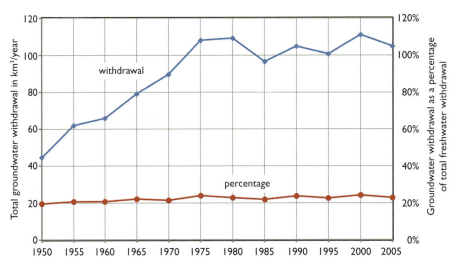

Figure 5.5 Evolution of total fresh groundwater withdrawals in the USA, 1950–2005. [*Data source*: USGS, 2009].

Periods of marked growth have also been observed in several European countries, although generally at lower exploitation levels. They include more than 54% of growth in 25 years in the UK (1950–1975), more than 70% in seven years in Denmark (1970–1977), more than 15% in ten years in Spain and in the former Soviet Union, and more than 12% in five years in The Netherlands (1971–1976). Groundwater abstraction has remained relatively stable over time in Germany, Belgium, France, Sweden and Canada. Since the beginning of the 21th century, almost all European countries have shown stabilisation or slightly declining groundwater abstraction trends. In Australia, groundwater abstraction has almost tripled in 30 years, during the period 1970–2000.

In developing countries, strong increases in groundwater abstraction have generally been observed from the period 1970–1980 onwards, especially in countries with strong demographic and economic growth, and where irrigation has expanded significantly, such as India and China. Large increases have occurred in countries of the arid and semi-arid zone where oil revenues facilitated deep groundwater (including non-renewable groundwater resources) to be withdrawn for irrigation. Total groundwater withdrawal has increased by a factor of 11 in Libya between 1970 and 2000; by 10 in Saudi Arabia between 1975 and 2000; by 6 in Egypt between 1972 and 2000; by 3.3 in Iran between 1965 and 1995; and by 3.2 in Tunisia between 1977 and 2000.

The share of groundwater in the total water withdrawals has often increased over time, even during periods when these total withdrawals tended to stabilise. In the United States, for example, this share increased from 20 to 24% between 1950 and 2000 (Figure 5.5).

The incredibly rapid intensification of groundwater exploitation in Asia and several countries in the arid and semi-arid zone elsewhere, mostly for irrigation, is one of today's major trends and constitutes a true *'Silent Revolution'* (Llamas and Martinez-Santos, 2005a, 2005b, 2006), full of consequences. In the vision of the cited authors, intensive exploitation of groundwater appears in water policies in most countries of the arid and semi-arid zone after a stage of *'hydroschizophrenia'* (see Figure 5.6). During the hydroschizophrenia stage, government water development plans are devoted exclusively to surface water and ignore the possible exploitation of groundwater. Intensive groundwater exploitation then develops as a largely market-driven phenomenon, often leading to conflict situations where increases in the value of the products of irrigated agriculture and in irrigation costs are accompanied by a disturbance of the hydrological equilibrium of the groundwater systems. This trend towards intensive exploitation happens in the short to medium term and may not be sustainable due to physical or socio-economic constraints and due to the increasing negative impacts of intensive exploitation.

5.3.2 Outlook

National projections for water demands and abstractions are almost never specified by source of water. Therefore, every single estimate of the quantities of groundwater to be developed in the future is based on assumptions.

One of the most plausible assumptions is that current trends will continue during the decennia to come, particularly when groundwater has a large share in total water

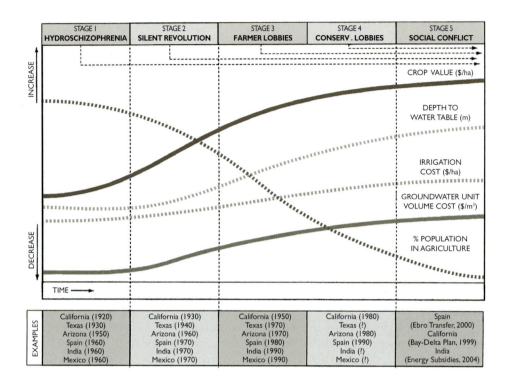

Figure 5.6 Groundwater policy trends in arid and semi-arid countries. [*Source*: R. Llamas & P Martínez-Santos (2005a)].

supply. Stabilisation or a slight decrease in groundwater abstraction is observed in most industrialised countries. This is in contrast to a still relatively strong but decelerating increase in countries where economic and demographic growth is considerable and where irrigation makes significant use of groundwater, particularly in Asia (India, China).

However, with some variations, a stabilisation or a decrease in abstraction could be anticipated in countries that, under a sustainable development policy, successfully stop excessive exploitation. A similar tendency would be observed in countries where a reduction in mining of non-renewable groundwater resources is forced by factors such as rising pumping costs, salinization risks and depletion of the reserves. This is expected to occur in several Arab countries.

On the other hand, a trend towards increasing groundwater withdrawal is to be expected in areas where the groundwater potential is still sparsely utilised, especially if options for developing surface water resources are exposed to constraints (irregular regime, silting-up of reservoirs, effects of climate change).

Nevertheless, the future global picture will be dominated by what will happen in those countries that currently withdraw the largest quantities of groundwater: India, USA, China and the other countries shown in Table 5.3.

5.4   THE CONTRIBUTION OF GROUNDWATER RESERVES

The exploitation of groundwater is mainly – and in many countries completely – a matter of capturing renewable resources. Abstracted groundwater is in such cases balanced by a reduction in the natural groundwater discharge and/or by induced additional recharge, which means that the groundwater reserves remain intact in the longer term. However, a proportion of the total groundwater withdrawal on Earth is balanced by groundwater storage depletion. In this regard, a distinction can be made between groundwater mining and over-abstraction of renewable groundwater, depending on the conditions of the exploited aquifer. Given the lack of an unambiguous definition of 'non-renewable groundwater resources' in general use among hydrogeologists (see Section 4.6), this distinction is in practice somewhat arbitrary, which may lead to inconsistencies in and between data sets. Countries for which such inconsistencies have been found include Australia, Niger, South Africa, Sudan, Oman and Yemen. In this book, groundwater resources are considered 'non-renewable' only in cases where the current groundwater recharge is negligible to the extent that local authorities consider it irrelevant as a criterion for planning and managing groundwater development.

5.4.1   Groundwater mining: exploiting fossil groundwater

Groundwater mining is abstracting groundwater from a non-renewable groundwater resource. Since the exploited aquifer in such cases is assumed to receive no significant contemporary recharge, the water stored in it is called *fossil groundwater*. Groundwater mining is comparable to any other form of mining: a finite volume is present and can be exploited. However, it should be kept in mind that slight discrepancies between the abstracted groundwater volumes and storage depletion are possible, because the aquifer may receive minor amounts of recharge and some natural discharge may still continue, especially if the aquifer system is very large.

How much water is currently being abstracted worldwide by mining non-renewable groundwater reserves? According to available statistics, undoubtedly incomplete, mining of groundwater at the end of the twentieth century may have been around 31 km^3/year worldwide (see Table 5.4 and Figure 5.7). Hence, the mass of groundwater mined annually in the world is nearly ten times that of oil (on the basis of data for 2001). Tentatively, it may be assumed that the annual volume of mined fossil groundwater by the year 2010 will have increased to some 35 km^3/year. This rate is only 3.5% of the estimated global groundwater withdrawal, and only 1.5% of total water withdrawal, thus a minimal share. However, this percentage is much higher and sometimes very high for some countries that rely on fossil groundwater (see Figure 5.8). One should bear in mind that the data inconsistencies mentioned before add some uncertainty to the contents of Table 5.4 and the Figures 5.7 and 5.8.

Groundwater mining is geographically concentrated. Four countries – Saudi Arabia, Algeria, Libya and the United Arab Emirates – have together a share of 86% of the total estimated global groundwater mining. These countries have an economy that depends largely on mining (oil), and hence they are familiar with non-renewable resources, even if this is water. Groundwater mining is a speciality of the arid and hyper-arid zone: almost all of it (96%) takes place in Arab countries.

Table 5.4 Mining of non-renewable groundwater resources around the world (*according to national statistics – countries listed in decreasing order of mining rate*).

Country	Year	Arabian Multilayer Aquifer System	NW Sahara Aquifer System	Nubian Aquifer System	Murzuk Basin	Great Artesian Basin	Disi	Maestrichtien	Taoudeni-Tanezrouft Basin	Karroo	Iullemeden	TOTAL	References
Saudi Arabia	2006	20.04										20.04	7
Libya	2000		0.55	0.90	1.75							3.20	1,2,4
Algeria	2000		1.68									1.68	2,4
UAE	2005	1.57										1.57	6
Egypt	2000			0.91								0.91	8
Tunisia	2000		0.68									0.68	2,4
Australia	2000					0.63						0.63	3,4
Kuwait	2002	0.42										0.42	6
Sudan	1998			0.41								0.41	4
Jordan	2005						0.21					0.21	6
Mali	2000								0.20		0.00	0.20	4,5
Senegal	2003							0.18				0.18	4,5
Qatar	2005	0.16										0.16	6
Bahrain	1999	0.10										0.10	6
South Africa										0.10		0.10	9
Mauritania	2003							0.09				0.09	4,5
Israel	2004			0.05								0.05	6
Botswana	2000									0.03		0.03	6
Niger	2000										0.01	0.01	4
TOTAL		22.29	2.91	2.27	1.75	0.63	0.21	0.27	0.20	0.13	0.01	30.67	

References:
1. Bakhbakhi, 2002
2. OSS, 2003
3. Habermehl, 2006
4. Foster and Loucks, 2006
5. OSS, 2005
6. AQUASTAT, 2008
7. AQUASTAT, 2009
8. Plan Bleu, 2008
9. Margat, 2008

In the countries with highest withdrawal of non-renewable groundwater, this resource is a significant, usually predominant source of water. It covers a major part of the total water demands (predominated by irrigation water demand): 85% in Saudi Arabia, 62% in Libya, 39% in the United Arab Emirates and 25% in Algeria (~ 100% in its Saharan part). Mining non-renewable groundwater resources supports economic development in the short and medium term, particularly as it enables

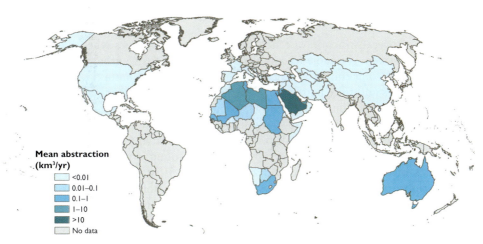

Figure 5.7 Abstraction of non-renewable groundwater by country (at the beginning of the 21st century). [*Data sources*: sources mentioned in Table 5.4 (19 countries) and selected other sources (mainly AQUASTAT and Plan Bleu) for 28 additional countries].

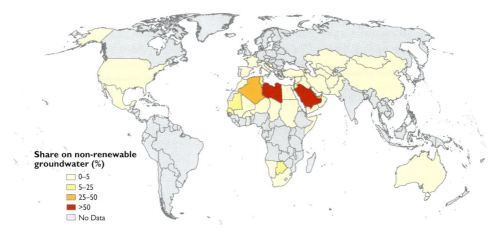

Figure 5.8 Percentage of total water withdrawal covered by non-renewable groundwater (at the beginning of the 21st century). [*Data sources*: those mentioned under Figure 5.7].

millions of hectares to be irrigated. However, it is clear that this development is not sustainable in the longer term.

It should be noted that there is much uncertainty and inconsistency in available data on non-renewable groundwater abstraction as found in different sources. This uncertainty and inconsistency results mainly from the lack of uniform and unambiguous criteria to define whether groundwater resources are renewable or non-renewable. Hence, groundwater resources classified by some as renewable may be classified by others as non-renewable. Consequently, the data presented in Table 5.4 and the patterns shown in the Figures 5.7 and 5.8 are also subject to uncertainty. Data has been accepted as referring to 'non-renewable groundwater' only if it was related to a specified aquifer system and if it was judged correct to classify groundwater in this aquifer system as non-renewable.

5.4.2 Hydrological overabstraction of renewable groundwater

Hydrological overabstraction of renewable groundwater occurs when the exploited groundwater system is unable to completely balance intensive groundwater abstraction rates in the long run by induced recharge and reduced natural discharge[2]. The abstraction rates and their pattern then are hydrologically unsustainable and cause a steady depletion in groundwater storage, eventually leading to complete exhaustion. The phenomenon is observed especially in areas where low to moderate groundwater recharge coincides with high demands for groundwater. Consequently, the most prominent areas of groundwater overabstraction are found in relatively densely populated parts of the arid and semi-arid zone. Table 5.5 lists areas in the world that are notorious for hydrological overabstraction of renewable groundwater resources. Most of them are located in the Western USA, Mexico, Southern Europe, the Middle East, and South and East Asia. The data on the depletion rates is of varying quality and particularly that based on calculation by difference (flux-based approach) is subject to considerable uncertainty. Nevertheless, the listed values give an idea of the order of magnitude of the depletion rates.

It is estimated that hydrological overabstraction of renewable groundwater in the world is currently between 100 and 150 km^3/year, which is significantly more than the annual volume of groundwater mined. Although the physical boundary conditions theoretically allow aquifers to recover from overabstraction, in practice this remains an illusion in most cases. In spite of some positive exceptions, strong demographic and economic pressures, combined with the absence of effective groundwater governance, often present insuperable obstacles to efforts intended to stop or reverse groundwater depletion trends. Therefore, like mined non-renewable aquifers, many aquifers with renewable groundwater affected by over-abstraction are doomed to become physically or economically exhausted during the twenty-first century. The more intensive the overabstraction, the sooner the exhaustion of the exploitable reserves will become reality.

5.4.3 Evolution of groundwater storage depletion around the world

Assessing global groundwater depletion accurately is difficult and it is even more difficult to estimate its evolution over time. Nevertheless, Konikow (2011) recently produced a well-documented attempt, the results of which are summarised in Figure 5.9. Due to the better data availability, groundwater depletion could be assessed with much more confidence for the USA than for elsewhere; for estimating depletion outside the USA, sufficient data was available for only five groundwater systems with large volumes of depletion. Estimates for the remainder of the world were made by assuming that the share of depletion in the total groundwater withdrawal would be equal to that observed in the USA; this component of the global depletion estimate is thus rather uncertain. As a result, the estimated cumulative global groundwater deple-

---

2 'Over-abstraction' is used here objectively (in a neutral sense), without expressing a judgement on whether it should be avoided or not.

Table 5.5 Areas around the world notorious for significant hydrological overabstraction of renewable groundwater.

Aquifer, region or country	Lateral extent (km²)	Estimated rate of depletion (in recent years)			Period or year	Reference
		(km³/yr)	Method*	mm/yr water**		
NORTH AMERICA						
High Plains, USA	483 844	12.4	1	26	2000–2007	McGuire, 2003; 2009
Central Valley, California, USA	58 000	3.7	3	64	2003–2009	Famiglietti et al., 2009
Western Alluvial Basins, USA		0.29	5		2001–2008	Konikow, 2011
Atlantic & Gulf Coast, USA		8.75	5		2001–2008	Konikow, 2011
Scattered basins, Mexico		5.4	6		2005?	Chávez Guillén, 2006
SOUTH AMERICA						
Ica Province, Peru		0.284	6		2008?	Bayer, 2009
EUROPE						
Upper Guadiana Basin, Spain	18 900	0.25	2	13	1998 (peak)	Villarroya Gil, 2008
Segura Basin, Spain	18 870	0.27	2	14	2009	Cabezas, 2011
Canary Islands, Spain		0.16	6		1993	Plan Hidrológico Nacional, 1993
ASIA						
Arid margins of Central Syria						Jaubert et al., 2009
Yemen (countrywide)		1.4	2		2001	ESCWA, 2007
Oman (countrywide)		0.6	2		2003	ESCWA, 2007
UAE (countrywide)		2.1	2		2002	ESCWA, 2007
Bahrain, Dammam aquifer		0.11	2		2001–2002	Zubari, 2005
Iran (countrywide)***		3.2	2		2004?	Rahmatian, 2005
N. India & adjacent areas	2 700 000	54	3	20	2003–2009	Tiwari et al., 2009
(including: NW India	438 000	17.7	3	40	2003–2009	Rodell et al., 2009)
North China Plain	136 000	3.52	4	26	2000–2008	Liu et al., 2011

* Methods used for estimating depletion:

1. Direct volumetric accounting based on terrestrial observations (groundwater levels)

2. Flux-based approach (depletion calculated by difference)

3. Volumetric accounting based on remote sensing (GRACE)

4. Groundwater flow modelling

5. Combination of methods, including volumetric accounting

6. Method not specified

** Expressed as depth of an equivalent layer of water over the total horizontal extent of the aquifer system (scale-independent depletion indicator)

*** Includes scattered mountain basins, such as Varamin, Kashmar, Masshad, Rafsanjan, Yazd, Zarand (Motagh et al., 2008)

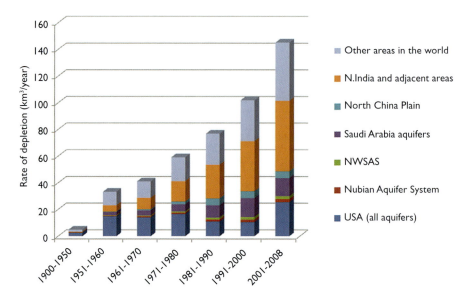

Figure 5.9 Estimated net global rates of groundwater depletion, period 1900–2008. [*Data source*: Konikow, 2011].

tion during the entire 20th century is around 3 400 km³ and for the period 1900–2008 it amounts to 4 500 km³. The estimated mean annual rate of global depletion during the period 2000–2008 is 145 km³/year. This is significantly lower than the rate of 256 km³/year estimated for the year 2000 on the basis of a global model by Wada *et al.* (2010), who used a flux-based approach (sensitive to error accumulation) and a rather simple algorithm.

Considering the results reported by Konikow as the best estimates currently available, a few observations can be made:

- Groundwater depletion started becoming a globally relevant phenomenon only after 1940 and global groundwater depletion rates have been accelerating ever since. The practical implications of this unprecedented storage depletion require further study.
- Current rates of depletion are rapidly accelerating especially in Northern India and adjacent areas.
- Over the period 1900–2008, cumulative groundwater depletion in Northern India and adjacent areas was around 30% of the global total and that in the USA was almost 23%.
- The cumulated global groundwater depletion between 1900 and 2008 would have contributed 12.6 mm to sea level rise, which is 6–7% of total sea level rise observed during that period.

It should be born in mind that the construction of surface water reservoirs has an effect on sea level rise that is opposite to that of groundwater depletion. Over the last

50 years, surface reservoirs have withheld a cumulative 10 800 km³, which would have reduced sea level rise by 30 mm (Jones, 2011).

5.5   PURPOSES FOR WHICH GROUNDWATER IS USED

Most – but not all – uses of groundwater require groundwater to be abstracted. Strictly speaking, there is a difference in volume between groundwater abstracted for a certain purpose and groundwater effectively used or even consumed for the same purpose. This is because a fraction of the abstracted water tends to be lost during conveyance or at the water use location (see Box 5.3). Numerical information on these losses is usually not available. Unless indicated otherwise, the fractions or quantities mentioned below for any specified type of use should therefore be interpreted as quantities or fractions of water abstracted for that purpose.

5.5.1   Main categories of groundwater use

Groundwater use is often subdivided into three main categories: irrigation water, domestic water and industrial water. Even if the available statistics follow this subdivision, there are many uncertainties and inconsistencies left; e.g., many irrigation wells are also used for rural domestic water supply (a minor fraction of the water pumped), while in other cases domestic water supply systems may contribute to small-scale irrigation. The domestic water sector includes not only water for households, but also for offices, public services and small industries, as far as served by public water supply systems. Industrial water use statistics are likely to be limited to water withdrawn by the industries themselves, but practices in preparing these statistics may vary between countries.

According to the processed information (Table 5.2), the global shares are as follows (see Table 5.6):

*   70% of all groundwater abstracted in the world (some 688 km³/year) is intended for irrigation;
*   21% (around 209 km³/year) is meant for domestic use;
*   the remainder, 9% or approximately 85 km³/year, is abstracted for industrial use or pumped for mining purposes.

These shares, however, vary substantially from one continent to another, as can be observed in Table 5.6 and Figure 5.10. In Europe as well as in Oceania and in Central America and the Caribbean, approximately half of all groundwater pumped is for the domestic use sector, compared to less than a quarter in Asia and North America. Approximately two-thirds to three-quarters of the groundwater withdrawals in North America, Asia and Africa are for irrigation – elsewhere the corresponding percentages are less.

The shares of the major water use sectors vary even more at the level of individual countries. This has already been shown in Table 5.1 but more information can be found in Appendix 5. Figure 5.11 shows the dominant use of abstracted groundwater by country. It shows simply the sector among the three considered that has the highest share in groundwater abstraction, which is only moderately informative since this share may still range from 33.4% to 100%. Some more details on groundwater abstraction and use in each sector will follow.

Box 5.3 What proportion of abstracted and used groundwater is really consumed?

Abstracted and used water – surface water and groundwater alike – is never fully consumed: part of it returns to the terrestrial water systems. The share that is consumed, i.e. either escaping to the atmosphere in the form of vapour or transformed into a constituent of a living organism or a manufactured product, varies enormously according to the type of use and local conditions.

Information on the proportion of abstracted groundwater really consumed is scarce, except for the irrigation sector. The really consumed share of groundwater abstracted for irrigation (known as 'consumptive use') depends on the overall irrigation efficiency of the irrigation units concerned and is on average in the order of 80%.

In the domestic and industrial water sectors, generally a much smaller share of abstracted groundwater is really consumed, but the fractions are highly variable. If water has to be fetched with considerable effort and carried over some distance before it is used at home, then it is likely that a rather large share, maybe even more than 50%, is consumed for drinking and cooking. Urban households in industrialised countries, on the other hand, consume only a few per cent of all supplied domestic water, while the large majority is used for cleaning and washing purposes. Where domestic water is also used for watering gardens, then the percentage consumed may increase somewhat. Nevertheless, only a minor fraction of the global quantity of groundwater abstracted for domestic purposes is really consumed, probably around 15%. This share is even less for groundwater used in the industrial sector: globally aggregated this is estimated to be in the order of only 5%.

Combining the estimated shares with the breakdown of global groundwater abstraction by sector (Table 5.2) leads to the conclusion that probably around 60% of all abstracted groundwater on the globe is really consumed. Part of the non-consumed 40% is lost during conveyance between the site where groundwater is withdrawn and the location where it is used, e.g. by seepage from irrigation canals or leaks from domestic water distribution networks. This water, usually of unchanged quality, often returns to the aquifers. The remainder, on the other hand, may become severely polluted during water use and more commonly is discharged into surface waters or eventually into the sea. In the case of domestic and industrial water use, this non-consumed surplus is called wastewater, which in an increasing number of countries gets treated before being discharged into open water.

In many countries, the *agricultural sector* (or *irrigation sector*) is by far the largest groundwater exploiter and user. This is the case in developed countries where irrigation is widespread, as well as in almost all developing countries outside the humid tropics. Table 5.7 lists the countries abstracting more than one cubic kilometre of groundwater per year, of which more than 50% is for irrigation.

Table 5.6 Breakdown of groundwater abstraction by main use category and by continent (Based on the data of Table 5.2).

Continent	Breakdown by type of use			Breakdown by continent	
	Irrigation	*Domestic*	*Industrial*	*Total*	
	(%)	*(%)*	*(%)*	*(km³/yr)*	*(%)*
North America	71.1	23.0	5.9	143	14.6
Central America and the Caribbean	30.2	56.9	12.9	14	1.4
South America	48.8	31.5	19.7	26	2.6
Europe (incl. Russian Federation)	36.4	45.0	18.6	72	7.3
Africa	64.7	31.7	3.6	41	4.1
Asia	75.6	16.3	8.1	680	69.3
Oceania	48.5	48.1	3.4	7	0.7
WORLD	70.1	21.2	8.7	982	100.0

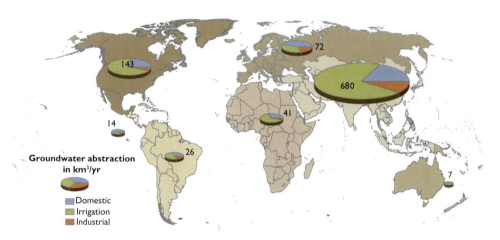

Figure 5.10 Groundwater withdrawal for each continent (in km³/year) and its breakdown by water use sector. [*Data and data sources:* see Appendix 5].

According to the most recent global irrigation inventory (Siebert *et al.*, 2010), there are currently almost one hundred million groundwater-irrigated hectares on our planet, which corresponds to 39% of the total irrigated area. The data processed in this inventory represents different periods of time for different territorial units, and hence it is impossible to tell exactly for which reference year the inventory is representative. However, the authors state that for many countries and other territorial units the situation around the year 2000 is represented. As Table 5.8 shows, approximately three quarters of the groundwater-irrigated land is located in Asia and another 16% in North America.

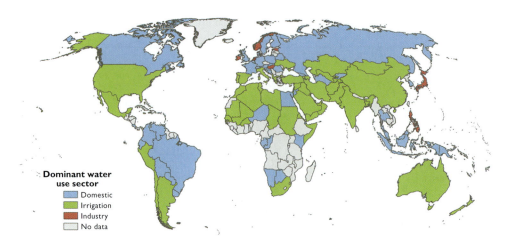

Figure 5.11 Dominant groundwater use sector by country. [*Data and data sources*: see Appendix 5].

Table 5.7 Countries abstracting at least 1.0 km³ of groundwater annually, of which 50 to 100% is used for irrigation *(in descending order of percentage)*.

Country	%	Country	%	Country	%
United Arab Emirates	100	Greece	86	Kazakhstan	71
Afghanistan	94	South Africa	84	Italy	67
Pakistan	94	Libya	83	Algeria	65
Saudi Arabia	92	Morocco	83	Turkey	60
Syria	90	Argentina	79	Peru	60
India	89	Tunisia	74	Uzbekistan	57
Portugal	89	Mexico	72	China	54
Iran	87	Spain	72	Ukraine	52
Yemen	86	Israel	71	Australia	52
Bangladesh	86	USA	71	Iraq	50

Table 5.8 Breakdown of groundwater-irrigated land by continent (after Siebert *et al.*, 2010).

Continent	Groundwater-irrigated area (x 1 000 ha)	Percentage of the continent's total irrigated area	Percentage of the world's groundwater irrigated area
North America	15 738	54.4	16.1
Central America and the Caribbean	330	30.8	0.3
South America	1 558	16.3	1.6
Europe (incl. Russian Federation)	4 817	36.2	4.9
Africa	2 153	18.7	2.2
Asia	72 531	39.1	74.1
Oceania	694	22.7	0.7
WORLD	97 821	38.7	100.0

Figure 5.12 Groundwater-irrigated date palm orchards in the arid Nizwa region, Oman. [*Photo*: Jac van der Gun].

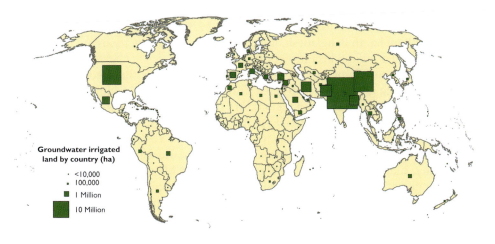

Figure 5.13 Groundwater-irrigated land by country. [*Data source*: Siebert *et al.*, 2010].

The bulk of agricultural groundwater used for irrigation is concentrated in a limited number of countries, as is illustrated by Figure 5.13 and 5.14. According to the most recent global irrigation inventory (Siebert *et al.*, 2010), more than 80% of the land irrigated by groundwater on Earth is found in the following six countries: India (37.7%), China (16.2%), USA (13.8%), Pakistan (4.9%), Iran (4.1%) and Bangladesh (3.5%). Five of these countries are located in Asia.

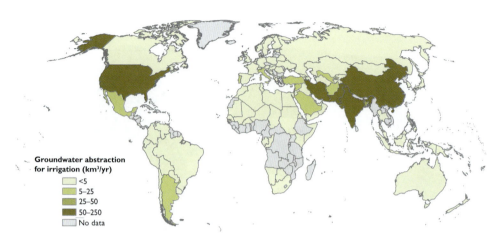

Figure 5.14 Current groundwater withdrawal rates for irrigation, by country. [*Data and data sources: see* Appendix 5].

The breakdown shown in Table 5.8 highlights the very prominent position of Asia with respect to groundwater-based irrigation, followed by North America. In Asia, irrigation by pumped groundwater has been expanding explosively during recent decades. For instance, the groundwater-irrigated area in India has doubled from the period 1970–1975 (13 million ha) to 1990–1993 (26.5 million ha) and is currently around 37 million ha. In Bangladesh, it has increased from a few thousand hectares in 1973 to almost 3.5 million hectares at present.

Public water supply or *private domestic water supply* is the predominant purpose of abstracting groundwater in most European countries (including Russia, but except those in the Mediterranean region), in Canada and in most countries in South America, tropical Africa and South-East Asia. Table 5.9 lists countries for which it is known that more than half of their total groundwater abstraction is for domestic use. The largest exploiters of groundwater for the domestic water sector are the USA, India, China, Indonesia, Russian Federation, Iran and Mexico; together they pump half of the global total. Well-developed public water supply networks usually contain a substantial technical infrastructure of physical components below and above ground surface (see Figure 5.15).

The *industrial sector* is the largest groundwater-using sector in some industrialised countries (Estonia, Norway, Hungary, Japan, The Netherlands and Ireland), while it is second in terms of the volume of groundwater used in countries such as Germany, Belgium, France, Austria, the UK and the Czech Republic. Countries using more than 50% of all pumped groundwater for industry are only Estonia (80%), Norway (73%) and the Philippines (64%), as far as is known. In absolute terms, the largest withdrawers of groundwater for industrial purposes are China, the USA, Japan, India, Thailand and Germany.

The Figures 5.14, 5.16 and 5.17 offer a visual impression of the quantities of groundwater pumped in each country to supply each of the three main water use

Table 5.9 Countries where 50 to 100% of the groundwater abstraction is for domestic use (in descending order of percentage).

Country	%	Country	%	Country	%
Gabon	100	Russian Federation	79	Bolivia	60
Montenegro	100	United Kingdom	77	Ecuador	60
Maldives	98	Congo	76	Thailand	60
Indonesia	93	Switzerland	72	Uruguay	60
Sweden	92	Costa Rica	71	Venezuela	60
Belize	90	Poland	70	Egypt	59
Papua New Guinea	90	Bosnia & Herzegovina	67	Niger	58
Djibouti	89	Georgia	66	Namibia	55
Croatia	86	Finland	65	Belgium	55
Niue	85	France	63	Turkmenistan	53
Slovakia	84	Malaysia	62	Austria	52
Republic of Korea	83	Botswana	61	Belarus	52
Slovenia	83	Romania	61	Jordan	51
Barbados	80	American Samoa	60	Kyrgyzstan	50
Kenya	79	Antigua and Barbuda	60	Paraguay	50

Figure 5.15 Water tower in Deventer, The Netherlands. Built in 1892 and still in use to maintain a constant pressure in the urban drinking water network. Groundwater is – after minor treatment – pumped into this tower's reservoir, from where it enters the public water supply distribution network. [Photo: Jac van der Gun].

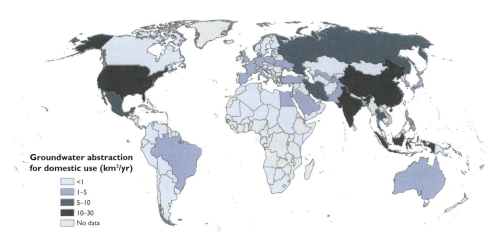

Figure 5.16 Current groundwater withdrawal rates for domestic water supply, by country. [*Data and data sources*: see Appendix 5].

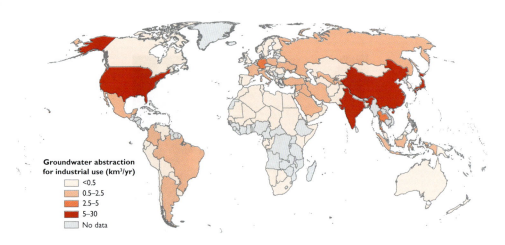

Figure 5.17 Current groundwater withdrawal rates for industrial water use, by country (that which is not provided by public water supply). [*Data and data sources*: see Appendix 5].

sectors. Figure 5.18, finally, provides an integrated picture of the relative shares of the quantities pumped for each sector, by using a pattern of colour gradations according to these proportions (see legend).

5.5.2 Other uses of groundwater

Other uses of groundwater, of more secondary importance and usually limited to small quantities, are often not included in the national statistics on water use, although they may be important for the local economy:

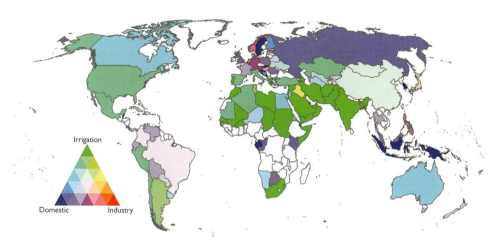

Figure 5.18 Groundwater abstraction proportions according to water use sectors. [*Source:* After an earlier version by IGRAC in WWDR2 (2004), updated on the basis of data shown in Appendix 5].

- *Thermal baths or spas*

Groundwater is the raw material for facilities and activities in this field, developed for therapeutic or touristic purposes. The undeniable importance of this type of use is growing in many countries.

- *Mineral water and bottled water*

Mineral water is natural water with a special taste or therapeutic value attributed to the presence and specific concentrations of certain dissolved substances. Traditionally, mineral water was consumed at its source (at spas or springs), such as at Vichy in France, at Spa in Belgium and at Karlovy Vary in the Czech Republic, but in modern times most of the exploited volume of mineral water is bottled for consumption elsewhere. Bottled water nowadays includes also ordinary pumped groundwater that does not qualify to be called 'mineral water'; it is especially in demand in areas where public water supply agencies distribute water of unreliable or poor quality. Most of the water used in the bottled-water industry is more than likely included in national statistics on industrial water use.

- *Geothermal energy*

Groundwater is essential in the exploitation of geothermal energy. Geothermal resources are commonly subdivided into low-temperature resources and high-temperature resources. The *low-temperature resources* category utilises warm to relatively hot groundwater (30° to 90°C) of deep aquifers for direct heating purposes (buildings, urban heating, greenhouses, fish farms, thermal baths). The standard technology uses pairs of wells (doublets), each consisting of a pumped well and an injection well. Because the water is re-injected, there is no significant consumptive use of groundwater. Geothermal energy development is significant in various countries rich in groundwater of anomalously high temperature (geothermal water), such as Iceland, Hungary, Japan, New Zealand, Germany and China. In Germany 14 million m³ of geothermal water was used in 2004 and in China 10, 12 and 18 million m³ are

Figure 5.19 Discharge of geothermal water at Hammam Dammt, Yemen [*Photo*: Jac van der Gun].

currently being exploited on an annual basis in Beijing, Tianjin and Fujian Province, respectively. Low-temperature geothermal resources are also developed in areas of deep hot groundwater with a normal temperature regime, for example in the Paris Basin in France, where approximately thirty doublets have already been in use for more than twenty-five years, with a total yield of nearly 8 000 m³/hour.

Geothermal resources in the *high-temperature* category consist of deep very hot water (> 90°C) or steam, used for the generation of electricity in geothermal power plants. The water used is usually saline and unfit for any other use. This type of geothermal resources has been developed especially in volcanic regions in several countries: Germany, Italy, Iceland, Russia, Mexico, USA (California), Indonesia, The Philippines, Japan, Australia, Papua New Guinea, New Zealand, the Azores and the French West Indies (Guadeloupe).

The total installed geothermal power capacity in the world has increased from 8.9 GW in 2005 to 10.7 GW in 2010. During the latter year, a total of 67 246 GWh was generated in 24 countries. The USA leads with the largest installed capacity, followed by the Philippines, Indonesia, Mexico, Italy, New Zealand, Iceland and Japan (Holm *et al.*, 2010). Geothermal energy currently corresponds to only 0.3% of the 20 181 TWh global electric power production capacity in 2008 (International Energy

Agency, 2010), but it is expected to boom among the renewable energy sources. Of the 39 countries identified as having the potential to meet 100% of their electricity needs through domestic geothermal resources, nine (including El Salvador, Iceland and the Philippines) have already developed it as a significant part of their energy production (Holm *et al.*, 2010).

• *Subsurface heat storage*

The subsurface storage of heat makes use of the heat storage capacity of groundwater, but in the very low energy domain. Shallow groundwater is used for temporary storage of heat, with the purpose of recovering it later when it is needed. During recovery, energy is exchanged with heat pumps designed for individual or collective heating or cooling systems. In addition to the heat that has been stored, minor quantities of heat originating in the subsoil may be intercepted. This is an emerging technology, used for heating buildings collectively or privately. It is rapidly expanding in many European countries, the United States, Canada and Japan. The 355 heat pumps identified in 2002 in the European Union use together probably 1–2 km³ of groundwater per year. Again, this is non-consumptive use, because the pumped groundwater is usually re-injected.

• *Contributions to the environment*

The 'integrated water resources management' paradigm (IWRM), now widely advocated and applied, implies that indirect uses and non-consumptive functions of groundwater are no longer overlooked or excluded. Indirect uses include the use of regular surface water generated by groundwater, starting with the water of springs. This role of groundwater is considered in only a few countries. In Spain, for example, the use of natural groundwater discharge, some 14 km³ per year on average, is explicitly taken into consideration.

It goes without saying that groundwater is not only beneficial to mankind. It also plays an important role in keeping surface water flows permanent (groundwater produces the bulk of the baseflows of streams) and many wetlands are dependent on groundwater seepage or groundwater levels. Therefore, groundwater is an important factor in ensuring the sustainability of aquatic ecosystems and wetlands. Whether to conserve these or not is a choice for human society (see Section 4.4).

Preserving the environmental functions of groundwater usually results in restrictions to groundwater exploitation, which has repercussions for the assessment of exploitable groundwater resources. In some cases an 'environmental demand' can be defined that has to be reconciled with economic uses of groundwater. In other cases this is a less suitable approach, if what really counts is maintaining a particular preferred state of the groundwater system.

As an example of the significance of environmental demands, Spain estimated in its *Plan Nacional Hidrológico* of 1993 that in addition to the direct use of groundwater of around 5 km³/year, another 2 km³/year is needed to satisfy the country's environmental demands. In The Netherlands, groundwater levels have a strong impact on the environment and keeping them within prescribed strict limits – optimally tuned to land use – is a major objective of water resources management across the country. The environmental functions of groundwater form a major constraint to groundwater abstraction in all Dutch groundwater management policies and plans.

5.6　CONTRIBUTION OF GROUNDWATER TO WATER SUPPLY IN DIFFERENT SECTORS

The share of groundwater in the supply of each water use sector is one of the indicators of the economic and social importance of the resource. The shares of groundwater withdrawal and use vary from region to region and from country to country. Groundwater is a significant source of supply in many countries and in some countries it constitutes the principal source of supply (see Appendix 5).

As shown in Table 5.2, groundwater withdrawal constitutes about one quarter of the total global water withdrawal, estimated to be in the order of 3 800 km³/year (excluding water losses from reservoirs by evaporation). Taking into account that 'water losses' (i.e. the difference between water withdrawn and water used) in groundwater irrigation tend to be smaller than in surface water irrigation (see below), it may be assumed that groundwater covers somewhat more than one quarter of today's global water use for all sectors combined, perhaps reaching as much as 30%.

Figure 5.20 shows the geographical variation of the share of groundwater in the national freshwater withdrawals. In half of all the countries for which corresponding data is available, groundwater withdrawals exceed 25% of the total water withdrawal for all uses (including cooling water for electric power plants), while in approximately 25% of these countries groundwater accounts for more than 50% of the total withdrawal.

Groundwater is obviously the predominant source of water supply in most arid countries not endowed with inflows of surface water from elsewhere (Mongolia, Oman, Yemen, etc.) and in a number of cases depending on non-renewable resources (Saudi Arabia, Libya). It is the principal source also in countries where surface water sources are difficult to tap (e.g., Bangladesh, Botswana, Congo), or where groundwater is preferred as a source of water supply, like in Denmark and Croatia. But a substantial portion of the spatially scattered water demands is also met by groundwater in other countries (most rural water supplies, private irrigation supplies). For example, in India, the world's largest groundwater user, where during 2010 approximately

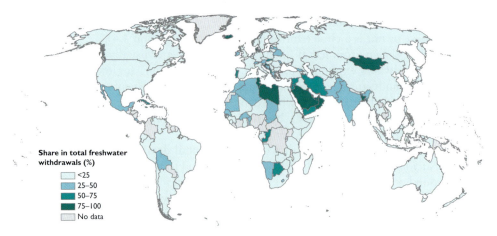

Figure 5.20　Share of groundwater in the total freshwater withdrawal in each country. [*Data and data sources:* See Appendix 5].

223 km³ of groundwater was pumped for irrigation; groundwater covers there 64% of the total irrigation consumptive water use (Siebert *et al.*, 2010).

Worldwide, groundwater withdrawal covers approximately 45% of the domestic water demands (mostly by public water supplies: households, public services and the publicly supplied commercial sector), 24% of the irrigation water demands and 15% of the industrial demands not served by public water systems. As mentioned before, these percentages vary considerably from country to country.

Groundwater use is negligible in the energy sector, in terms of quantities of freshwater used. The special applications mentioned in Section 5.5 (geothermal energy and seasonal subsurface heat storage) usually re-inject the abstracted fresh groundwater.

5.6.1   Domestic water sector

The term 'domestic water' is used here to indicate drinking water and other municipal water supply services, such as water for cooking and cleaning, public services, offices, shops and small industry connected to public water supply. In many countries, groundwater (including water from springs) is the main and sometimes almost the only source of domestic water (see Appendix 5 and Figure 5.21).

For example, the share of groundwater in Europe is close to 100% in Austria, Denmark, Montenegro and Croatia, more than 90% in Italy and Hungary, and between 90 and 60% in Bosnia-Herzegovina, Switzerland, Russia, Germany, Poland, Belgium, The Netherlands and France. European countries with a relatively small share of groundwater in domestic water are Ireland, Spain, Greece and the United Kingdom. In the entire European Union, groundwater covers around 70% of domestic water supplies. Outside Europe, this share is almost 100% in Pakistan, Botswana, Iran and Yemen, around 64% in India, but only 33% in the United States of America and 29% in China. Reliable data is scarce, but groundwater is undoubtedly also the main source of domestic water supply in several countries in Africa, the Middle East and Central Asia. Significant use of desalinated water has resulted in certain dry countries in a relatively modest share of groundwater in the total abstraction of fresh

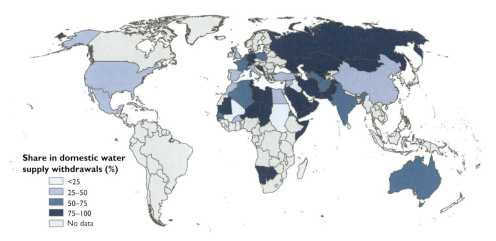

Figure 5.21 Share of groundwater in the total freshwater withdrawal for domestic water supply in each country. [*Data and data sources*: See Appendix 5].

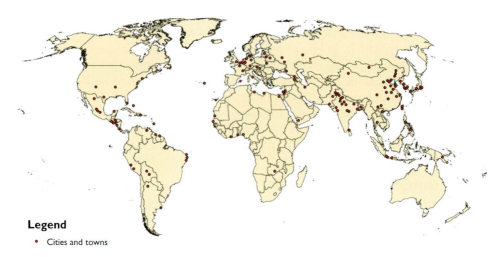

Legend

• Cities and towns

Figure 5.22   Cities and towns around the world supplied exclusively or mainly by groundwater. [*Data sources*: Howard, 1997; World Bank, 1998; H. Zaisheng/WHYMAP, 2006; Zektser *et al.*, 2004].

and saline water combined (Cyprus, Bahrain, Saudi Arabia) or even a very low share (United Arab Emirates, Qatar, Kuwait, Malta, Oman). A complicating factor in the available data sets is that in some cases springs are apparently included in the statistics on groundwater abstraction and in other cases not (e.g. Italy and Israel: some information sources mention percentages twice to triple those mentioned in Appendix 5, probably due to this fact).

Groundwater is usually the most appropriate local source of domestic water for rural areas, but also for a significant proportion of the urban population. The public water supply of numerous cities around the world is entirely or for a substantial part based on abstracted groundwater (see Figure 5.22). Groundwater is also by far the largest source used to cover the domestic water demands of sparsely populated areas where no public water supply system is available. In the Unites States of America, for example, 98% of the corresponding domestic demands (in 2000 amounting to a total of 4.95 km³, or 8% of the national drinking water demand) are met by private groundwater abstractions, in addition to 25 km³ of groundwater pumped by public water supply companies.

5.6.2   Agricultural sector

Groundwater is for many farmers the most easily and individually accessible source of irrigation water. It is also often the cheapest source to exploit – by pumping from dug or drilled wells, powered by electricity or diesel – and the most flexible one in daily practice. Therefore, although in a smaller proportion than for drinking water, it contributes in significant percentages to meeting irrigation water demands. Table 5.10 shows a number of irrigation statistics for selected countries where groundwater-based irrigation is relatively well developed and for which sufficient data is available.

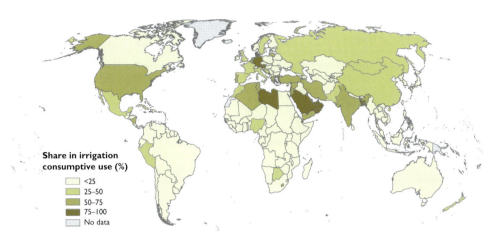

Figure 5.23 Share of groundwater in irrigation consumptive use in each country. [*Data source:* Siebert *et al.,* 2010].

The table reveals a number of interesting features, such as:

- Even within this selection of countries, the total area of the groundwater-irrigated land varies by more than two orders of magnitude.
- The share of groundwater in irrigated land is not significantly different from its share in irrigation consumptive use.
- The share of groundwater in the total irrigation water abstraction, on the other hand, is on average significantly lower than the share of groundwater in irrigation consumptive use. The explanation lies in a usually lower percentage of irrigation water losses in groundwater irrigation compared to surface water irrigation.

Figure 5.23 presents a global picture of the proportion of groundwater in irrigation consumptive use, at the country level. It is significant in the first place in countries with predominantly arid climates (99% in Libya, 97% in Saudi Arabia, 66% in Algeria, 64% in Iran, 63% in Yemen, 48% in Morocco, 33% in Pakistan, 20% in Australia), or where surface water resources are difficult to control (90% in Bangladesh), but also in several other countries (69% in Syria, 64% in India, 57% in the USA, 51% in Greece, 50% in Turkey, 49% in Israel, 42% in Mexico, 40% in France and Italy, 39% in Spain and 19% in Argentina). The proportion of groundwater in irrigation water may be even greater than these numbers indicate, because the statistics do not usually include irrigation water from springs, a traditional and valued source in many countries, especially in dry climates.

This proportion, however, is not by itself an indicator of the importance of groundwater-based irrigation, because it does not say anything about the absolute quantity of groundwater involved. Furthermore, a high proportion of groundwater in some countries where irrigation is supplemental (e.g. Germany) has a completely different meaning than equally high proportions in arid and semi-arid countries where groundwater resources serve as the nearly exclusive source of water. In the more humid regions, where rain-fed agriculture is dominant and where irrigation is only supplementary, groundwater is increasingly used to comply with increasing regulatory constraints on production, quantities and qualities and on timing set by the market.

Table 5.10 Irrigation on the basis of groundwater. *Statistics for a selection of countries where irrigation is relatively developed.*

Country	Groundwater irrigation parameters as estimated by Siebert et al. (2010)			Groundwater abstraction for irrigation (estimates for 2010)		
	Groundwater irrigated area		Groundwater share in total consumptive use for irrigation (%)	in millions of m³/year	% of total groundwater abstraction	% of total abstraction for irrigation
	x 1 000 ha	% of total irrigated area				
AFRICA						
Algeria	290	64	66	1 878	94	43
Egypt	332	10	9	3 203	41	2
Libya	312	99	99	3 915	83	97
Morocco	661	46	48	2 533	83	28
South Africa	127	9	9	2 632	84	–
Tunisia	224	59	62	1 501	74	73
NORTH & CENTRAL AMERICA						
Cuba	82	45	45	1 582	–	–
Mexico	2 191	39	42	21 281	72	33
USA (conterminous)	13 469		57	79 597	71	41
SOUTH AMERICA						
Argentina	323	24	19	5 260	79	–
Brazil	591	19	22	3 822	38	–
Peru	426	28	27	1 734	60	–
ASIA						
Afghanistan	282	16	19	6 720	94	22
Bangladesh	3 459	74	76	25 984	86	–
China	15 847	30	35	60 455	54	15
India	36 850	64	64	223 390	89	–
Iran	3 988	62	64	54 926	87	53
Iraq	146	6	6	1 344	50	–
Israel	89	49	49	896	71	66
Japan	232	9	9	2 516	23	–
Kazakhstan	111	5	5	2 285	71	6
Myanmar	100	5	5	2 406	–	–
Pakistan	4 837	36	33	60 932	94	32
Saudi Arabia	1 156	97	97	22 201	92	94
Syria	950	68	69	10 146	90	56
Tajikistan	66	9	9	5 712	39	–
Thailand	481	9	10	1 534	14	–
Turkey	1 730	49	50	7 973	60	13
Un. Arab Emirates	227	100	100	3 533	100	84
Uzbekistan	268	6	6	5 644	57	8
Yemen	483	66	63	2 788	86	68
EUROPE						
France	854	44	40	790	14	20
Germany	185	79	76	233	4	–

(Continued)

Table 5.10 (Continued)

Country	Groundwater irrigation parameters as estimated by Siebert et al. (2010)			Groundwater abstraction for irrigation (estimates for 2010)		
	Groundwater irrigated area		Groundwater share in total consumptive use for irrigation (%)	in millions of m³/year	% of total groundwater abstraction	% of total abstraction for irrigation
	x 1 000 ha	% of total irrigated area				
Greece	623	48	51	3 131	86	41
Italy	894	33	40	7 000	67	35
Portugal	136	55	55	5 598	89	–
Russian Federation	338	36	36	349	3	–
Spain	1 276	37	39	4 100	72	19
OCEANIA						
Australia	537	21	20	2 581	52	14

The information in Table 5.10 regarding groundwater-irrigated hectares and ground-water abstracted for irrigation, together with the information provided in Section 5.5, do not leave any doubt that groundwater irrigation is a major factor in global agricultural production and thus in food security. The global proportion of groundwater (39%) in the world's irrigated area (see Table 5.8) is higher than its share in total irrigation water withdrawal, because the irrigation efficiency in the case of groundwater tends to be significantly higher than that of irrigation by surface water, as is generally recognised. The share of groundwater in the global economic profit from irrigation may even be higher than the 39% quoted in the table, as a result of on-average higher economic returns per unit of irrigation water used than in the case of surface water irrigation. This is because groundwater is usually a more reliable source; hence, water shortage risks are lower, which encourages farmers to grow and irrigate high-value cash crops.

Groundwater is also an essential source of water for livestock farming and in several countries also a substantial source of water for aquaculture, in some cases as a source of heat rather than as a source of water. In the United States, for example, it covered 29% of the aquaculture water demands in 2000 (1.46 km³).

5.6.3 Industrial sector

In many industrialised countries, groundwater often covers a significant portion of the water demand of industries that are not connected to public water supplies: 40% in Japan, 27% in India, 41% in France, 21% in the United States and 20% in China. This share used to be 15% in the former USSR (the 2010 figure for the corresponding territory is 18%). It can be very high – close to 100% – in dry countries such as Libya, Burkina Faso, Mongolia, Namibia, Oman, Saudi Arabia, Tunisia and Yemen (see Appendix 5 and Figure 5.24). Again, this rate is largely dependent on the relative importance of groundwater resources in each country and it is normal that it is high in arid and semi-arid countries. In some, water use for energy development (such as

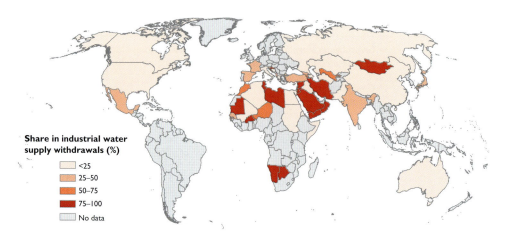

Figure 5.24 Share of groundwater in the industrial water withdrawals in each country.

geothermal use in Iceland) or mine drainage (e.g., in Niger) is included in the statistics for the industrial sector, leading to higher figures for the share of groundwater.

The drainage performed by mining industries to dewater the underground is rarely taken into account in statistics on groundwater abstractions, or they include only that part that is actually used for processing the materials or minerals. In the United States, for example, 2.8 km^3 of water – 62% of which is saline – was withdrawn during 2000 as mining water, but not included in the statistics on industrial water withdrawal.

5.6.4 Other benefits from aquifers

The economic benefit from aquifers is not limited to the use of abstracted ground-water. First of all, aquifers can be used as buffers to complement or as alternatives for surface water storage dams, with the advantage that a groundwater reservoir is not vulnerable to siltation and usually suffers much less from evaporation losses. The exploitation strategies then become more comprehensive than only abstracting groundwater, and may also include artificial recharge, sometimes supplemented by subsurface dams to increase the stored volume of groundwater. Next, aquifers are also used to improve water quality by natural purification processes, especially in cases where wastewater is reused. Furthermore, in many industrialised countries (Europe, USA) they are used for underground storage of other useful fluids (gas), under differ-ent environmental constraints in order to prevent adverse effects on the resources.

Aquifers are also used for the permanent storage of certain types of waste at great depth, under conditions of assured safety and strict control. A prominent application is the sequestration of carbon dioxide (CO_2) in deep saline aquifers (deeper than 800 m, to facilitate the transition into the supercritical state of CO_2), where groundwater is unfit for any use and disconnected from the active hydrological cycle. Known by the name 'carbon capture and sequestration' (CCS), its purpose is to reduce the accu-mulation of carbon dioxide in the atmosphere. In this way, artificial deposits of CO_2

are created similar to natural CO_2 deposits. Often, depleted oil or gas fields are used. The storage capacities in the large sedimentary basins around the world are enormous and could accommodate about 10 000 billion tons of CO_2 (800 billion in Europe), which is the equivalent of several centuries of global CO_2 emission.

CCS seems an excellent contribution to the implementation of the Kyoto Protocol aiming at reducing the greenhouse effect. It is under consideration and already tested in several countries – notably in Denmark and Norway (North Sea), the United Kingdom, Germany and Canada – and it is the subject of study by several groups in Europe and the United States.

5.7   SOCIO-ECONOMIC IMPLICATIONS

5.7.1   Main groundwater exploiters

Privately supplied water – be it drinking water, irrigation water or industrial water – is much more often groundwater than surface water, partly because groundwater is more generally present. The number and identity of groundwater exploiters thus are related to the proportion of private water withdrawal in each water use sector.

In the irrigated agriculture sector, there is no doubt that there are far more self-supplying farmers using groundwater than those using surface water. Farmers who use surface water for irrigation are in most cases served by public or co-operative irrigation water supply networks. Irrigating farmers form the largest group of groundwater exploiters and their number has increased in recent times more quickly than that of any other group of groundwater exploiters. There are millions of them in countries of massive use of groundwater for irrigation: India ranks first, followed by China, Pakistan, Bangladesh and the United States.

Urban or rural agencies in charge of or commissioning the withdrawal and distribution of drinking water come next. In many countries the majority of them tap groundwater or capture springs for this purpose, rather than surface water. This includes rural communities that merely create water points, as was done in the framework of the rural water supply development in the African Sahel-Sudan zone during the years 1980–1990, when tens of thousands of wells and boreholes were constructed. It should be mentioned that there are tendencies in many countries to merge water supply authorities or companies into larger ones, belonging either to the public or to the private sector. This leads to a smaller number of independent groundwater exploiters in the drinking water sector.

5.7.2   Specific economic aspects

The diversity of uses of abstracted groundwater results in significant differences in value attributed to its use. Similarly, differences in the share of groundwater as a source of supply cause the relative importance of groundwater to vary from one economic sector to another.

The drinking water sector represents economically and socially high-valued water use and can be considered as a public service, usually available on commercial terms. Predominant are intermediaries in the production-distribution chain, according to country and to local conditions organised in very different ways under public agencies (state, municipalities, syndicates) or private companies (delegated tasks). The frequent predominance of this sector in the exploitation and use of groundwater globally boosts the role and the market value of groundwater.

In the self-supplying industrial sector, groundwater exploiters are companies (private or even state-owned) that use their pumped water directly. The abstracted groundwater is not traded and can only be subject to charges by the government.

In the groundwater irrigation sector, the large majority of the groundwater exploiters is also the user – the irrigating farmer – of the withdrawn groundwater. They pay the direct cost of groundwater exploitation (more often subsidised than subject to taxes) and usually they pump groundwater by virtue of formal or assumed water rights attached to land ownership. The abstractions, however, may be subject to charges. The direct exploitation costs vary considerably depending on local conditions and energy costs. Llamas *et al.* (2004) mention a range of 0.01 to 0.2 US$/m^3 and a corresponding annual cost of groundwater irrigation between 20 and 1 000 US dollars per ha.

Because irrigation farmers are directly confronted with the cost of groundwater, they tend to strike a balance between the cost of water and their agricultural production, which means that they go for the highest economic profit rather than for highest production. In this connection, one has to realise that the relationship between the level of satisfying the water needs of crops and the corresponding production level is far from linear. It is often observed that the supply of irrigation water originating from groundwater remains well below the theoretical water demand. The reason behind it is the following: each cubic metre of water has the same cost, but the final inputs of water contribute comparatively less to increasing the production, as described by the *'law of diminishing returns'* in economics.

Therefore, groundwater is generally more efficiently used and more profitable in irrigation than surface water distributed by shared networks. Physical efficiency, related to the number of cubic metres of irrigation water used per hectare, is on average twice that of irrigation by surface water, according to Burke (2003) and Shah (2004), while the socio-economic efficiency, in terms of the value of the production per cubic metre of water ($US/m^3) or jobs created per million of cubic metres used, can be three to ten times higher.

According to Llamas (1996), groundwater irrigation in Andalucía (Spain) is five times more productive – in terms of economic efficiency (value of production per cubic metre of water used) – than irrigation by surface water. The explanation is twofold: (1) it has on average a much higher water use efficiency, which permits a reduction in the gross water demands per hectare (average of 5 000 m^3/year for groundwater, compared to 10 000 m^3/year for surface water) and (2) it allows crops to be grown that are more sensitive to water shortage but produce higher added value. In the studied area, groundwater-fed irrigation contributes to 30–40% of the total production of irrigated agriculture, while it uses only 20% of the total volume of irrigation water. This does not necessarily hold for other regions.

According to the Indian Water Resources Society (IWRS, 1999; cited by Llamas *et al.*, 2003), groundwater irrigation in India uses on average four times less water per hectare than surface water irrigation, or 4 000 versus 16 000 m^3/year. IWMI (2005) estimated the output of the world's agriculture irrigated by groundwater to have a value of US$ 150 to 170 billion some years ago, which at that time would have corresponded to an average productivity of around US$ 0.2 per cubic metre of water used.

Altogether, groundwater use belongs more to the private than to the public sector of the water economy. The share of groundwater in the wealth created by water use is much higher than its share in the total volume of water used. In some cases, e.g. on Gran Canaria and Tenerife, privately controlled groundwater markets function, most likely contributing to the most profitable allocation of scarce water.

Planet Earth (2005) estimated the total value of the world production of groundwater in 2001 to be some 300 billion Euros – 37% of the value of oil production. However, the calculated amount is based on an average water price of 0.5 Euros per cubic metre, which is probably overestimated, and the price is based on production and distribution costs minus subsidies, and thus does not necessarily reflect the economic value of groundwater. Furthermore, this price refers only to domestic water, a little more than one-fifth of all groundwater used around the world, sold more or less on a commercial basis but not fully exposed to market forces.

In conclusion, abstracted groundwater is generally better used than surface water: both more efficiently and often for higher-valued purposes. Groundwater also usually has greater economic value. Therefore, conserving its quantity and preserving its quality should become priorities in water resources management.

REFERENCES

Abdurrahman, W., 2002. *Development and management of groundwater in Saudi Arabia.* GW-MATE/UNESCO Expert Group Meeting, Socially sustainable management of groundwater mining from aquifer storage. Paris, 6 p.

Alcamo, J., P. Döll, Th. Henrichs, F. Kaspar, B. Lehner, Th. Rösch & S. Siebert, 2003. Global estimates of water withdrawals and availability under current and future "business-as-usual" conditions. *Hydrological Sciences Journal,* Vol. 48, No 3, pp. 339–348.

AQUASTAT, 2011, 2012. *FAO's Global Information System on Water and Agriculture.* Available at: http://www.fao.org/nr/water/aquastat/main/index.stm [Accessed in 2011 and 2012, respectively (updates refer generally to the years 2009 and 2010, respectively)].

Bakhbakhi, M., 2002. *Hydrogeological framework of the Nubian Sandstone Aquifer System.* GW-MATE/UNESCO Expert Group Meeting, Socially sustainable management of groundwater mining from aquifer storage. Paris, 6 p.

Bayer, D., 2009. *La crisis del agua en Ica y cómo resolverla.* Unpublished article on Internet, dated 22 November 2009.

Bocanegra, E., M. Hernandez & E. Usinoff (eds), 2005. *Groundwater and Human Development.* IAH selected papers on Hydrogeology 6, Heise, Hannover, 278 p.

Burke, J.J., 2003. Groundwater for irrigation: Productivity gains and the need to manage hydro-environmental risk. In: *Intensive use of groundwater. Challenges and opportunities,* M.R. Llamas and E. Custodio, eds., Balkema, Lisse (Allemagne): 59–92.

Cabezas Calvo-Rubio, F., 2011. *Las aguas subterráneas en la cuenca del Segura.* Presention at the II° Seminario Nacional "El papel de las aguas subterráneas en la política del agua en Espana", Observatorio del Agua de la Fundación Botín, Madrid, 1 February 2011.

Carruthers, I., & R. Stoner, 1981. *Economic Aspects and Policy issues in Groundwater Development.* Worldbank Staff Working Paper No. 496. Washington DC, World Bank, 110 p.

Chávez Guillén, R., 2006. *Disponibilidad de auga subtránea en Mexico.* Presentation on Session 5.33 'The manifold dimensions of groundwater sustainability', 4th World Water Forum, Mexico, March 2006.

Conf. Rome, 1992. *Deuxième Conférence Méditerranéenne de l'Eau*, Rome, 28–30 Oct. 1992, organised by the Government of Italy and the ECE.

Deb Roy, A., & T. Shah, 2001. Socio-ecology of groundwater irrigation in India. In: Llamas, M.R., & E. Custodio (eds), 2003. *Intensive use of groundwater: challenges and opportunities.Intensive use of groundwater*. Balkema, Lisse, The Netherlands, pp. 307–335.

DGRE, 1996. *Direction Générale des Ressources en Eau, Ministère de l'Agriculture de Tunesie, Tunis*.

ESCWA, 2007. *ESCWA Water Development Report 2: State of Water Resources in the ESCWA Region*. United Nations Publication, Sales No. E.08.II.L.1, United Nations, New York.

EUROSTAT, 2011. *The European Commission's information system providing statistics on EU and candidate EU countries*. Available at: http://epp.eurostat.ec.europa.eu/portal/page/portal/eurostat/home/ [Accessed in 2011].

Famiglietti, J., S. Swenson & M. Rodell, 2009. *Water storage changes in California's Sacramento and San Joaquin river basins, including groundwater depletion in the Central Valley*. PowerPoint presentation, American Geophysical Union Press Conference, December 14, 2009, CSR, GFZ, DLR and JPL.

FAO, 1997. *Irrigation in the Near East region in figures*. FAO Water Reports 9, Saudi Arabia 205212, Rome.

FAO, 1998. *Water resources in the former Soviet Union: a review*. Rome.

FAO, 1999. *Irrigation in Asia in figures*. Water Report No 18, Rome.

FAO, 2000. *Irrigation in Latin America in figures*. Water Report No 20, Rome.

FAO, 2001. *La situation mondiale de l'alimentation et de l'agriculture 2001*. FAO, Rome.

FAO, 2003. *Rethinking the Approach to Groundwater and Food Security*. Water Report No 24. Rome, FAO, 49 p.

FAO, 2005. L'irrigation en Afrique en chiffres. Enquête AQUASTAT, Water Report No 29, Rome.

Foster, S., & D. Loucks, 2006. *Non-renewable groundwater resources. A guidebook on socially-sustainable management for water policy-makers*. Paris, UNESCO-IHP, IHPVI, Series on Groundwater No. 10.

GEF, 2004. *Draft Review and Synthesis Document (R&Sd)*. STAP/GEF Technical Review Workshop on Strategic Options and Priorities in Groundwater Resources, 5–7 April 2004, Paris.

Giordano, M., 2006. Agricultural groundwater use and rural livelihoods in Sub-Saharan Africa: A first-cut assessment. *Hydrogeology Journal*, Vol. 14, No 3, pp. 310–318.

GWA/Salem, O.M., 1992. The Great Manmade River Project. *Water Resources Development*, Vol. 8, No 4, pp. 270–278.

Habermehl, M.A., 2006. *Development and management of groundwater in the Great Artesian Basin, Australia*.

Holm, A., L. Blodgett, D. Jennejohn & K. Gawell, 2010. *Geothermal Energy: International Market Update*. Geothermal Association, 77 p.

IGRAC, 2010. *The International Groundwater Resources Assessment Centre's Global Groundwater Information information System (GGIS)*. Available at: http://www.un-igrac.org/ [Accessed in 2010].

Indian Water Resources Society, 1999. *Cited in:* Llamas, M.R., & E. Custodio (eds), 2003.

Jaubert, R., M. Al-Dbiyat, F. Debaine & F. Zwahlen, 2009. *Exploitation des eaux souterraines en Syrie centrale: enjeux politiques & réalités locales*. Available from: http://ifpo.revues.org/1332

Jones, J.A.A., 2011. Groundwater in peril. In: J.A.J. Jones (ed.): *Sustaining Groundwater Resources*. The International Year of Planet Earth. Springer, Books on Environmental Sciences, pp. 1–19.

Konikow, L., 2011. Contribution of global groundwater depletion since 1900 to sea-level rise. *Geophysical Research Letters*, Vol. 38, Paper and Auxiliary Materials. Available from: L17401, doi: 10.1029/2011GL048604, 2011.

Liu, Ch., J. Yu & E. Kendy, 2001. Groundwater exploitation and its impact on the environment in the North China Plain. *Water International*, Vol. 26, No 2, pp. 265–272.

Llamas, 1996. Eaux souterraines et agriculture en Espagne. *Vida Rural*, No 41.

Llamas, M.R., W. Back & J. Margat, 1992. Groundwater use: equilibrium between social benefits and potential environmental costs. *Applied Hydrogeology*, 1992, No 2, Heise, Hannover, pp. 3–14.

Llamas, M.R., & E. Custodio (eds), 2003. *Intensive use of groundwater: challenges and opportunities*. Balkema, Rotterdam, 478 p.

Llamas, M.R. & Martinez-Santos, P., 2005a. Intensive groundwater use: Silent revolution and potential source of social conflicts. In: *Journal of Water Resources Planning and Management*, ASCE., Sept.–Oct., pp. 337–341.

Llamas, M. & P. Martínez-Santos, 2005b. Intensive groundwater use: a silent revolution that cannot be ignored. In: International Water Association: *Water Science and Technology Series*, Vol. 51, No 8, pp. 167–174.

Llamas, M.R., & P. Martinez-Santos, 2006. Significance of the Silent Revolution of intensive groundwater use in world water policy. In: Rogers, P., M.R. Llamas & L. Martínez Cortina (eds), *Water crisis: myth or reality?* Chapter 10. Balkema, Rotterdam and London, Taylor & Francis, pp. 163–180.

Margat, J., 2008. *Les eaux souterraines dans le monde*. UNESCO & BRGM, 187 p.

McGuire, V., 2003. *Water-level changes in the High Plains aquifer, predevelopment to 2001, 1999 to 2000, and 2000 to 2001*. USGS Fact Sheet FS-078-03, 4p, USGS, Reston, Virginia.

McGuire, V., 2009. *Water-level changes in the High Plains aquifer, predevelopment to 2007, 2005–06 and 2006–07*. Scientific Investigations Report 2009–5019, USGS, Reston, Virginia.

Motagh, M., T Walter, M. Sahrifi, E. Fielding, A. Schenk, J. Anderssohn & J. Zschau, 2008. Land subsidence in Iran caused by widespread water reservoir overexploitation. *Geoph. Res. Lett.*, 35, L16403. Available from: doi: 10.1029/ 2008GL033814.

OSS, 2003. *Système aquifère du Sahara septentrional. Rapport de synthèse*. Observatoire du Sahara et du Sahe, Tunis, 129 p.

OSS, 2005. = UNESCO/OSS, 2005. *Ressources en eau et gestion des aquifères transfrontaliers de l'Afrique du Nord et du Sahel*. ISARM-Africa, UNESCO-IHP IV, Series on Groundwater No 11, Paris.

Plan Blue/J. Margat, 2008. *L'eau des Méditerranéens. Situation et perspectives*. Paris, L'Harmattan.

Plan Hidrológico Nacional, 1993. National hydrological plan of the Government of Spain, as defined in 1993.

Rahmatian, M., 2005. *Iran: Cost assessment of environmental degradation*. Presentation at the Caspian EVE 2005, Ashgabad, Nov. 2005, UNDP and World Bank.

Rodell, M., I. Velicogna & J. Famiglietti, 2009. *Satellite-based estimates of groundwater depletion in India*. Nature, Vol 460. Available from: doi: 10.1038/nature08238.

Sahuquillo, A., J. Capilla, L. Martinez-Cortina & X. Sanchez-Vila (eds), 2004. *Selected papers of the International Symposium on Intensive use of Groundwater: Challenges and Opportunities, Valencia, Spain, 10–14 December 2004*. Balkema, Dordrecht, The Netherlands.

Shah, 2004. *Groundwater and Human Development: Challenges and Opportunities in Livelihoods and Environment*. Stockholm, World Water Week 2004, Water Science and Technology.

Shah, T., O. Singh & A. Mukherji, 2006. Some aspects of South Asia's groundwater irrigation economy: analyses from a survey in India, Pakistan, Nepal and Bangladesh. *Hydrogeology Journal*, Vol. 14, No 3, pp. 286–309.

Shah, T., J. Burke & K. Villholth, 2007. Groundwater: a global assessment of scale and significance. In: David Molden (ed.), *Water for food, water for life. A comprehensive assessment of water management in agriculture*. IWMI, Colombo, Earthscan, London, pp. 395–423.

Siebert, S., J. Burke, J. Faures, K. Frenken, J. Hoogeveen, P. Döll & T. Portmann, 2010. Groundwater use for irrigation – a global inventory. *Hydrol. Earth Syst. Sci.,* Vol. 14, pp. 1863–1880.

Tiwari, V., J. Wahr & S. Swenson, 2009. Dwindling groundwater resources in northern India, from satellite gravity observations. *Geophysical Research Letters*, Vol. 36, L18401. Available at: doi: 10.1029/2009GL039401.

USGS, *Estimated Use of Water in the United States in 2005*.USGS Circular 1344, 2009, Washington.

Villarroya Gil, F., 2008. *El acuífero 23 en el marco hidrológico de la cuenca alta del Guadiana. (Historia reciente de su evolución hídrica).*

Wada, Y., L. van Beek, C. van Kempen, J. Reckman, S. Vasak & M. Bierkens, 2010. Global depletion of groundwater resources. *Geophysical Research Letters*, Vol 37, L20402. Available from: doi: 1029/2010GL044571.

Wang, R., H. Ren & Z. Ouyang (eds), 2000. *China Water Vision: The ecosphere of water, life, environment and development.* China Meteorological Press, Beijing, 178 p.

WRI, 2000. *World Water Resources 2000–2001.* UNEP, UNDP, World Bank, World Resources Institute, Washington.

WRI, 2004. *World Water Resources 2002–2004.* UNEP, UNDP, World Bank, World Resources Institute, Washington.

Zubari, W.K., 2005. Spatial and temporal trends in groundwater resources in Bahrain, 1992–2002. *Emirates Journal for Engineering Research,* Vol. 10, No 1, pp. 57–67.

Chapter 6

Growing needs for groundwater resources management interventions

- *Rationale behind groundwater resources management*
- *What impacts do groundwater exploitation and use have on groundwater availability and on the environment?*
- *Current pressures on groundwater quantity and their variation around the world*
- *What will be the impact of climate change?*
- *How and where does groundwater quality degradation occur?*

6.1 WHY SHOULD GROUNDWATER RESOURCES BE MANAGED?

For centuries, groundwater has been exploited and used by a variety of actors without previous assessments of groundwater resources and without any attempt to manage these resources rationally. But access to groundwater has gradually become easier and its exploitation has dramatically increased – as documented in the previous chapter. As a result, wells started interfering with each other and aggregated pumping rates became large enough to significantly modify the groundwater regimes. Side effects of individual pumping operations – such as falling water levels, wetland degradation and reduction of spring discharge and baseflows – are to a large extent spontaneously transferred to third parties, not merely to other groundwater exploiters, but to the entire local community and the local environment. In other words, groundwater pumping produces externalities. And vice versa, numerous activities of individuals and local communities unrelated to groundwater withdrawal also produce considerable externalities to groundwater systems and their exploiters – especially by polluting groundwater.

In short, the state of groundwater systems is continuously changing in a rapidly changing world, groundwater systems are dynamically interacting with other physical and socio-economic systems, and the many actors involved are pursuing their goals relatively independently. Observed or expected changes in the groundwater state often diverge from what is most desirable, but it is evidently far beyond the reach of single individuals to control these changes. Furthermore, groundwater is a common pool resource and there may be large discrepancies between the interests and preferences of individuals and those of the local community as a whole. All these reasons form a motivation and justification for groundwater resources management interventions.

Evolution from an initial stage where the focus is predominantly on exploiting groundwater to the stage of groundwater resources management is a logical response to the observed increases in complexity and number of perceived problems. It creates opportunities to address emerging problems by actions such as:

- Resolving water use conflicts between groundwater exploiters.
- Slowing down or stopping non-sustainable abstraction.
- Mediating conflicts between those who benefit from groundwater development and those who do not abstract but suffer from harmful externalities of groundwater withdrawal.
- Identifying, reducing and eliminating threats to the groundwater resources.

Box 6.1: Groundwater resources management

Unlike surface water, groundwater is freely accessible to numerous people scattered over an area, who may exploit it and may influence its quality. These individuals have their own objectives on a different scale from the scale of an aquifer and in general they have no knowledge of the aquifer's properties and potential.

Managing groundwater is first of all tuning individual actions to shared societal goals and constraints as defined for each physical management unit – usually a certain aquifer system or part of it. It also means taking into account the specific advantages of groundwater resources (buffer capacity, large areal extent, stable water quality) in the choices on allocation, by allocating groundwater to the most essential and most rewarding uses, as well as by making use of the buffer and purification capacity of aquifers. Finally, it includes protection and conservation of the groundwater resources while maintaining their positive role and functions in the water cycle and the environment.

In general, groundwater and surface water cannot be managed independently from each other. According to how both are linked and interrelated with respect to quantity and quality and how intensely (see Chapter 4, Box 4.1), groundwater management should be integrated into overall water resources management. The river basin has been adopted as the fundamental spatial unit for application of the integrated water resources management process (IWRM). However, some specific hydrogeological settings will require a modified approach, with aquifers or groundwater bodies as the appropriate spatial unit within which to address groundwater management and protection. Extensive deep aquifer systems in arid regions are an example of such a setting (Foster and Ait-Kadi, 2012).

Groundwater resources management implies establishing appropriate institutions for each spatial management unit*, including the creation of an authority with adequate mandate, powers and resources. It also requires defining how users participate in defining objectives and selecting management instruments, as well as an effective control of management interventions.

* 'Aquifer management organization' (AMOR), in the terminology of the World Bank; or 'Aquifer community' in Shah's terminology.

But groundwater resources management is more than simply reacting to observed or anticipated problems. It should ensure that the potential of groundwater as a resource is not overlooked or underestimated and that full advantage is taken of the opportunities offered by groundwater. Groundwater management has to be guided by a strategic vision on the role of groundwater in water supply, in the economy and in the natural environment of the country or area concerned. This implies that a balance has to be struck between exploiting and conserving groundwater, in harmony with political preferences regarding water, socio-economic development and the environment.

Although groundwater resources management can and should not be planned or implemented in isolation from overall water resources management, it is nevertheless characterised by some very specific objectives, features and instruments, as illustrated in Box 6.1.

A framework of analysis may assist in identifying, overviewing and understanding the many and often complex change processes in groundwater systems and their repercussions. The DPSIR framework, shown in Figure 6.1, is a convenient tool for this purpose. It makes a distinction between five interconnected classes of variables that together describe the dynamics and the interrelations, in a cyclical setting.

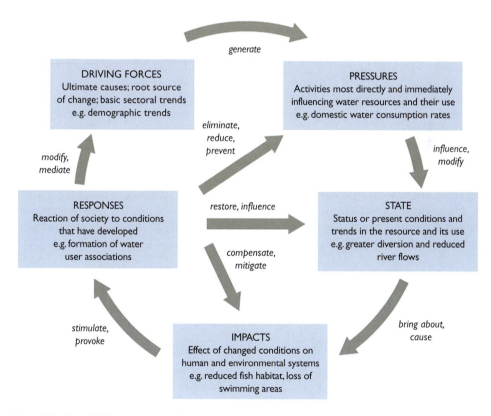

Figure 6.1 The DPSIR framework of analysis. [*Source:* WWAP (2006), adapted from Costantino *et al.* (2003)].

In the terminology of this framework, changes in groundwater conditions or *state* (water levels, fluxes, stored groundwater volume, water quality) are produced by *pressures* (or 'stresses') on the corresponding groundwater system (withdrawal, recharge modifications, emission of pollutants). Behind these pressures, in turn, are so-called *driving forces* or root causes of change. Driving forces relevant in the context of groundwater resources management are demographic trends, socio-economic conditions and aspirations, scientific and technological innovation, conditions of policy/law/finance, climate variability and change, and sudden disasters. The changes in state have their *impacts* on human society, ecosystems and the environment, which triggers *responses* in terms of groundwater resources management interventions and behaviour of individuals alike.

In the following sections, attention will be paid to the most important types of changes of state in groundwater systems around the world and their impacts. The focus will be partly on observed changes and partly on changes expected to occur in the future. Especially to underpin the latter category, the linkages with pressures – and to some extent with driving forces – are described. This may contribute to a better understanding of what groundwater resources management, addressed in Chapter 7, is about.

6.2 HUMAN ACTIVITIES MODIFY GROUNDWATER LEVELS AND FLUXES

As may be concluded from the previous chapter, groundwater is withdrawn and used in large quantities around the world because of its immense value for human life and socio-economic development. However, groundwater withdrawal inevitably modifies the state and regime of the exploited groundwater systems, because it intercepts part of the flux through an aquifer and it may deplete stored groundwater. This is revealed by internal physical side effects (related to the groundwater systems themselves) as well as by external ones (related to the interconnected surface water and environmental systems). These side effects produce impacts that are usually considered to be negative, but in some cases they may be positive, e.g. if waterlogged zones are drained by groundwater abstraction. Groundwater withdrawal is not the only category of human activities with the potential to significantly modify groundwater regimes: land use also produces important pressures.

6.2.1 Internal side effects of groundwater withdrawal

Groundwater withdrawal produces declines in the groundwater levels. These levels tend to stabilise after some time, unless groundwater withdrawals continue increasing. Declining groundwater levels have their repercussions for the use of individual wells, leading to higher water lifts and often to lower well productivity. These effects result in higher pumping costs and sometimes require wells to be deepened. Wells that were originally artesian may stop flowing and then need to be pumped.

In addition to their impact on the conditions for withdrawal, groundwater level declines may trigger other internal effects. They may enhance groundwater recharge, either by reducing 'rejected recharge' (water that runs off to surface water bodies when insufficient subsurface storage capacity is available above the phreatic level) or by inducing recharge from surface water bodies that are hydraulically connected to the aquifer. Often this enhanced recharge is considered a positive effect, except when the induced recharge consists of poor quality water, such as intruding saline water, which is the case in many coastal areas, e.g. along the Mediterranean Sea, in North America and in North China. Modified flow patterns associated with groundwater level declines may also cause undesired migration of poor quality groundwater.

A distinction has to be made between groundwater withdrawal for the purpose of water use and groundwater withdrawal for drainage purposes. In the latter, groundwater level declines are not a side effect, but an intended effect, with the aim of facilitating mining activities, preventing waterlogging in irrigation districts or creating terrain conditions that are optimal for the envisaged type of land use.

In many areas around the world, groundwater levels are known to have declined by several tens of metres, sometimes nearly one hundred metres or even more. This is mostly observed in areas where groundwater is intensively exploited for urban or agricultural water use purposes, occasionally also in areas where massive drainage of mines takes place. Box 6.2 and Figure 6.2 give some examples.

6.2.2   External side effects of groundwater withdrawal

Groundwater abstraction comes initially at the expense of groundwater storage, which – in turn – modifies local hydraulic gradients and the corresponding fluxes, until eventually a new dynamic equilibrium is reached. During the transitional phase, the abstracted groundwater is gradually balanced less by storage depletion and increasingly more by induced recharge and/or reduced natural groundwater outflow. When equilibrium has been reached, the withdrawal is completely balanced by the induced recharge and reduced natural outflow combined. Only if the long-term abstraction rate is higher than can be balanced by these components, does progressive depletion of stored groundwater occur.

Consequently, groundwater withdrawal around the world is to a large extent at the expense of natural groundwater outflows and thus leads to a reduction in spring flows and baseflows, degeneration of wetlands, increased salinity levels in brackish coastal waters and a decrease in the natural discharge by evaporation in groundwater outcrop zones such as salt flats. Even traditional groundwater exploitation systems by infiltration galleries or qanats (see Chapter 4) owe their sustainability as 'artificial springs' to the reduction in more or less visible and diffuse natural groundwater outflows, including evapotranspiration in zones where phreatophytes are present.

Declining spring discharges are among the most easily visible side effects (see Figure 6.3). Examples of affected large springs are those of Jeffara in Tunisia (declining by 2 m³/s between 1900 and 1990), those of Bahrain (by 3 m³/s between 1970 and 1990), and several other ones in Saudi Arabia and China.

Reduction or disappearance of baseflows is easily noticed as well. Figure 6.4 shows significant flow reduction in the Fuyang River as a side effect of groundwater depletion on the North China Plain. In Spain, the baseflow of the Júcar River decreased between 1970 and 1990 from 200 to 50 million m³ per year, as a result of increasing exploitation of the La Mancha Oriental aquifer in the upper part of the basin.

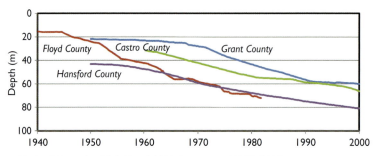

(a) Ogallala aquifer (High Plains), USA. *Data source*: USGS (2000).

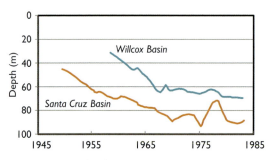

(b) Alluvial aquifers, Arizona, USA. *Data source*: USGS (1985).

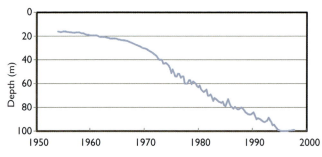

(c) Jeffara aquifer at Bin Gahir, South of Tripoli, Libya. *Data sources*: GWA (1992), Salem (1999).

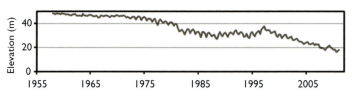

(d) China. Beijing Plain Aquifer. *Data source*: Zhou et al. (2012).

Figure 6.2 Examples of groundwater level declines in intensively exploited aquifers. (*Continued*).

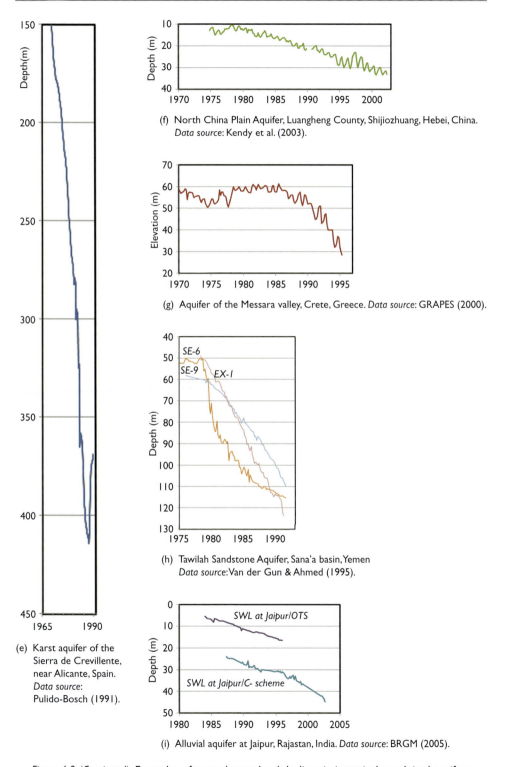

(f) North China Plain Aquifer, Luangheng County, Shijiozhuang, Hebei, China.
     Data source: Kendy et al. (2003).

(g) Aquifer of the Messara valley, Crete, Greece. Data source: GRAPES (2000).

(h) Tawilah Sandstone Aquifer, Sana'a basin, Yemen
     Data source: Van der Gun & Ahmed (1995).

(e) Karst aquifer of the
     Sierra de Crevillente,
     near Alicante, Spain.
     Data source:
     Pulido-Bosch (1991).

(i) Alluvial aquifer at Jaipur, Rajastan, India. Data source: BRGM (2005).

Figure 6.2 (Continued) Examples of groundwater level declines in intensively exploited aquifers.

Box 6.2: Examples of large declines in groundwater level as a result of intensive groundwater exploitation

The most significant declines in groundwater level are the result of intensive groundwater exploitation over a long period of time (a century or more), such as in California in the USA, the regions of Lille and Paris in France, and the basins of the Dniepr-Donetz and of Azov-Kouban in Russia. During the second half of the twentieth century, continuous declines in the levels at a rate of several metres per year have become common in many aquifers, reflecting the rapid growth of urban and agricultural groundwater withdrawals.

Since the early twentieth century, intensive exploitation of the Ogallala Aquifer in the High Plains of the United States has caused declines of up to 60 m in Texas and 50 m in Kansas. The exploitation of 'basin and range' alluvial aquifers has reduced the groundwater levels in Arizona during the same period by 40 to 70 m, with rates of decline of 1 to 6 m per year.

Intensive and increasing exploitation of the Po Valley aquifer in Italy has drawn down the levels in Milan more than 25 m (up to 40 m) in 80 years. In Ravenna, declines in the coastal aquifer have exceeded 40 m, with a cone of depression under the city reaching 40 m below sea level.

In Spain, exploiting the karst aquifer of the Sierra de Crevillente, near Alicante, has reduced groundwater levels by 250 m in 20 years. The groundwater level of the Campo de Dalias aquifer, near Almeria, fell by 15 m in 20 years, and reached 8 m below sea level in 1984. On the islands of Tenerife and Gran Canaria, groundwater level drops exceeding 300 m have been observed within the time span of a century.

In the aquifer of the plain of Argolis, Greece, exploited by 6 000 dug and drilled wells, groundwater levels fell by 80 to 150 m, while the water level was drawn under sea level over an area of some 100 km^2 (1983).

In Mexico, declines of 1 to 3 m per year were observed in many aquifers between 1950 and 1990, with cumulative declines of 70 to 130 m.

The increase of pumping on the Jeffara plain in Libya resulted in water level declines of several tens of meters between 1960 and 1990, to 70 metres south of Tripoli.

In Saudi Arabia, intensive exploitation of the Saq aquifer since the 1980s caused declines of 100 to almost 200 m in several areas, including Qasim, in a few years.

In China, 56 depressions in the groundwater piezometric surface have been reported with declines greater than 100 m, currently scattered over an area of 87 000 km^2. In the Northern China Plain (Huang Huai Hai), groundwater levels are falling at rates of 1.4 m per year in Shijiazhuang and 3.3 m per year in Soshun (Wang et al., 2000).

In the Philippines, declines of 5 to 12 m per year have been observed in Manila, while groundwater levels have reached depths of 70 to 80 m below sea level.

Figure 6.3 Discharge from a spring at the village of Hadda near Sana'a (Yemen) in 1982. [*Photo*: Jac van der Gun]. *This spring, emanating from volcanic rocks, had kept the hill-slope part of the village green and provided water for its fruit orchards from time immemorial. It dried up once and for all during the 1980s, as a side effect of many new pumped wells in the surrounding area.*

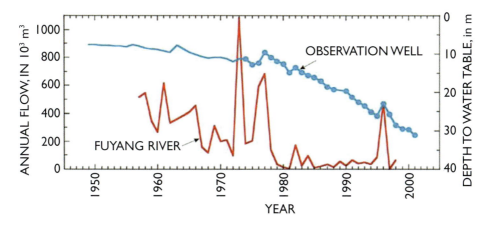

Figure 6.4 Stream and well hydrographs from North China Plain showing evidence of reduced stream flow caused by groundwater depletion. [*Source*: Konikow and Kendy (2005)].

Box 6.3: Land subsidence as a side effect of intensive groundwater exploitation

An international UNESCO working group inventoried 42 cases of land subsidence due to intensive groundwater exploitation around the world during the early 1980s, mainly in urban agglomerations, of which 35 in Europe, the USA and Japan (UNESCO, 1984). The following cases were identified:

- In Italy: Milan (subsidence of 0.2 m from 1952 to 1972); Venice (more than 0.2 m since the 1930s, which is particularly harmful at this location susceptible to flooding by the sea). In the Po Delta, abstractions lowered ground level by more than 3 m during the 1950s.
- In the United Kingdom: London.
- In Mexico: Mexico City (since the 1920s up to 0.4 m per year in the centre, accumulating to 10 m).
- In the United States: Denver, Houston, Las Vegas, San Francisco, Tucson; up to 2–9 m of subsidence in several cities in California, where the area of subsidence exceeding 0.3 m, as observed in 1984 in the Central Valley, extended over 13 000 km².
- In China: 700 cases of land subsidence, affecting more than 50 cities, including Tianjin (2.6 m), Beijing (0.6 m), part of the North China Plain (maximum 1.1 m in Changzhou), Shanghai and Taipei.
- In India: Calcutta.
- In Thailand: Bangkok.
- In Japan: Tokyo (starting in 1910; subsidence up to 4 m; land surface fell to 1 m below sea level), Osaka (up to 2.5 m) and another 62 cases reported in 1998.
- In Indonesia: Jakarta.

In more recent years, several of these cases have been studied and monitored in more detail, while additional cases of land subsidence have been identified, such as Hanoi in Vietnam (0.10–0.15 m of subsidence), the Bolsón del Hueco basin around El Paso in Texas (0.3 m of subsidence since the 1950s), the Sagamigawa alluvial plain in Japan (up to 0.32 m during 1975–1995) and several areas in Iran such as the Yazd-Ardakan basin (0.5–1.2 m during 1985–2010). Other examples, listed by Galloway and Burbey (2011), are: Coachella Valley, California (70 mm/yr during 2003–2009); Jakarta (250 mm/yr during 1997–2008); Mashad Valley, Iran (280 mm/yr during 2003–2005); Mexico city (300 mm/yr during 2004–2006); Saga Plain (160 mm/yr during 1994); Semarang, Indonesia (80 mm/yr during 2007–2009); Tehran Basin, Iran (205–250 mm/yr during 2004–2008); Toluca Valley, Mexico (90 mm/yr during 2003–2008) and Yunlin, China (100 mm/yr during 2002–2007). A series of international symposia convened by IAHS/UNESCO (e.g. FISOLS 95, SISOLS 2000, SISOLS 2005 and EISOLS 2010) have contributed significantly to the international exchange of knowledge on land subsidence.

Withdrawal of oil and gas may produce similar effects (e.g. in Ravenna, Italy) and may be accompanied by the abstraction of large quantities of water, usually saline. In the case of dewatering mines, subsidence due to the abstraction of groundwater may be aggravated by collapsing rock masses.

Not all land subsidence around the world is caused by abstraction of groundwater or other fluids (oil and gas) or by drainage. Land subsidence may also result from the natural consolidation of thick unconsolidated young deposits (due to the weight of the overburden), from the weight of newly built constructions, from the oxidation of peat, or from the extraction of minerals and other solid raw material.

As an example of the degradation of groundwater-supported wetlands it may be mentioned that in Spain 440 km² of wetlands (35% of the original extent) has disappeared, according to the National Hydrological Plan (Plan Hidrológico Nacional, 1993) mainly due to intensive groundwater withdrawal. In a more general sense, groundwater level declines may lead to dryer soils, because they cause the contribution of groundwater to soil moisture by capillary flow to become more difficult. This results not only in damage to wetlands and natural vegetation, but also to loss of agricultural production or increased soil erosion. In this context, in The Netherlands it has been common practice for more than half a century already to make a forecast of such losses of agricultural production for each planned major groundwater abstraction, as a requirement for obtaining an abstraction permit and as the basis for compensation to be paid for possible damage.

Declining groundwater levels may also cause local terrain collapses or relatively widespread land subsidence. Such terrain deformations are particularly harmful in urban areas, where they undermine the stability of buildings and roads, damage water supply and sanitation networks, affect water courses and increase flooding risks. These phenomena render the land less suitable for construction and thus reduce its value. Land subsidence in the order of decimetres or even metres affected many cities in the world during the twentieth century (Box 6.3).

The observed side effects and impacts of groundwater exploitation as described above have heavily motivated groundwater resources assessment efforts around the world, usually with some emphasis on determining the intensity of pressures imposed upon the resources and analysing how these are related to driving forces. Population growth and agricultural development, facilitated by improved access to modern technologies and funds, have been and still are major driving forces behind the increasing pressures of groundwater abstraction in many countries. Higher water demands in response to climate change are likely to aggravate the situation in several regions in the near future.

6.2.3 The influence of land use on the groundwater regime

Man's influence on groundwater levels and fluxes is not restricted to groundwater withdrawal, but can also be very significant as a result of land use and related practices. These may modify both groundwater recharge and discharge, as compared to natural conditions.

Important land use features that influence recharge include the following:

- *Increase in the impermeable surface* in urban areas, including industrial zones, road networks and airports. For example, a 36% decrease in recharge due to urbanisation was estimated for Ontario, Canada (OECD, 1991). Urbanisation in the northern region of France is assumed to have reduced the recharge of the chalk aquifers by 100 000 m^3 per year per km^2, on average. Under certain conditions, however, opposite effects are observed and urbanisation does result in increased recharge, compensating for or even exceeding the earlier mentioned losses of natural recharge. Sometimes this is a problem rather than a benefit (see Box 6.4).

- *Changes in vegetation cover or in cropping patterns*. Afforestation will usually reduce groundwater recharge because of the relatively high rate of evapotranspiration of forest canopies. Different crops have different water demands; consequently a particular rainfall pattern may produce different rates of recharge on agricultural land, depending on the type of crop.

- *In-stream hydraulic works*. Some hydraulic works in water courses that recharge their adjoining alluvial aquifers lead to a reduction in the recharge volumes. This is particularly the case if reservoirs or thresholds produce clogging of the river bed, such as in the case of the lower Var valley, near Nice in France. In semi-arid zones, certain flood control dams may result in a reduction or even nullification of the recharge to hydraulically connected alluvial aquifers, such as occurred in the Ziz valley in the Moroccan Tafilalt. Elsewhere, an opposite effect may be observed, e.g. at the large recharge dams in Oman and the numerous small recharge dams constructed in Africa, India and elsewhere. It is obvious that each system has to be analysed individually for a correct evaluation of its effect on recharge.

- *Irrigation*. Under most irrigation methods, a significant share of the applied irrigation water passes through the soil zone and percolates downwards, thus contributing to groundwater recharge. Part of the percolated water is even intended as a means to control the salt balance. The increase in recharge may be considered as a benefit, but the additional water often leads to water quality deterioration (salinisation and pollution) and sometimes – especially in the case of surface water irrigation – to waterlogging, as observed in parts of the Indian state Haryana, in the Sindh province of Pakistan and in Uzbekistan.

The most prominent effect of land use on groundwater discharge is produced by *artificial drainage*. This is particularly relevant in flat lowlands with predominantly shallow groundwater tables where groundwater levels are controlled artificially (*polders*), such as in most of The Netherlands. Drainage works there have the objective of controlling the groundwater levels, but as a side effect they trigger a massive artificial discharge of groundwater, in volumes often exceeding groundwater abstraction. Once the drainage provisions are there (drains, ditches, canals, pumping stations, etc.), it is virtually impossible to make a distinction between natural groundwater discharge components and 'artificial' ones induced by drainage works.

Although the influence of land use on the groundwater regimes has been identified and analysed in several countries, no detailed global inventories have yet been undertaken, as far as is known. It is therefore impossible to specify to what extent

Box 6.4: Excessive groundwater replenishment

In contrast to intensive exploitation that depletes groundwater and unlike artificial recharge operations, certain land use activities unintentionally cause an excessive replenishment of the local groundwater resources. The associated negative impacts may outstrip the advantage of increasing these resources.

In urban areas it is not uncommon that the rate of recharge is greater than that under natural conditions, in spite of the recharge-reducing effect of impermeable surfaces. This is because of contributions to recharge by often significant losses of clean water originating from outside and distributed by public water supply networks (up to 50%), by percolation losses from irrigation of urban green zones and by losses of effluent from sewerage systems. These contributions may more than offset the reduction of natural recharge caused by impermeable surfaces, with the result that groundwater levels start rising – even more if there are tendencies to reduce local groundwater abstraction due to pollution. This has produced problems in cities such as Caracas, Lima, Los Angeles, Merida (Mexico), Moscow, Perth, Mar del Plata, Barcelona and Seoul (Morris et al., 2003).

Irrigation using surface water often causes groundwater levels to rise (see Figure 6.5), which eventually produces waterlogging and in dry areas salinisation of soil and water. These phenomena, that have to be controlled by drainage, reduce land productivity. Large areas of irrigated land have been affected all over the world: in China, Egypt, United States, India, Iran, Iraq, Pakistan and many other countries.

It is estimated that salinisation, alkalinisation and waterlogging due to rising groundwater levels have heavily affected about 22 million hectares around the world and another 55 million hectares moderately (Morris et al., 2003).

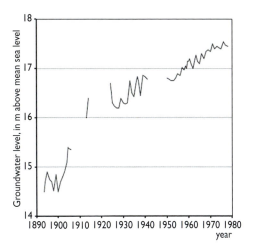

Figure 6.5 Example of rising groundwater levels caused by irrigation, Nile Delta, since the beginning of the 20th century. [*Source:* Shahin (1987)].

land use has already affected the groundwater resources, either at the country level or worldwide. However, information on the presence of artificially drained agricultural areas around the world – by far the largest share of all artificially drained areas –, has been compiled by researchers at Frankfurt University (Feick *et al.*, 2005) and is depicted in Figure 6.6. This compilation has identified 167 million hectares of artificially drained agricultural land on our planet, of which 28.5% is located in the

(a)

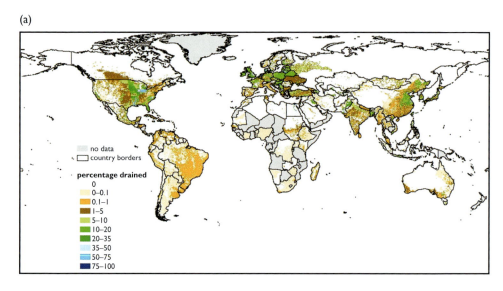

(b)

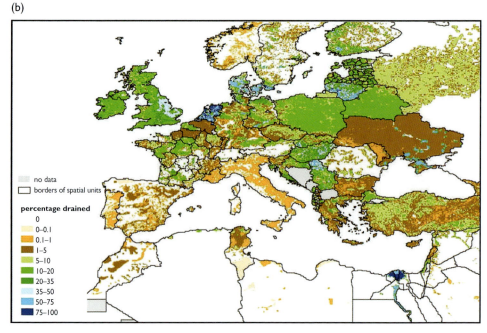

Figure 6.6 Artificially drained agricultural areas, in percentage drained per 0.5 ° × 0.5 ° grid cell: (a) World map; (b) Zoom to Europe and North Africa; (c) Zoom to North/Central America [*Source*: Feick, Siebert and Döll (2005)]. (*Continued*).

(c)

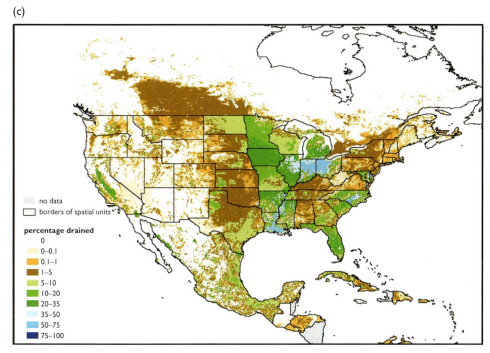

Figure 6.6 (Continued)   Artificially drained agricultural areas, in percentage drained per 0.5°×0.5° grid cell: (a) World map; (b) Zoom to Europe and North Africa; (c) Zoom to North/Central America [*Source*: Feick, Siebert and Döll (2005)].

USA, 27.2% in Europe, 12.0% in China, 3.7% in Pakistan, 3.5% in India and 3.1% in Mexico. Artificially drained agricultural land occupies a large percentage of the national territory in some countries, e.g. in The Netherlands (55%), Lithuania (40%), Denmark (35%), Serbia (27%) and Hungary (25%).

6.3   CURRENT STATE OF THE PRESSURES PRODUCED BY GROUNDWATER WITHDRAWAL

6.3.1   Overall groundwater development stress

An indicator of the degree of pressure or stress imposed by groundwater withdrawal upon the groundwater resources is the so-called '*groundwater development stress indicator*' *(GDS)*. This is the ratio of groundwater abstraction for a given year to the mean annual groundwater recharge (including induced and artificial recharge). This indicator, usually expressed as a percentage, is most appropriately used at the level of a single aquifer of limited size and containing renewable groundwater. Applying it to large compartmentalised aquifers systems, to regions that encompass several aquifer systems or even to entire countries leads progressively to smoothing, which tends to mask local critical conditions. In addition, a mix of poorly and intensely recharged aquifers within one single area undermines the indicator as a diagnostic tool even more. Nevertheless, even if applied to large and complex spatial units, the indicator is not meaningless.

In such cases, it expresses in a single number to what degree the area's groundwater regimes are on average likely to be modified compared to natural conditions, before pumping started. Its value therefore is a measure for the probability of occurrence of negative side effects of groundwater exploitation, such as storage depletion and decreased natural discharge, in at least part of the area concerned. Applied at the level of countries, values of the groundwater development stress indicator above 50% certainly point to significant groundwater regime changes and the likely occurrence of substantial negative side effects of withdrawal. However, stressed aquifers may also be present in countries – especially in large ones – that have a much lower score.

The geographical variation in the groundwater development stress indicator around the world can be assessed by examining its values presented in Appendix 5, based on groundwater abstraction estimates for 2010 by country. There is a large range of values, from almost 0% to more than 100%. The highest values, exceeding 100%, are found in semi-arid and arid zones, in particular in most Arab countries, Pakistan, Uzbekistan and Iran. One should be aware that the calculated values are far from accurate, due to uncertainties in both the national abstraction data and the estimated mean rate of recharge. For the latter, the estimates of mean diffuse recharge over the period 1961–1990 as obtained by the WaterGap model (Döll and Fiedler, 2008) were used as a proxy.

Although the indicator is more meaningful at the individual aquifer level than at the national level, data is lacking to prepare a fairly complete global picture of it for the world's countless aquifers. Very few countries carry out the inventories in their territory that would be required for this purpose. Table 6.1, nevertheless, presents indicator values for a number of aquifers around the world. The selection shown in this table is limited to examples of intensively exploited aquifers with renewable groundwater resources. The values of the groundwater development stress indicator (and the groundwater exploitation index) are above 100% for nearly all the cases shown, which indicates that the aquifers are severely stressed. The values given are subject to large uncertainty margins and comparing the values should be done with caution, because methods to estimate recharge and abstraction may vary widely from one case to another. Unlike the scare-mongering message sometimes spread that 'groundwater is excessively exploited every-where', one should not draw general conclusions from Table 6.1, because it is limited to intensively exploited aquifers. There are also thousands of aquifer systems around the world that are still hardly exploited. At the global level, current abstractions from renewable groundwater add up to only 6–7% of the total global groundwater flux (see Chapters 2 and 5) and to perhaps one-fifth of the exploitable renewable resources. However, the global scale is obviously not the appropriate level to evaluate the pressures, because problems arise and resources are managed rather at a local level.

6.3.2   Differentiating between renewable and non-renewable groundwater

The groundwater development stress indicator introduced above does not make a distinction between withdrawals from renewable and those from non-renewable groundwater resources. If conditions allow a reliable subdivision of the available groundwater abstraction data into these two categories, then a more differentiated use of indicators for the pressures due to groundwater withdrawal is possible. In other

Table 6.1 Renewable groundwater development stress indicator (RGDS) for a selection of intensively exploited aquifer systems around the world with renewable resources.

Country	Aquifer system	Year	Withdrawal km³/year	RGDS indicator* %	References
Algeria	Mitidja	2000	0.273	107	Min. NSRE 2003
	Mostaganem Plateau	2000	0.036	155	Min. NSRE 2003
	Hodna	2000	0.131	107	Min. NSRE 2003
	Ghriss Plain	2000	0.08	125	Meddi & Hubert 2003
	Set of 20 overexploited aquifers	2000	0.7	121	Min. NSRE 2003
China	Huang Huai Hai Plain (NCP)	1997	19.8	169	Wang et al. 2000
	Hebei Plain (part of HHH Plain)	1997	16.5	183	
Gaza and Israel	Coastal Aquifer	1999–2000	0.55	178	Israel Water Comm. 2002
India	Alluvial aquifer Σ Punjab	1995	11.4	120	CGWB India
	Indus basin	2000	46.7	~76	Kahlown 2004
Israel and Palestine	Mountain Aquifer	1999–2000	0.76 (incl. springs)	106	Israel Water Comm. 2002
Libya	Jeffara	1995	0.225	113	Salem 1998, Al-Miludi 2001
Malta	MSL Aquifer	1997–1998	0.025	250	Riolo 1999
Mexico	Valley of Mexico	1988	0.63	~200	Ragone 2003
	Baja California	1980	0.12	150	UNEP 1985
	Set of 100 overexploited aquifers	2000	13.9	160	Marin 2006
Morocco	Souss-Massa	1998	0.64	267	FAO 2005
		2005	0.59	158	Oubalkace 2006
	Haouz-Ksob	2005	0.59	132	Oubalkace 2006
Pakistan	Indus Basin	2000	46.7	~76	Kahlown 2004
Spain	Mancha Occidental	1989	0.58	171	Lopez-Camacho et al. 1992;
	Campo de Cartagena	1989	0.075	231	Anonymous 1994 (Libro
	Sierra de Crevillente	1989	0.015	750	blanco de las aguas subt.)
	Campo de Dalias	1989	0.11	123	
	Balearic Islands	1989	0.285	109	
	Canary Islands	1989	0.34	110	
Tunisia	Sahel/Ras el Jebel	1990	0.013	153	Khater 2003
	Jeffara	1990	0.126	64	DGRE 1995 and 2005
	Set of 32 overexploited aquifers	2000	0.504	168	Hamdane 2005
USA	Arizona (alluvial aquifers)	1990	3.78	1 022	USGS
	Central Valley of California	1990	20	286	USGS
	Ogallala Aquifer	2000	21.5	~300	USGS
Yemen	Sana'a Basin	1994	0.18	190	Foppen 1996
	Tihama Coastal Plain	1994	0.81	107	Van der Gun & Ahmed 1995
	Highland Plains	1994	0.5	220	Van der Gun & Ahmed 1995

* In these cases, where all withdrawn groundwater belongs to the 'renewable groundwater' category, the values of the RGDS indicator and the GDS indicator are the same.

words: separate indicators may then be used for the renewable groundwater resources and for the non-renewable groundwater resources of the areas concerned.

An indicator expressing the degree of pressure on renewable groundwater as a consequence of withdrawal is the *'renewable groundwater development stress indicator' (RGDS)*.[1] This is the ratio of the annual groundwater abstraction from renewable groundwater resources over the mean annual recharge in the area concerned, expressed as a percentage. It will therefore become clear that the numerical value of the RGDS is identical to that of the GDS indicator in those areas where all groundwater withdrawal is from renewable groundwater (see also Appendix 5). Figure 6.7 shows the global pattern of the renewable groundwater development indicator, at the country level.

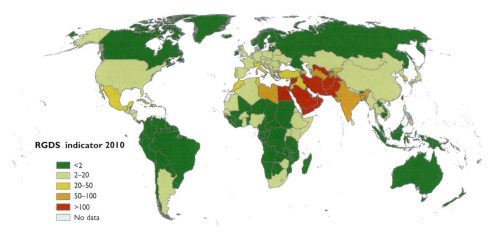

Figure 6.7 Renewable groundwater development stress indicator at country level, year 2010. [*Data and data sources:* see Appendix 5].

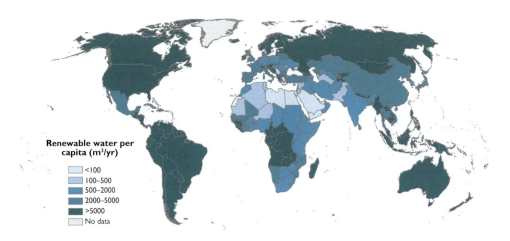

Figure 6.8 Mean annual blue water renewal per capita, by country, year 2010. [*Data sources:* AQUASTAT (2011) and UNDESA (2011)].

---

1 Sometimes also called: *'groundwater exploitation index'* or *'indice d'exploitation'*.

Another indicator of the stress produced by groundwater abstraction from an aquifer with renewable groundwater is the recently proposed 'groundwater footprint'. This is calculated by multiplying the aquifer's area by the dimensionless ratio of total annual groundwater abstraction over the difference between mean annual recharge and annual groundwater contribution to environmental streamflow. It thus attempts to explicitly incorporate the ecosystem requirements of groundwater (Gleeson et al., 2012).

The pressure imposed upon non-renewable groundwater resources by withdrawals ('mining'; see Chapter 5, Table 5.4) should be assessed in relation to the groundwater reserves (stock), because recharge is irrelevant and insignificant in such cases. The assessment should take into account only the exploitable part of the reserves. The results can be expressed in terms of either annual storage depletion (for a given year) or accumulated storage depletion since predevelopment, in both cases as a percentage of the estimated exploitable reserves. However, estimates of the exploitable groundwater reserves are much more inaccurate than those of the abstracted volumes, which undermines the value of such an indicator. In practice, in the most frequent case of confined groundwater, the comparison of the drawdown (drop of the groundwater level compared to the initial level) measured at a given moment with the maximum permissible drawdown may be a better indicator of 'pressure' than a volumetric ratio.

6.3.3   Interpreting these pressures in the context of total water stress (scarcity or abundancy)

The messages conveyed by the indicators defined above become more meaningful by interpreting them in the context of total water resources, i.e. groundwater and surface water combined ('blue water'). A useful indicator of water scarcity or abundance of water resources to satisfy human water supply needs is the *mean annual renewal of blue water per capita*. Figure 6.8 depicts the global pattern of this indicator.

If a high value of the GDS or RGDS indicator coincides with a relatively high mean annual renewal of blue water per capita, then options may exist to reallocate withdrawals from groundwater to surface water, which would relieve the pressure on groundwater. Countries in this category are Iran, Afghanistan, Armenia, India, Lebanon and Tajikistan. Countries where this does not seem to be an option include Saudi Arabia, Yemen, Kuwait, Bahrain, Qatar, United Arab Emirates and Libya.

Mean annual blue water renewal per capita is also known as the *Falkenmark water stress indicator*. Tentatively, values above 1 700 m^3/year are considered to indicate the likely absence of significant water stress. Below this level, water stress is expected to appear regularly, while below 1 000 m^3/year they become a limitation to economic development and human well-being, and below 500 m^3/year even a major constraint to life (severe water stress). It needs to be emphasised that the class boundary values mentioned enable only a very crude initial interpretation. The diagnosis becomes more reliable if other factors relevant for water stress are also taken into account, e.g. climate and water quality. Another way of looking at Figure 6.8 is relating it to the average *water footprint* of national consumption per capita (Mekonnen & Hoekstra, 2011). This footprint can be split into green, blue and grey water components. If the sum of the blue and grey components of this footprint exceeds the mean annual blue water renewal per capita, then the difference is covered by non-conventional water sources, storage depletion and/or imported *virtual water*.[2] Countries where this is the case, according to the available data, include again Saudi Arabia, Yemen, Kuwait, United Arab Emirates and Libya, The deficits in these countries range from a few hundreds to more than one thousand cubic metres per capita per year.

6.3.4   Hydrological overabstraction and aquifer overexploitation

Terms like overexploitation, overdraft, overabstraction, overpumping and overdevelopment are commonly used to indicate a critical or excessive intensity of groundwater abstraction from an aquifer with renewable groundwater resources. The different perceptions and interpretations of 'critical or excessive intensity' are often not explicitly defined or explained, which easily leads to confusion. However, as explained in more detail in Box 6.5 and discussed in several papers (e.g. Custodio, 1992 and 2002; Margat, 1992), they boil down to the following two conceptually different types of critical conditions:

- Permanent disruption of the dynamic hydrological equilibrium of the aquifer (permanent imbalance): for this category the term *'hydrological overabstraction'* is introduced here.
- Unfavourable balance of benefits from groundwater and its negative side effects (including the occurrence of unacceptable side effects): in this category the term *'aquifer overexploitation'* or simply *'overexploitation'* is used.

Unlike the common practice in the international hydrogeological community, it seems only logical and useful to use two different terms to distinguish between these two conceptually different types of critical conditions.

Hydrological overabstraction is a purely scientific-technical qualification, dealing with physical aspects only. It is a neutral concept, with an impartial criterion for diagnosis (namely: under the current abstraction regime, storage depletion will continue in the medium and long term). Data like that shown in Table 6.1 and Table 5.5 gives an impression of the degree of hydrological overabstraction as observed in a number of intensively exploited aquifers around the world. Estimates by country are shown in Table 6.2. It should be emphasized, however, that data for various important countries is missing from this table (e.g. India and China) and that available data for several of the listed countries often conflicts or is estimated only in a very tentative way. Therefore, the presented data should be used with caution.

The current globally aggregated rate of hydrological overabstraction from renewable groundwater may be estimated by subtracting the estimated rate of non-renewable groundwater depletion (31 km³/year; see Section 5.4) from the global rate of groundwater depletion as estimated by Konikow (2011) for the period 2001–2008 (145 km³/year). Hence, it is probably of the order of 115 km³/year, or around 13% of global groundwater withdrawal. Nevertheless, for individual countries this percentage may be considerably higher, as Table 6.2 shows.

---

2 For definitions of the different terms and concepts mentioned above, see Appendix 1.

Box 6.5: What is groundwater overexploitation?

The term 'overexploitation' is widely used to indicate that the rate of groundwater exploitation from a certain aquifer is considered to be excessive. But excessive in relation to what? If this question is not addressed and answered, then the qualification 'overexploitation' remains ambiguous and may even lead to confusion and controversy. There is a need to define the concept more clearly than is currently common practice.

Usually, one of the following two criteria is used to judge whether an aquifer is excessively exploited, each of them leading to a distinct type of diagnosis:

a The aquifer's water budget: long-term dynamic equilibrium or permanent state of non-equilibrium?
This is a physical criterion, allowing an objective and reproducible diagnosis to be made, provided that sufficient information is available. In response to pumping, groundwater levels will initially drop and this will induce an adjustment of the aquifer's hydrological regime by a reduction of natural outflows and by induced recharge. If these two components can no longer balance the withdrawal rate, then the dynamic hydrological equilibrium will not be restored and storage depletion will continue as long as the groundwater abstraction rates – unsustainable rates in this case – are not reduced. As already introduced in Section 5.4, such abstraction conditions may be called '*hydrological overabstraction*'.

Many groundwater professionals overlook the described dynamic response of the aquifer to pumping and suggest that hydrological overabstraction occurs if groundwater withdrawal exceeds natural groundwater recharge. Although such a simplification of the criterion does not preclude a reasonable diagnosis for many aquifer systems in the world, it may produce significant errors in individual cases (e.g. in coastal aquifers) and therefore has been baptised 'the water budget myth' (Bredehoeft, 1997; 2002).

b The balance of benefits and side effects of groundwater exploitation
In contrast to the previous criterion, this is subjective, referring to undesirable conditions produced by intensive groundwater withdrawal. The word 'undesirable' implies that the diagnosis made on the basis of this criterion involves preferences and thus it varies according to persons, area and time. A major role is played by the negative side effects of groundwater abstraction that tend to increase if groundwater withdrawal becomes more intensive. These side effects include worsening of conditions for groundwater developers in the area (falling groundwater levels, depletion, wells running dry, change of groundwater quality) but also external effects, such as reduced natural outflows (springs, baseflow), degradation of wetlands, land subsidence and reduction of soil moisture on agricultural land. The term '*groundwater overexploitation*' applies if the balance of benefits from groundwater abstraction and the negative side effects is considered unsatisfactory, that means it is either negative or likely to be improved if the abstraction were reduced.

In short, the term 'groundwater overexploitation' has a negative connotation in practice and should only be used to indicate exploitation of an undesired intensity, which produces more negative side effects than is acceptable. It does not necessarily require a permanent lack of equilibrium in the groundwater regime, because certain negative impacts (such as land subsidence and wetland degradation) may also become prohibitive under long-term hydrological equilibrium conditions. On the other hand, most cases where hydrological overabstraction occurs will probably also be classified as overexploitation cases.

Finally, the planned mining of non-renewable groundwater and the planned temporary exploitation of the reserves of renewable groundwater – making use of the groundwater buffer – should not be considered as groundwater overexploitation. The decision to deplete groundwater storage in these cases is based on the judgment that the benefits of the planned depletion outstrip its negative side effects.

Table 6.2 Estimates of overabstraction of renewable groundwater for different countries.

| Country | Reference year | Withdrawal from renewable groundwater resources | | | References |
| | | Total km³/year | Overabstraction | | |
			km³/year	% of total	
Algeria	2000	~ 1.2	0.145*	12	MNSRE 2003
China	1997	96.8	30	31	Wang et al. 2000
of which in North China		55.7	30	54	
Cyprus	1998	0.145	0.04	28	Tsiourtis 1999
Gaza	1995	0.13	0.06	46	Al Jamal & al. 1996
Iran	1993	57	11	19	ICID 2000
	1995	53	16	30	FAO 1995
Israel	2000	1.2	0.187	16	Israel Wat. Comm. 2002
Jordan	2000	0.5	0.225	45	Min. Water & Irrig. 2002
Lebanon	2003	0.24	0.09**	38	ESCWA 2007
Libya	1999–2000	1.28	0.765	60	Salem 2000
Malta	1997–1998	0.025	0.015	60	Riolo 1999; WSC 1998
Mexico	2000	27.2	13.9	51	Marin 2006
Morocco	2005	3.68	0.54	15	Oubalkace, 2006
Oman	2003	1.15	0.60**	52	ESCWA 2007
Portugal	2000	4.75	0.75	16	INAG 2001
Spain	1995	5.53	1.05	19	Anonymous 1998
Syria	2005	8.34	3.44**	41	ESCWA 2007
Tunisia	2000	1.21	0.175	14	DGRE 1995, 2005
	2000	1.21	0.204	17	Hamdane 2005
United States	2000	115	17.7	15	Konikow 2011
Yemen	2001	2.40	1.40**	58	ESCWA 2007

* On the basis of 40 aquifers in North Algeria for which water balances have been established
** Proxy data obtained by difference between total groundwater withdrawal and mean annual groundwater recharge aggregated over the entire country. Such estimates are relatively unbiased if virtually all the aquifers are overabstracted and have negligible submarine discharge, but otherwise are an underestimate.

Note that 'abstraction exceeding mean recharge' is not a correct criterion for the diagnosis of hydrological overabstraction, as explained in Box 6.5. It may lead to erroneous conclusions.

Groundwater overexploitation is a more complex concept, especially because it includes subjective elements. In the absence of rigorous definitions and uniform assessment procedures, individual cases of overexploitation are not truly comparable, which hampers any attempt at a worldwide inventory. Assigning the qualifier 'overexploitation' is usually triggered by observed or anticipated negative side effects of groundwater abstraction rather than by perceived changes in the aquifer's groundwater regime. Certain types of side effect – e.g. land subsidence and wetland degradation – are more sensitive to local changes in groundwater state variables than to aggregated groundwater budget changes. In such cases, the identification or prediction of the corresponding threats should take place at a more local level than the aquifer scale. Overexploitation often implies hydrological overabstraction of the aquifer, but – as also mentioned in Box 6.5 – this is not a necessary condition, because side effects like land subsidence and ecosystem degradation may become problematic even if the rate of groundwater withdrawal does not preclude a long-term hydrological equilibrium.

Few countries have inventoried and mapped the aquifers they consider as overexploited. One of these countries is Spain, where by the year 1997 seventy-seven hydrogeological units had been identified as being overexploited or suffering from increasing salinity; among these, fifteen units have been legally declared overexploited (MMA, 2000). RIVM and RIZA inventoried seventy cases of overexploitation in the then twelve countries of the European Community in 1991. In Algeria, out of forty aquifers for which a hydrological balance was established, twenty-five were overexploited by the year 2000, according to the *Agence National des Ressources Hydrauliques* (ANRH). In Tunisia, thirty-two out of 239 inventoried aquifers are overexploited, according to Hamdane (2005). Among the 653 aquifers identified in Mexico, the number of those considered overexploited has increased from thirty-five in 1975 to one hundred and two in 2005 (Marin, 2006).

Early efforts to avoid aquifer overexploitation – dating from many years before this term came into use – focused on concepts such as safe yield and sustainable yield (see Appendix 1). Van der Gun and Lipponen (2010) review how the interpretation of these concepts and related paradigms have evolved over time and how these have been and can be used as a guide towards sustainable groundwater resources management.

6.3.5 Depletion of groundwater reserves

As a cumulative side effect of hydrological overabstraction of renewable groundwater and mining of non-renewable groundwater during the 20th and the beginning of the 21th century, significant groundwater depletion has been produced in several large aquifers around the world. This depletion is inseparable from prominent water level declines as discussed in Section 6.2, and adds up to hundreds of cubic kilometres or more (see Figure 6.9 and Table 6.3). On a human time scale, much of this depletion is virtually irreversible and certainly the loss of storage capacity is permanent in those cases where soft earth layers have been compressed and consolidated. According to estimates made by Konikow (2011), the groundwater reserves of the

Table 6.3 Principal cases of groundwater depletion caused by intensive groundwater exploitation during the 20th century (*Overabstraction of renewable resources plus mining of non-renewable ones*).

Country	Aquifer system or region	Reference period	Duration (years)	Cumulative depletion of the reserves (km³)	References
Australia	Great Artesian Basin	1880–1973	94	25	Habermehl 1980, 2002
Algeria,	North Western Sahara	1950–2000	50	9.35	OSS 2005 (modelling)
Tunisia, Libya	Aquifer system (CT + CI)	1900–2000	100	53	Konikow 2011
		1900–2008	108	70	Konikow 2011
China	Hebei Plain	1960–1980	20	15 to 20	Wang et al. 2000
	(part of North China Plain)	1960–2000	40	70	
	North China Plain	1900–2000	100	130	Konikow 2011
		1900–2008	108	170	Konikow 2011
Egypt, Libya	Nubian Sandstone Aquifer System + Post Nubian	1960–2000	40	30	Bakhbakhi 2002 (modelling)
	Nubian Aquifer System	1900–2000	100	80	Konikow 2011
		1900–2008	108	98	Konikow 2011
India, Pakistan,	Northern India and	1900–2000	100	938	Konikow 2011
Nepal and Bangladesh	adjacent areas	1900–2008	108	1 361	Konikow 2011
Saudi Arabia	Multi-layered Arabian Platform Aquifer	1975–2000	25	~ 380	Abdurrahman 2002
	Arabian aquifers	1900–2000	100	359	Konikow 2011
		1900–2008	108	468	Konikow 2011
Spain	Volcanic aquifers of Gran Canaria and Tenerife (Canary Islands)	1900–1990	100	8 to 10	
United States	Atlantic Coastal Plain	1900–2000	100	14.4	Konikow 2011
		1900–2008	108	17.2	Konikow 2011
	Gulf Coastal Plain	1900–2000	100	199	Konikow 2011
		1900–2008	108	266	Konikow 2011
	High Plains Aquifer	1900–2000	100	259	Konikow 2011
	(Ogallala)	1900–2008	108	353	Konikow 2011
	Central Valley, California	1900–2000	100	113	Konikow 2011
		1900–2008	108	145	Konikow 2011
	Western USA Alluvial Basins	1900–2000	100	175	Konikow 2011
		1900–2008	108	178	Konikow 2011
	Western volcanic aquifer system	1900–2000	100	− 48	Konikow 2011
		1900–2008	108	− 45	Konikow 2011
	Deep confined bedrock aquifers	1900–2000	100	33	Konikow 2011
		1900–2008	108	36	Konikow 2011

Note: Negative values for the cumulative storage depletion as reported for the Western Volcanic system in the United States indicate an increase in storage.

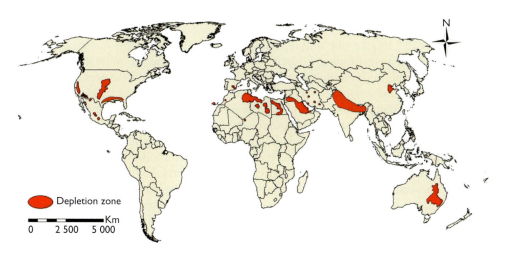

Figure 6.9 Zones of very significant groundwater depletion during the twentieth century as a result of overabstracting renewable or mining non-renewable groundwater resources.

Earth's continents have been depleted by almost 3 400 km³ during the 20th century and even by some 4 500 km³ during the period 1900–2008. Part of this volume corresponds to a reduced thickness of phreatic aquifers (by falling water tables), and another part to decompression of confined aquifers (reduced volume of voids and expansion of water inside these voids).

Although the globally aggregated depleted volume of groundwater is very small compared to the volume of water stored on the planet (Table 2.1) or present in the subsurface of some large countries (Table 2.2), its impact is not necessarily negligible because most of the depletion is concentrated in a few areas only, where it has removed relatively accessible water. These areas are particularly located in the dry zones of the Northern hemisphere and in Australia. The groundwater reserves there have lost contact with today's active water cycle and abstraction occasionally causes external side effects, such as land subsidence.

Depletion threatens long-term sustainability of the groundwater resources. Available data suggest that the High Plains aquifer system has lost some 8 to 9% of its stored volume since predevelopment (McGuire, 2009), but the depletion is likely to represent a much higher share of the exploitable groundwater resources. The Alluvial Basins of Arizona, responsible for almost 60% of the cumulative depletion of the Western Alluvial Basins in the USA, show how quickly conditions may change. Depletion there had increased to some 3 km³/year during the period 1960–1980, but serious problems of subsidence and water-table declines caused water management and water use practices to change drastically, resulting in an average recovery of groundwater storage of 0.41 km³/year after 1980 (Konikow, 2011). The exploitable resources of numerous other aquifers in the dry zones around the world, especially the smaller aquifers, may have been already partially or entirely lost as a result of depletion, but it is usually difficult to find reliable data to prove and document this objectively.

6.3.6   Are the world's groundwater resources being exploited excessively or insufficiently?

It is not possible to give one simple, generally valid answer to this question. This is because there is an enormous diversity in pressures on the groundwater resources around the world, in the responses of groundwater systems to these pressures and in the impacts of these responses. It is clear that there are numerous cases of intensive groundwater exploitation where the qualification 'overexploitation' seems appropriate, but these cases are scattered around the world and do not represent the entire category of intensely exploited aquifers. It would also be worthwhile to inventorize and analyse the large number of aquifers where the opposite conditions prevail: aquifers that are hardly being exploited (*'underexploited' aquifers*). Unlike the overexploited aquifers, these are not in a critical state, but it is useful to better understand the reasons behind the low intensity of their exploitation and to assess the forgone benefits.

Cases of hydrological overabstraction or mining represent only a minor but not negligible fraction of global groundwater abstraction. They pose nevertheless serious problems to countries where the groundwater resources in question play an important or even dominant role in their water supply, because such supplies are not sustainable. The higher the contribution of the overabstracted or mined source of water to the current provision of water and the projected water supply for the near future, the more critical are the consequences. One should be aware that these supplies necessarily have to diminish and most likely will be phased out during the 21st century. Governments and water management agencies should identify and monitor the unsustainably exploited water resources within the territory of their mandates and prepare to replace them in due course with alternative water sources or, if these are not available, to reform water use practices. The latter may require a transition to a completely different regional economy, much less dependent on water. This constitutes a major challenge in regions where the economy is dominated by irrigated agriculture that is largely based on non-sustainable exploitation of groundwater resources, such as on the Highland Plains in the USA or in many regions in the Middle East.

There is no doubt that in many areas around the world there is still scope for gaining considerable benefits or profits by making more intense use of scarcely exploited groundwater resources. This is especially evident in the context of rural water supply. Rural settlements are dispersed over large areas, which often makes groundwater a more suitable source than surface water to satisfy domestic water demands. This is because groundwater tends to be easily accessible, reliable, of good quality (little or no treatment needed) and available where it is needed (minimal transport cost). Groundwater therefore is expected to play an important role in achieving Millennium Goal No. 7 that includes the target 'halving by 2015 the proportion of people without access to safe drinking water' (UNDP, 2010) and in achieving more ambitious drinking water supply targets during the period beyond 2015.

6.4   PRESSURES PRODUCED BY CLIMATE CHANGE AND SEA LEVEL RISE

6.4.1   Climate change

Global climate change as a result of the greenhouse gas effect is expected to affect groundwater systems during the 21st century and beyond considerably. As a driver of change it will produce or modify stresses, notably by changing groundwater recharge volumes and their distribution in space and time, by modifying demands for groundwater and by its contribution to sea level rise. The contrasts known today in climatic and groundwater conditions are likely to become more pronounced, although it is still difficult to quantify the changes and their timing. The climate change predictions of the Intergovernmental Panel on Climate Change (IPCC) are based on four families of greenhouse gas emission scenarios and use different climate models, which leads to a diversity of alternative climate change projections. In spite of a high degree of uncertainty about climate change, the different projections share certain robust findings. These include that changes in the global climate during the 21th century will very likely be larger than those observed during the 20th century, that land will warm more than the adjacent oceans, that frequencies and intensities of some extreme weather events are very likely to increase and that anthropogenic warming and sea level rise will continue for centuries even if greenhouse gas concentrations were to stabilise due to a reduction in emissions (IPCC, 2007).

Various studies have attempted to predict the effects of climate change on groundwater in the basin of the Mediterranean Sea (in Spain, France and Greece), focusing on special, most sensitive local conditions of reduced groundwater recharge combined with increased groundwater exploitation for irrigation. The investigations showed that the groundwater budgets would be reduced on average by some 100 mm/year or more, mainly due to an increase in potential evapotranspiration (over 50%), resulting in a significant decrease in the mean groundwater level (Vachaud *et al.*, 1997).

On a global scale, the IPCC reports provide useful information. The third IPCC report published two world maps depicting projected changes in average annual runoff by the year 2050, relative to average runoff for 1961–1990 (IPCC, 2001). These changes were calculated with a hydrological model using as inputs two versions of the Hadley Centre atmosphere-ocean general circulation model for a scenario of 1% per annum increase in effective CO_2 concentration in the atmosphere. According to these and other model projections, runoff will increase in the high latitudes and Southeast Asia, and decrease in central Asia, the area around the Mediterranean, southern Africa and Australia. For other areas of the world, changes in precipitation and runoff are much more scenario- and model-dependent. The maps do not differentiate between groundwater and surface water. Except for a few very general comments, the report does not address groundwater.

The fourth IPCC report (2007) confirms the global pattern of contrasting changes in rainfall expected to occur between now and the end of this century, with decreases in rainfall compared to the 1980–2000 averages that may exceed 20% in arid and semi-arid zones. Again, this report is not very explicit about the changes that may be expected in the mean surface water or groundwater fluxes and in their variability. Even in a special Technical Paper on Climate Change and Water issued by IPCC (Bates *et al.*, 2008), the attention paid to groundwater is mainly in the form of outlining potential impacts of climate change on groundwater systems in a generic way rather than predicting region-specific impacts.

The projections of changes in groundwater recharge made on the basis of the Water GAP Global Hydrology Model, developed and used at the Universities of Kassel and Frankfurt, in Germany, offer much more area-specific information. Döll (2009) describes such projections on the basis of four climate change scenarios of the IPCC and compares the model outcomes with those of the reference period 1961–1990. The model projections suggest that by the 2050s groundwater recharge is likely to have increased in the northern latitudes, but heavily decreased (by 30–70% or even more) in some currently semi-arid zones, including the Mediterranean, north-eastern Brazil and south-western Africa (see Figure 6.10). Simulations for ten other climate scenarios produced different trends for some regions, except for the Mediterranean region and the high northern latitudes. The four simulated projections shown in Figure 6.10 suggest a decrease in the long-term mean global groundwater recharge by more than 10% and a tendency to increasing contrast in recharge rates between humid and arid/semi-arid zones. Climate change is difficult to predict and at the scale of tens of years it is hard to distinguish it from climate variability as produced by El Niño Southern Oscillation (ENSO), Pacific Decadal Oscillation (PDO), Atlantic Multidecadal Oscillation (AMO) and other inter-annual to multi-decadal climate oscillations (Gurdak et al., 2009).

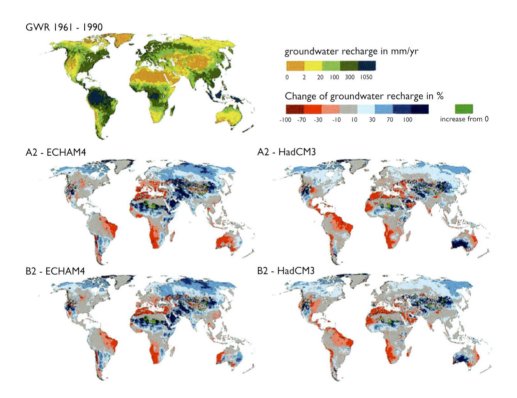

Figure 6.10 Global pattern of estimated mean groundwater recharge during 1961–1990 and percentage change between 1961–1990 and 2041–2070 for four IPCC climate change scenarios. [Source: Döll, 2009].
Note: The climate scenarios were computed by the climate models ECHAM4 (Röckner et al., 1996) and HadCM3 (Gordon et al., 2000), respectively, on the basis of IPCC's greenhouse gas emission scenarios A2 (emissions increase during 1990–2050 from 11 to 25 Gt C yr⁻¹) and B2 (increase from 11 to 16 Gt C yr⁻¹).

Climate change will not only change mean annual groundwater recharge and mean annual surface water flow, it is expected to affect also their *distribution in time*. Wet episodes may become more concentrated in time in many regions, while dry periods will tend to become longer. However, this will not significantly affect the water supply capacity of most groundwater systems, due to their buffer capacity. The buffers cannot prevent long-term groundwater availability from becoming less if climate change reduces the mean recharge rates, but they facilitate a gradual adaptation to new conditions.

Climate change will also modify *water demands and water use*, which – in turn – may affect groundwater systems and their exploitation. As the patterns of change in mean groundwater recharge shown in Figure 6.10 show positive correlation with patterns of change in mean precipitation and mean runoff as predicted by the IPCC (Bates *et al*., 2008), it may be concluded that higher water demands will particularly affect areas where mean groundwater recharge is expected to decrease. This may produce severe problems – in the numerous small and shallow alluvial aquifer systems scattered over arid and semi-arid regions ('wadi aquifers') probably more than anywhere else (Van der Gun, 2009). Nevertheless, it is expected that in many increasingly water-scarce areas around the world, dependency on groundwater will increase since the storage buffer renders groundwater more resilient than dwindling surface water sources. This is one more reason to manage groundwater carefully in such regions.

These changing conditions inspired UNESCO-IHP to launch the project *Groundwater Resources Assessment under the Pressures of Humanity and Climate Change (GRAPHIC)* in 2006. It aims to better understand how groundwater systems around the world will change under various population and climate change scenarios. Subjects addressed are groundwater recharge, discharge, storage, quality and management. A variety of methods is used and pilot studies are carried out in different parts of the world (UNESCO, 2006; Treidel *et al*., 2012).

6.4.2   Sea level rise

The global-mean sea level rose during the twentieth century by 17 to 18 cm, at a steadily increasing rate (Church and White, 2006). Ongoing and predicted sea level rise is caused to a large extent by climate change, but progressive groundwater depletion contributes to it as well. Konikow and Kendy (2005) argue that the ultimate sink for groundwater removed from the aquifers by depletion is the oceans. Accordingly, Konikow (2011) calculates that groundwater depletion in the USA would have contributed 2.2 mm to sea level rise during the 20th century, while the contribution of total global groundwater depletion during the same period would have been 9.3 mm. For the period 1900–2008, the corresponding estimates are 2.8 mm and 12.6 mm, respectively (see Section 5.4.3). The estimated long-term global groundwater depletion thus balances 6 to 7 per cent of the observed sea level change since 1900. The corresponding mean annual rate currently is some 0.403 mm (period 2001–2008), or about 12% of the current rate of sea level rise estimated by the IPCC. Although all these estimates are subject to considerable uncertainty, it may be concluded that groundwater storage depletion contributes significantly to sea level rise. It should be noted, however, that another human intervention in the hydrological cycle – augmenting surface water storage by the construction of dams – has an opposite effect on sea level. The same is true for augmenting groundwater storage by artificial recharge.

The position of sea level relative to land surface in coastal areas is not only affected by climate change and human interactions with surface water and groundwater. Geological processes such as tectonic uplift or subsidence, erosion and sedimentation also impact on the relative position of sea level.

In relation to groundwater, the main impact of sea level rise is intrusion of saline water into coastal aquifers. Worldwide, sea water intrusion is a real threat to coastal aquifers in flat areas and it may have huge repercussions since a large percentage of the world's population lives in coastal zones. A series of papers in the Hydrogeology Journal provides a geographic overview of saltwater-freshwater interactions in coastal aquifers (Barlow, 2010; Bocanegra *et al.*, 2010; Custodio, 2010; Steyl and Dennis, 2010; White and Falkland, 2010). A study on the impact of sea level rise on coastal groundwater in The Netherlands (Oude Essink *et al.*, 2010) concluded that the expected sea level rise will affect the Dutch coastal groundwater systems and trigger saline water intrusion, but only in a narrow zone within ten kilometres from the coastline and the main lowland rivers.

Another impact of sea level rise on groundwater is an increase in the volume of both saline and fresh groundwater stored in coastal aquifers, because it moves the drainage base level to a topographically higher position.

6.5   GROUNDWATER QUALITY DEGRADATION

As pointed out already in Section 6.2, human influences on groundwater systems are not restricted to groundwater withdrawal and artificial drainage, but are also associated with other human activities. This applies in particular to groundwater pollution, a side effect of human activities of which the majority has no intention at all to interfere with groundwater. It is also important to note that those who pollute groundwater are not necessarily identical to those who use the same resource and benefit from it. Many polluters thus may not get direct feedback from their polluting activities, which does not help to create or increase their motivation for improving their practices and behaviour in this respect.

Broadly, three main categories of groundwater quality degradation can be distinguished:

- Groundwater quality degradation caused either by injection of contaminants into the aquifer or – more commonly – by movement of contaminants through the soil into the groundwater domain. This form of groundwater quality degradation, characterised by *substances being added* to the volumes of water, is generally called groundwater pollution. As already mentioned, it results from activities of people who are either not aware of this side effect or not motivated to prevent it.
- Groundwater quality degradation induced by changes in the groundwater regime (in particular triggered by withdrawal), causing the movement of water masses of different quality. This form of groundwater quality degradation is characterised by *migration* of substances already present in the water masses

or in the subsoil, rather than by additional substances entering these domains. Therefore, there is a tendency in part of the groundwater community to avoid using the term 'groundwater pollution' for this category of water quality degradation in aquifers. The most common example in this category is the local increase in groundwater salinity due to either sea water intrusion or upconing of saline water underlying fresh groundwater, both triggered by groundwater abstraction.

- A mixed form of groundwater quality degradation, where modification of the groundwater regime and the influx of contaminants both play a significant role. An example is the increase in salinity in aquifers receiving irrigation return flows. The latter not only return water in which the minerals contained in the abstracted local groundwater have become more concentrated, but they also flush down part of the applied fertilizers and pesticides. Another example is the induced increase of the arsenic content in shallow groundwater in Bangladesh, triggered by intensive groundwater exploitation.

Box 6.6: Groundwater pollution

Pollution of groundwater or surface water is revealed by man-made changes in their chemical composition or other characteristics (physical, biotic), usually resulting in a less favourable water quality. Like water quality itself, water quality degradation is assessed on the basis of water quality standards as defined for different uses. The concept pollution leaves room for different interpretations, as it may refer to multiple uses and functions of water, including its contribution to the living conditions for aquatic ecosystems.

As a direct or induced result of human action, pollution should not be confused with natural water quality deficiencies such as salinity or excessive levels of various solutes (sulphate, fluoride, arsenic, etc – see Section 4.3). The term pollution is not unambiguous, because it may refer to both the process (producing threats) and its effects (observed or predicted water quality degradation).

Being less directly exposed, groundwater is usually better protected against pollution than surface water. Compared to surface water with its much smaller turnover time, however, groundwater is less resilient. This means that pollutants remain for a much longer time once they have entered, even if their ingress took place only during a short period. A certain degree of natural attenuation and decay by microbiological and chemical processes may ease groundwater pollution conditions to some extent.

Quality degradation makes groundwater unsuitable for certain – or sometimes even all – uses, unless it is treated. In particular it will reduce or undermine its suitability for use as drinking water.

Although not as often as in cases of surface water pollution, impacts of groundwater pollution on the state of aquatic ecosystems do occur, provided that the latter are dependent on groundwater.

6.5.1   Groundwater pollution

The main threats to groundwater quality come from pollution risks. There are many kinds of pollution, with a wide range of sources, processes and substances involved (see Box 6.6). The threats are either that components naturally present in groundwater (salt, nitrate, iron, manganese) reach concentrations beyond the accepted standards for a certain type of water use, or that other substances are introduced, often harmful even at very low doses (hydrocarbons, pesticides, or organic contaminants). The list of these man-made pollutants is long and continues to grow along with the introduction of new chemicals and with the refinement of standards.

Given a certain level of pollutants present at the surface, groundwater pollution risks are greater if the exposed groundwater systems are more vulnerable and less resilient, as has already been discussed in Section 4.8. Obviously, taking stock of groundwater pollution risks is different from making an inventory of cases where groundwater quality degradation has already caused impairment or even loss of the resources. The latter is also much more difficult.

According to process and spatial distribution, observed or predicted groundwater pollution is commonly subdivided as follows:

* *Diffuse pollution*, spread over relatively large areas. Its most widespread example is groundwater pollution due to a surplus of fertilizers and pesticides used in intensive agriculture; excessive levels of nitrates or phosphates are the most common result. The use of only partially treated wastewater for irrigation is another source of diffuse pollution. Diffuse pollution can also result from airborne pollution transmitted by precipitation, especially in the form of acid rain. Rather acid water (especially at pH < 4.5) can be enriched with metal ions when flowing through aquifers or, after being abstracted, by corrosion of metal pipes (aluminium, cadmium, copper, lead, zinc), as has been observed in several countries where groundwater has significant natural acidity (Canada, USA, Sweden).

* *Local pollution (point source pollution and line source pollution)*. This has many potential sources, often related to malfunctioning or accidents at or near factories, mines, utilities or installations at fixed locations; hence these hazardous sites may be inventoried and mapped. Examples are poorly isolated waste dumps such as sanitary landfills and deposits of mining waste and tailings (e.g. salt heaps of the potassium mines in the Alsace, France, causing salinisation of the Rhine Valley aquifer), inadequate subsurface storage of dangerous substances (hydrocarbons), industrial accidents, leaks from sewerage networks, and losses from pipelines. The local pollution sources are sometimes subdivided into *point sources* and *line sources*. The latter are related to linear sources like highways (emission of pollutants produced by traffic), railways (spills and pesticides) and also water courses from which polluted water (often originating from industrial and urban waste) may percolate downward to underlying alluvial aquifers.

6.5.2   Groundwater quality degradation induced
by groundwater abstraction

The most prominent type of induced groundwater quality degradation – seawater intrusion and saline water upconing in coastal zones – has been observed during the twentieth century in many countries, e.g. in China, along the Mediterranean Sea and the North Sea, and in the United States (Atlantic Coast, Florida and California). It may be controlled by reducing groundwater abstraction rates, applying artificial recharge (The Netherlands) or by pumping barriers (Israel, Orange County in California, Barcelona).

Induced groundwater quality degradation may also result from poorly performed drilling operations, inadequate well construction and corrosion of well casing, causing hydraulic short-cuts between aquifers with groundwater of different quality.

6.5.3   State of knowledge of the threats to groundwater quality

An *ad hoc* UNESCO – IHP V working group (Theme 3.1: Groundwater contamination inventory) conducted a comprehensive review of the current state of affairs related to the identification and inventory of the potential factors relevant for groundwater pollution. The results of such inventories are generally laid down in specific maps:

- Maps of identified potential local pollution sites, such as different classes of residential and industrial buildings, waste dumps, deposits of mining waste or tailings, oil pipelines, etc.
- Maps of zones characterised by special conditions related to diffuse pollution, particularly from agriculture, such as areas of inputs of nitrate and pesticides (e.g. maps for the European Community drawn in 1991 by RIVM/RIZA), or industrial zones.

Such inventories and mapping programmes have been carried out in only a few countries, mainly in industrialised countries. The outcomes are usually accompanied by maps of intrinsic vulnerability (see Section 4.8), but prepared on the basis of a diversity of criteria, which forms an obstacle to any attempt to derive a worldwide picture from the collection of these examples.

Only overlaying the maps of pollution threats and those of intrinsic vulnerability can lead to reliable maps of the groundwater pollution risk. This has been done in a very few countries, with the objective to trigger and guide efforts for groundwater protection, giving priority to preventing pollution. The approaches used vary and the outcomes are therefore of limited comparability. A complete overview of all risks to ground quality around the world should seek a more universally joint effort. As the next step, efforts should move from the inventory of sources and risks to the inventory of pollution that really has occurred.

6.5.4   Does a clear picture exist of the current groundwater pollution around the world?

Groundwater pollution is more often addressed in general terms or described for single cases by scientific journalists, environmentalists and movie producers than identified and assessed by groundwater professionals on the basis of exhaustive inventories. Inventories of actual cases of pollution and the presentation of their outcomes in the form of maps are usually less systematic and standardised than those of potential pollution sources and they tend to get much less attention. This is reflected in the UNESCO report (Zaporozec, 2002) mentioned earlier that despite its title deals mainly with the inventory of potential sources and processes of groundwater pollution.

It is true that the existence of so many types of pollution (different in terms of pollutants, degree of harmfulness and persistence) is not helpful for the development of one uniform cartographic methodology for presenting the findings. In addition, assessing certain types of diffuse pollution is hampered by the difficulty of defining natural background concentrations in areas where man-made changes to the natural environment started long ago. It is also true that maps of observed pollution may not be universally welcome because these can more easily be perceived as a tool for allocating blame than maps of potential pollution sources. But the relative scarcity of inventories and maps of existing groundwater pollution is undoubtedly also due to the fact that they require time-consuming field work and expensive laboratory activities.

Groundwater quality maps, usually limited to the upper aquifer and using standardised classifications, often include pollution but without clearly distinguishing it from natural water quality deficiencies. Maps of pollution should ideally delineate zones or 'black spots' where at a specified moment in time groundwater quality was found to be degraded, as compared to its natural state.

In several countries, inventories of local pollution sites and the associated damage allow statistical processing (classification by origin, type of contaminants and setting) and analysis of their spatial distribution. For example, 674 cases of groundwater pollution were identified in Italy during the period 1995–1996, affecting 594 sources of drinking water (Giuliano, 1998). Repeated groundwater pollution inventories in the State of South Carolina (USA), slightly more than a quarter the size of Italy, show an overall increase in identified pollution sites from 60 in 1980 to 4 527 in 2008 (Horton and Devlin, 2008). The majority of the sites is concentrated around population and industrial centres. Table 6.4 presents statistics on the pollution sources and types of pollutants at these sites. In The Netherlands, half the size of South Carolina, 6 500 to 7 500 polluted sites require clean-up operations (Noordhoff, 2010).

Table 6.4 Some statistics on the 4 527 identified groundwater pollution sites in South Carolina, USA (after Horton and Devlin, 2008).

Pollution sources	# of sites	Pollution types	# of sites
Underground storage tank	3 766	Petroleum products	3 941
Spills and leaks	319	Volatile organic compounds	479
Pits, ponds and lagoons	250	Metals	179
Aboveground storage tank	135	Base, neutral & acid extractables	33
Landfill	125	Nitrates	31
Septic tank/tile field	20	Pesticides/herbicides	22
Unpermitted disposal	15	Radionuclides	18
Spray irrigation	13	Polychlorinated biphenyls	10
Single-event spill	6	Phenols	7
Other	66	Other	28
Unknown	119		

In addition to the generally known agricultural and urban pollution sources and substances of which Table 6.4 gives an impression, in recent years there has been growing attention to micro-pollutants ('emerging pollutants'), in particular for *pharmaceuticals and personal care products (PPCPs)* and for *endocrine disruptive compounds (EDCs)* (Schmoll *et al.*, 2006; Musolff, 2009; Stockholm World Water Week, 2010). Disseminated by sewage, landfills and manure, these substances occur in natural waters in only very low concentrations (at the level of picograms to nanograms per litre) and are not removed by conventional wastewater treatment plants. They are relatively ubiquitous in surface water and shallow groundwater in densely populated areas, and progress in analytical methods is likely to reveal them more widely in future years. There is still much uncertainty on their possible impacts.

Maps showing the concentrations of one single parameter allow a more unambiguous interpretation, but they present information on only one of the relevant aspects. Under this category come the maps of nitrate concentrations in groundwater as established in the European Union, of which Figure 6.11 is an example. It refers mainly to diffuse pollution, which – unlike local pollution – can be identified by regional monitoring networks. Other maps portray the affected aquifer zones and permit their share in the total horizontal area occupied by the aquifers to be assessed. For example,

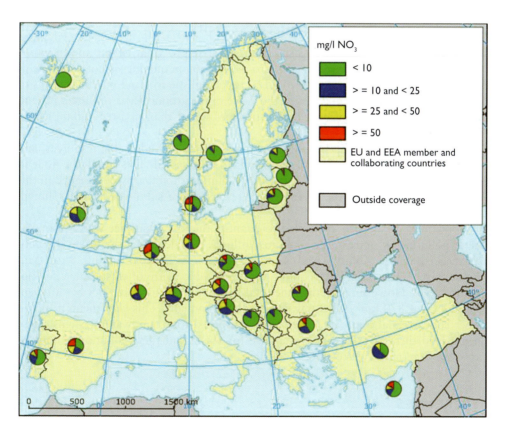

Figure 6.11   Average annual groundwater nitrate concentrations in groundwater, by country and by concentration class. [*Source*: European Environmental Agency, 2010].

the zone of seawater intrusion induced by intensive groundwater exploitation in North China spread in the 1990s over 1 430 km^2 (Wang *et al.*, 2000). The number of polluted wells is another indicator for the expansion of pollution or sea water intrusion in groundwater (more than 5 000 affected wells in North China).

To avoid difficulties in interpolating between points for which water quality analysis data is available, it is often preferred to present only the groundwater analysis results at the location of the sampling points on the maps. These may then be classified for a specific water quality variable on the basis of how much they diverge from natural conditions, or by comparison to water use quality standards. Counting the number of data points in each class may be a useful statistical indicator. For example, in 2002 around 62% of the 1 048 monitoring sites of nitrates in groundwater in France had concentrations above 10 mg/l, a threshold marking the beginning of water quality deterioration. However, only 10% of the sampled groundwater was reported to require treatment to make it potable (nitrate concentrations of 50 to 100 mg/l), while 1% (with nitrate concentrations above 100 mg/l) was considered unsuitable for drinking water production (Blum, 2004).

Such statistical indicators, based on sampling, however, cannot be converted directly to a degree of resource degradation.

6.5.5   Does groundwater pollution contribute to groundwater resources scarcity?

Giving an affirmative answer is easier than specifying to what extent and in which cases. Translating quality degradation into a reduction in quantity is not as simple as one might assume. In any case, it would be too simplistic to think purely in terms of quantity. Pollution results in a loss of the value of the groundwater resources (due to the additional costs of treatment or substitution) rather than in a reduction of their quantity.

Above all, pollution has an impact on the exploitable and usable resources, and it is the damage to these resources that needs to be assessed. This should be done in relation to categories of potential water use: pollution should only be assessed in this context, rather than that the groundwater resources are subdivided into 'polluted' (supposedly useless) and 'clean'.

Sometimes it has been attempted to quantify the loss of surface water resources due to pollution by untreated wastewater inflows, by calculating the fluxes theoretically needed for their dilution and 'self-purification'. As a rule of thumb, 8 to 10 m^3 of clean river water would be required for every cubic metre of wastewater (Shiklomanov, 1990). However, even then the total flux in the world would not be sufficient to dilute the 1 870 km^3 of wastewater produced annually. This very theoretical and rather controversial approach cannot be transposed to groundwater systems; these do not lend themselves to be a receptacle for waste and their pollution cannot be expressed in terms of wastewater fluxes.

REFERENCES

Abdurrahman, W., 2002. *Development and management of groundwater in Saudi Arabia.* GW-MATE/UNESCO Expert Group Meeting, Socially sustainable management of groundwater mining from aquifer storage. Paris, 6 p.

Abderrahman W.A., 2003. Should intensive use of non-renewable groundwater resources always be rejected? In: *Intensive Use of Groundwater*, Balkema, Lisse, The Netherlands, pp 191–203.

Al Jamal, K., 1996. Gestion des ressources en eau dans la bande de Gaza. Conf. euro-méditerranéenne sur la gestion de l'eau, Marseille, 25–26 Nov. 1996.

Alley, W.A., 1999. *Sustainability of groundwater resources.* US Geological Survey Circular 1186, Reston-Virginia, USA.

Anonymous, 1994. *Libro Blanco de las aguas subterráneas.* Ministerios de Industria y Energía, Obras Públicas, Transportes y Medio Ambiente, Madrid, 135 p.

Anonymous, 2007. Groundwater and Groundwater Monitoring. WRD Technical Bulletin, Volume 13.

AQUASTAT, 2011. *FAO's Global Information System on Water and Agriculture.* Accessed in 2011; retrieved data are 2009 updates.

Bakhbakhi, M., 2002. *Hydrogeological framework of the Nubian Sandstone Aquifer System.* GW-MATE/UNESCO Expert Group Meeting, Socially sustainable management of groundwater mining from aquifer storage. Paris, 6 p.

Barlow, P. & E. Reichard, 2010. Saltwater intrusion in coastal regions of North America. *Hydrogeology Journal,* Vol. 18, No. 1, pp. 247–260.

Bates, B., Z. Kundzewicz, S. Wu & J. Palutikof (ed.), 2008. *Climate change and water.* IPCC Technical Paper VI, WMO & UNEP.

Blum, A., 2004. *L'état des eaux souterraines en France. Aspects quantitatifs et qualitatifs.* Etudes et Traveaux No 43, *Institut francais de l'environnement (IFEN)*, Orleans, France, 38 p.

Bocanegra, E., G. Cardoso da Silva, E. Custodio, M. Manzano & S. Montenegro, 2010. State of knowledge of coastal aquifer management in South America. *Hydrogeology Journal,* Vol. 18, No. 1, pp. 261–268.

Bredehoeft, J., 1997. Safe yield and the water budget myth. Ground Water, 1997, 35, p. 929.

Bredehoeft, J., 2002. The water budget myth revisited: why hydrogeologists model. *Ground Water,* Vol. 40, pp. 340–345.

BRGM, 2005. *Données communiqués sur l'exploitation d'eau souterraine au Rajastan.* Centre Franco-Indien de Recherche sur les Eaux Souterraines.

Carbognin, L., L. Tosi & P. Teatini, 1995. Analysis of actual land subsidence in Venice and its hinterland (Italy). *Proceedings of the Fifth International Symposium on Land Subsidence,* Balkema, Rotterdam, The Netherlands, pp. 129–138.

Chai, J.C., S.L. Shen, H.H. Zhu & X.L. Zhang, 2004. Land subsidence due to groundwater drawdown in Shanghai. *Géotechnique* Vol. 54, No 2, pp. 143–147.

Church, J.A., & N.J. White, 2006. A 20th century acceleration in global sea-level rise. *Geophysical Research Letters*, Vol. 33, LO1602. Available from: doi: 10.1029/2005GL025826, 4 pages.

Costantino, C., F. Falcitelli, A. Femia & A. Tudini, 2003. Integrated environmental and economic accounting in Italy. In: *Measuring Sustainable Development – Integrated economic, environmental and social frameworks.* OECD, 2004, pp. 209–225.

Custodio, E., 1992. Hydrogeological and hydrochemical aspects of aquifer overexploitation. *Selected Papers on Aquifer Overexploitation from the 23th International Congress of the IAH,* Puerto de la Cruz, Tenerife, Spain, 15–19 April 1991, pp. 3–28.

Custodio, 2002. Aquifer overexploitation: what does it mean? *Hydrogeology Journal,* Vol. 10, No 2, pp. 254–277.

Custodio, E., 2010. Coastal aquifers of Europe: an overview. *Hydrogeology Journal,* Vol. 18, No. 1, pp. 269–280.

DGRE, 1995. *Situation de l'exploitation des nappes phréatiques.* Tunis, Min. Agr., DGRE, 263 p.

DGRE, 2005. *Annuaire d'exploitation des nappes profondes.* Tunis, Min. Agr., DGRE, 459 p.

Döll, P., & Fiedler, 2008. Global-scale modelling of groundwater recharge. *Hydrol. Earth Syst. Sci.,* 12, pp. 863–885.

Döll, P., 2009. Vulnerability to the impact of climate change on renewable groundwater resources: a global-scale assessment. *Environm. Res. Lett.* 4 (2009) 035006, 12 p.

ESCWA, 2007. *ESCWA Water Development Report 2: State of Water Resources in the ESCWA Region.* United Nations Publication, Sales No. E.08.II.L.1, United Nations, New York.

European Environmental Agency, 2010. *The European Environment, State and Outlook 2010 – Freshwater quality.* EEA, Copenhagen, Denmark.

FAO, 1995. *Water resources and irrigation in Iran (Report of W.Klemm, in preparation of Water Report 9).*

FAO, 2005. *L'irrigation en Afrique en chiffres.* AQUASTAT, Water Report No 29, Rome.

Feick, S., S. Siebert & P. Döll, 2005. *A digital global map of artificially drained agricultural areas.* Frankfurt Hydrology Paper 04, Institute of Physical Geography, Frankfurt University, Germany.

Foppen, J.W., 1996. *Evaluation of the effects of groundwater use on groundwater availability in the Sana'a basin.* SAWAS Technical Report No 5, 6 volumes, NWRA/Yemen & TNO, Delft, The Netherlands.

Foster, S., J Chilton, M. Moench, F. Cardy & M. Schiffler, 2000. *Groundwater in rural development: facing the challenges of supply and resource sustainability.* World Bank Technical Paper 463. World Bank, Washington DC.

Foster, S., & M. Ait-Kadi, 2012. Integrated Water Resources Management (IWRM): How does groundwater fit in? *Hydrogeology Journal,* Vol. 20, pp. 415–418.

Frans, R.P., D. Kalf & R. Woolley, 2005. Applicability and methodology of determining sustainable yield in groundwater systems. *Hydrogeol. Journal,* Vol. 13, pp. 295–312.

Galloway, D.L., & T.J. Burbey, 2011. Review: Regional land subsidence accompanying groundwater extraction. *Hydrogeology Journal,* Vol. 19, No 8, pp. 1459–1486.

Gleeson, T., Y. Wada, M. Bierkens & L. Van Beek, 2012. Water balance of global aquifers revealed by groundwater footprint. *Nature,* Vol. 488, pp. 197–200.

Gordon, C, C. Cooper, C.A. Senior, H. Banks, J.M. Gregory, T.C. Johns, J.F.B. Mitchell & R.A. Wood, 2000. The simulation of SST, sea ice extents and ocean heat transports in a version of the Hadley centre coupled model without flux adjustments. *Clim. Dyn.* 16, pp. 147–168.

GRAPES, 2000. *Πρόγραμμα δράσης για τους φυσικούς πορους των υπόγεων υδάτων και ποταμών σε Ευρωπαϊκή Κλίμακα. Ευρωπαϊκή Επιτροπή* (Action plan at European scale for the natural resources of groundwater and rivers. European Commission).

Gurdak, J., R. Hanson & T. Green, 2009. *Effects of climate variability and change on groundwater resources in the United States.* Factsheet 2009–3074. USGS, Office of Global Change, Idaho Falls, Idaho.

GWA/Salem, O.M., 1992. The Great Manmade River Project. *Water Resources Development,* Vol. 8, No 4, Dec 1992.

Haest, P.J., R. Lookman, I. Van Keer, J. Patyn, J. Bronders, M. Joris, J. Bellon & F. De Smedt, 2010. Containment of groundwater pollution (methyl tertiary butyl ether and benzene) to protect a drinking-water production site in Belgium. *Hydrology Journal,* Vol. 18, pp. 1917–1925.

Habermehl, M. A., 1980. The Great Artesian Basin, Australia. *BMR Journal of Australian Geology & Geophysics,* Vol. 5, pp. 9–38.

Habermehl, M., 2002. *Groundwater development in the Great Artesian Basin, Australia.* GW–MATE UNESCO Expert Group meeting, Socially Sustainable Management of Groundwater Mining from Aquifer Storage, Paris, 35 p.

Hamdane, A, 2005. *Gestion des eaux souterraines pour l'agriculture. Cas de Tunesie.* Plan Blue document, Min. of Agriculture of Tunesia.

Hernandez-Mora, N., M.R. Llamas & L. Martinez Cortina, 2001. Misconceptions in aquifer over-exploitation. Implications for water policy in Southern Europe. In: C. Dosi (ed), *Agricultural Use of Groundwater. Towards Integration between Agricultural Policy and Water Resources Management,* San Diego, Kluwer, pp. 107–125.

Horton, J., & R. Devlin, 2008. *South Carolina Groundwater Contamination Inventory*. Bureau of Water, South Carolina Dep. of Health and Environmental Control. Technical Report # 002–08.

Hu Ruilin, 2006. Urban land subsidence in China. *IAEG 2006*, Paper number 786.

ICID, 2000. *Irrigation and drainage in the world, Chapter Iran*. International Commission for Irrigation and Drainage.

INAG, 2001. *Plano Nacional da Água, Portugal*. Ministério do Ambiente e do Ordenamento do Território, Instituto da Água (INAG).

International Energy Agency, 2010. Key world energy statistics. IEA, Paris.

IPCC, 2001. *Climate Change 2001: Synthesis Report*. Published for the Intergovernmental Panel on Climate Change, Cambridge University Press.

IPCC, 2007. *Climate Change 2007: Synthesis Report*. An Assessment of the Intergovernmental Panel on Climate Change. Adopted in Valencia, Spain, 12–17 November 2007.

Israel Water Commission, 2002. *Water in Israel. Consumption and production 2001*. State of Israel, Ministry of National Infrastructures, Water Commission, Demand Management Division.

Israel Water Commission, 2002a. *Report on the water status in Israel*. The Israeli Parliament, June 2002 (in Hebrew).

Johnson, T., 2007. Battling Seawater Intrusion in the Central and West Coast Basins. Technical documents on Groundwater and Groundwater Monitoring. WRD Technical Bulletin Volume 13, Fall 2007.

Kahlown, M.A., & A. Majeed, 2002. Water resources situation in Pakistan: challenges and future strategies. In: *Science Vision*, Vol. 7, No 3, Islamabad, Pakistan, pp. 33–45.

Kahlown, M.A., & A. Majeed, 2004. *Water-resources development and management*. Pakistan Council of Research in Water Resources, Ministry of Science and Technology, Government of Pakistan.

Khater, A. R., 2003. Intensive groundwater use in the Middle-East and North Africa. In: Llamas and Custodio (eds), *Intensive use of Groundwater*. Balkema Publishers, The Netherlands, pp. 355–386.

Kendy, E., D.J. Molden, T. S. Steenhuis & C. Liu, 2003. *Policies drain the North China Plain: agricultural policy and groundwater depletion in Luancheng County, 1949–2000*. International Water Management Institute (IWMI), Colombo, Sri Lanka, IWMI Research Report 71, 45p. Available from: doi: 10.3910/2009.074

Konikow L. & Kendy, L., 2005. Groundwater depletion: a global problem. *Hydrogeology Journal*, Vol. 13, pp. 317–320.

Konikow, L., 2011. Contribution of global groundwater depletion since 1900 to sea-level rise. *Geophysical Research Letters*, Vol 38, L17401, doi: 10.1029/2011GL048604, 2011. Paper and Auxiliary Materials.

López – Camacho, B., A. Sánchez-González & A. Battle, 1992. Over-exploitation problems in Spain. In: *Selected Papers on Aquifer Overexploitation*, 23rd International Congress of the IAH, Puerto de la Cruz, Tenerife, April 1991, pp. 363–371.

Margat, J., 1992. The overexploitation of aquifers. *Selected Papers on Aquifer Overexploitation from the 23th International Congress of the IAH*, Puerto de la Cruz, Tenerife, Spain, 15–19 April 1991, pp. 29–40.

Marin, L.E., 2006. *El agua en Mexico: retos y oportunidades*. Real Academia de Ciencias Exactas, Físicas y Naturales de España, Vol. 98, No 2.

McGuire, V., 2009. *Water-level changes in the High Plains aquifer, predevelopment to 2007, 2005–06 and 2006–07*. Scientific Investigations Report 2009–5019, USGS, Reston, Virginia.

Meddi, M., & P. Hubert, 2003. Impact de la modification du régime pluviométrique sur les ressources en eau du Nord Ouest de l'Algérie. *Hydrology of the Mediterranean and Semiarid Regions*, IAHS Publication No 278, pp. 229–235.

Mekonnen, M.M., & A.Y. Hoekstra, 2011. *National Water Footprint Accounts: The Green, Blue and Grey Footprint of Production and Consumption. Volume 1: Main Report and Volume 2: Annexes.* Value of Water Research Report Series No. 50, UNESCO-IHE, Delft, The Netherlands.

Miludi, H. & O. M. Salem, 2001. Water policies in the Libyan Arab Jamahiriya. In: *Proceedings of the WHO/UNEP First Regional Conference on Water Demand Management, Conservation and Pollution Control*, Amman, Jordan, 7–10 Oct. 2011.

Ministerio de Medio Ambiente (MMA), 2000. *El libro blanco del agua en España.* MMA, Secretaría de Estado de Aguas y Costas, Dirección General de Obras Hidráulicas y Calidad de las Aguas, 637 p.

Min NSRE, Algeria, 2003. *Secteur de l'eau – Élements d'une stratégie sectorielle.* Internal report of the Ministry of Water Resources, in cooperation with World Bank and FAO.

Morris, B., A. Lawrence, J. Chilton, B. Adams, R. Calow & B. Klinck, 2003. *Groundwater and its susceptibility to degradation.* Early Warning and Assessment Report Series, RS 03–03, UNEP, Nairobi, Kenya.

Musolff, A., 2009. Micropollutants: challenges in hydrogeology. *Hydrogeology Journal*, Vol. 17, No. 4, pp. 763–766.

Noordhof, 2010. De Bosatlas van Nederland Waterland. 104 p.

OECD, 1991. *L'état de l'environnement.* Paris.

OSS/UNESCO, 2005. *Ressources en eau et gestion des aquifères transfrontaliers de l'Afrique du Nord et du Sahel.* OSS/UNESCO, ISARM-Africa, IHP-IV, Series on groundwater No 11, Paris.

Oubalkace, M., 2006. *Rapport final du Maroc à l'Atelier du Plan Bleu et de la Commission Méditerranéenne pour le developpemnt durable.* Saragoza, Spain.

Oude Essink, G.H.P., E.S. van Baaren & P.G.B. de Louw, 2010. Effects of climate change on coastal groundwater systems: A modelling study in the Netherlands. *Water Resources Research*, Vol. 46. Avaialble from: doi: 10.1029/2009 WR008719.

Pallas, Ph., & O. Salem, 1999. *Water Resources Utilisation and Management of the Socialist People Arab Jamahiriya.* Intern. Conf. 'Regional Aquifer Systems in Arid zones – Managing Non Renewable Resources', Tripoli, Nov. 1999, Paris, UNESCO, 65 p.

Plan Hidrológico Nacional, 1993. *National hydrological plan of the Government of Spain*, as defined in 1993.

Pulido-Bosch, A., 1991. *Sobreexplotación de acuíferos y desarrollo sostenible.* Paper initially prepared for "II Taller Internacional sobre Gestión y Technologías de Suministro de Agua Potable y Saneamiento Ambiental", La Habana, Cuba.

Pulido-Bosch, A., 1991. The overexploitation of some karstic aquifers in the province of Alicante (Spain). In: *XXIII Congress AIH Sobreexplotación de acuíferos, Santa Cruz de Tenerife*, pp. 557–561.

Riolo, A., 1999. *Annual report 1998/99 of the Water Services Corporation Malta.*

Röckner, E., K. Arpe, L. Bengtsson, M. Christoph, M. Clausen, L. Dümenil, M. Esch, M. Giorgetta, U. Schlese & U. Schulzweida, 1996. The atmosphetic general circulation model ECHAM-4: model description and simulation of present day climate. *MPI-Report No 218*, Hamburg, MPI für Meteorologie.

Salem, O., 1992. *Hydrogeology of the major groundwater basins of Libya.* Observatoire du Sahara et du Sahel, Project launch workshop 'Aquifères des grands bassins', Cairo, 22–25 Nov. 1992, 15 p.

Schmoll, O., G. Howard, J. Chilton & I. Chorus, 2006. *Protecting groundwater for Health – Managing the quality of drinking-water sources.* WHO Drinking-water Quality Series, World Health Organization, Geneva.

Shah, T., A. Deb Roy, A. Qureshi & J. Wang, 2001. Sustaining Asia's Groundwater Boom: An Overview of Issues and Evidence. *Natural Resources Forum*, Vol 27, pp. 130–140.

Shahin, M., 1987. Groundwater resources in Egypt: potentials and limitations. IAHS, *Proceed. Intern. Symp. Water for the Future, Rome, IAHS Publ. No 164, Hydrology in Perspective*, Wallingford, pp. 179–192. Available from: http://www.iahs.info/redbooks/164.htm.

Shiklomanov, I.A., 1990. Global water resources. UNESCO, *Nature and Resources*, Vol. 26, No 3.

Steyl, G. & I.Dennis, 2010. Review of coastal-area aquifers in Africa. In: *Hydrogeology Journal*, Vol. 18, No 1, pp. 217–226.

Stockholm World Water Week, 2010. The Malin Falkenmark Seminar: *Emerging Pollutants in Water Resources – A New Challenge to Water Quality*. Available from: http://www.world-waterweek.org/sa/node.asp?node = 750&sa _content_url = %2Fplugins%2FEventFinder%2Fevent%2Fasp&id = 3&event = 239 [Accessed 4 March 2011]

Treidel, H., J.L.Martin-Bordes & J.J. Gurdak (eds), 2012. *Climate change effects on groundwater resources. A Global synthesis of findings and recommendations.* IAH, International Contributions to Hydrogeology 27, CRC Press/Balkema, 401 p.

Tsiourtis, 1999. *Personal communication.*

UNDESA, 2011. *World population statistics for 2010.* Available from: http://www.un.org/esa/population/ [Accessed in 2011]

UNDP, 2010. *The path to achieving the Millennium Development Goals: a synthesis of MDG evidence from around the world.* United Nations Development Programme, One United Nations Plaza, New York, NY 10017, USA.

UNESCO, 1984. *Guidebook to studies of land subsidence due to ground-water withdrawal.* Edited by J.F. Poland, UNESCO – IHP, Paris.

UNESCO, 2006. *Groundwater Resources Assessment under the pressure of Humanity and Climate changes.* Paris, UNESCO/IHP, 19 p.

USGS, 1985. US National Water Summary 1984.

Vachaud, G., *et al.*, 1997. *Evolution des ressources en eau souterraine en zone de culture irriguée du Bassin méditerranéen en cas de changements climatiques.* Séminaire international sur l'agriculture et le développement durable en Méditerranée. Montpellier – Agropolis, 5 p.

Van der Gun, 2009. Climate change and alluvial aquifers in arid regions: examples from Yemen. Chapter 11 in: Ludwig, Kabatt, Van Schaik and Van der Valk (eds): *Climate change adaptation in the water sector.* Co-operative Programme on Water and Climate, Earthscan, London, pp. 159–176.

Van der Gun, J.A.M., & A.A. Ahmed, 1995. *The Water Resources of Yemen. A summary and digest of available information.* WRAY-project, Delft and Sana'a, 108 p. (text) + 115 p. (appendices).

Van der Gun, J. & Lipponen, A., 2010. Reconciling storage depletion due to groundwater pumping with sustainability. In: *Sustainability, Special Issue Sustainability of Groundwater*, November 2010. Available from: http://www.mdpi.com/2071–1050/2/11/3418.

Wang, R., H. Ren & Z. Ouyang (eds), 2000. *China Water Vision: The ecosphere of water, life, environment and development.* China Meteorological Press, Beijing, 178 pp.

White, I. & T. Falkland, 2010. Management of fresh-water lenses on small Pacific Islands. *Hydrogeology Journal*, Vol. 18, No 1, pp. 227–246.

World Bank, 1998. *India Water Resources Management Sector Review: Groundwater Regulation and Management Support.* Report prepared in collaboration with the Central Ground Water Board, Ministry of Water Resources, Government of India.

WWAP, 2006. *The United Nations World Water Development Report 2: Water, a shared responsibility.*

Zhou, Y., L. Wang, J. Liu & C. Ye, 2012. Impacts of drought on groundwater depletion in the Beijing Plain, China. Chapter 16 in: Treidel, Martin-Bordes and Gurdak (eds), 2012, *Climate change effects on groundwater resources. A Global synthesis of findings and recommendations.* IAH, International Contributions to Hydrogeology 27, CRC Press/Balkema, pp. 281–303.

Zaporozec, A. (ed.), 2002. *Groundwater contamination inventory. A methodological guide.* UNESCO, IHP-VI, Series on Groundwater No 2, 161 p.

Chapter 7

Groundwater resources management

- *How to manage the groundwater resources: information, objectives, instruments, process and other aspects*
- *Experiences in groundwater resources management in different parts of the world: how to enhance benefits from groundwater, avoid or reduce overexploitation, and control groundwater quality*
- *Transboundary aquifers*
- *Current state of affairs worldwide*

7.1 ELEMENTS OF GROUNDWATER RESOURCES MANAGEMENT

7.1.1 Area-specific information

Managing the groundwater resources of an area properly is not possible without sufficient and reliable area-specific information. In the first place, those in charge of managing the resources should have a correct and sufficiently accurate picture of the local groundwater resources, their current functions and use, their socio-economic relevance and their interactions with surface water, ecosystems and the environment. Secondly, they should be adequately informed about current groundwater–related problems and have sufficient information at their disposal to identify potential future problems, to assess the scope (if any) for expanding the use of the groundwater resources and to predict the likely effects of specific groundwater resources management interventions.

If sufficient and reliable area-specific information is lacking, then a *groundwater resources assessment study* needs to be carried out as an essential first step towards groundwater resources management. Such an assessment – of which a few examples are shown in Figure 7.1 – may include hydrological, geological, geophysical, hydrogeological, water use, socio-economic, stakeholder perception, institutional and other types of surveys. Most outputs of these surveys are of a descriptive nature, but together they also usually produce a diagnostic component. The area to be assessed should be the same as the chosen spatial management unit, which is often an aquifer system or part of it (if the aquifer is very large), but its delineation is also dependent

(a) Aquifer exploration by refraction seismics on the Chilean Altiplano.

(b) Gravimetric survey of the Tihama Quaternary aquifer system, Yemen.

(c) Exploratory borehole on the Altiplano, Chile.

Figure 7.1 Examples of groundwater assessment activities
[*Photos*: Ronnie van Overmeeren].

on the territorial boundaries of the institution that has the mandate for groundwater resources management. The components of the assessment activities should be in tune with the groundwater resources management objectives (see below) and thus may vary considerably from case to case.

Collecting area-specific information also includes the identification of potential actors in groundwater resources management. The role of each category of actors will have to be formalised in the initial stages of the management process.

A significant part of the area-specific information is time-dependent (groundwater quantity and quality, water demands, water use, profits from groundwater, groundwater-related problems, etc.). *Monitoring systems* are needed to observe changes over time in these variables and to provide feedback to the water management institutions.

7.1.2 Objectives

In general terms, groundwater resources management has multiple objectives. It pursues maximum benefits from groundwater resources to human society, in a good balance with and between the objectives of meeting water demands, making profits, maintaining resource sustainability, allocating benefits equitably and conserving the groundwater-related environment and ecosystems. These objectives partly conflict, and thus striking a balance is a subjective activity depending on political and societal preferences. The results of the choices made are laid down in a groundwater resources management strategy.

More specifically, groundwater resources management deals – preferably in a proactive way – with tasks or goals like the following:

- Ensuring that water supplies are safe in terms of quantity and quality, by allocating groundwater with priority to uses that would suffer most from supply failures (drinking water supply, irrigation), by regulating withdrawals, and especially by land use control to protect groundwater quality.
- Resolving groundwater use conflicts and preventing new ones from developing.
- Preventing or reducing overexploitation of renewable groundwater.
- Ensuring that the exploitation of non-renewable groundwater takes place according to planning, in order to avoid unexpected water supply failures and to consider the needs of future generations.
- Promoting groundwater to be used for purposes of highest economic or social value.
- Reconciling profitable use of groundwater with lowest possible negative impacts from groundwater abstraction.
- Protecting the resources by combating threats to their quantity or quality.
- Safeguarding the state of perennial water courses and wetlands that form protected natural environments and aquatic ecosystems, paying special attention to the conservation of groundwater that discharges into these water courses and thus contributes to a certain minimum flow and water quality.

Although the general objectives of groundwater resources management and surface water management are similar, at the more specific level their goals may diverge or have different degrees of relevance. For instance, ensuring safety for human survival by preventing disastrous floods is often a major goal in surface water management, but not in groundwater resources management. On the other hand, removing water locally from subsurface domains to facilitate mining and other uses of the subsurface is exclusively groundwater management. The same is true for defining an optimal balance between duration and intensity of exploiting water in the case of non-renewable resources, a problem shared with the entire mining industry.

7.1.3　Instruments

Many instruments can be used in groundwater resources management. Some of them are technical and direct, whilst others are either indirect and aim to influence people's behaviour or are needed for the development of strategies and plans, and for assessing their impacts. Collective technical works or provisions, that play such an important role in surface water management (e.g. dams and diversion works), are less ubiquitous and of secondary importance in groundwater resources management. In general, groundwater resources management depends predominantly on the implementation of indirect management tools, through which the mandated authorities try to ensure that people align their private activities optimally with the chosen management objectives that serve the public interest. These indirect management tools are of a regulatory, economic and informative nature.

The main categories of instruments can be briefly described as follows:

- *Technical instruments*
The most common instruments in this category are:

　　－　*Subsurface dams* to intercept or retain groundwater.
　　－　*Infiltration galleries* (preferably equipped with provisions to prevent wasting groundwater).
　　－　*Artificial recharge.* Usually local and often in combination with abstraction wells where increased productivity is sought; sometimes however spread over a larger area to enhance the aquifer's resources, as in the case of flood water spreading in California, USA. Artificial recharge is often implemented to compensate for groundwater overabstraction or overexploitation, but it has the potential to increase the groundwater resources by making more effective use of the regulatory capacity of certain aquifers (see Section 7.2).
　　－　*Soil and terrain shaping* to promote the infiltration of storm water (rainwater harvesting).
　　－　*Well drilling programmes* to stimulate beneficial use of underexploited aquifers. Such programmes often include objectives of regional economic development or improving drinking water supply.
　　－　*Remediation techniques* for polluted sites.

- *Legal and regulatory instruments*
Under the water-related laws of the country concerned (Box 7.1), individuals, companies or organisations may be obliged to comply with certain regulations, in particular related to the construction of new wells and to groundwater

Box 7.1: Groundwater rights and legislation

The fact that groundwater is at the same time part of the underground and part of the natural water cycle is reflected in groundwater rights and legislation. Groundwater legislation finds its roots in legislation on the subsurface (especially in mining law) as well as in general water legislation (mostly focussing on surface water, the most visible component). For a long time, the relation of groundwater with the underground was better understood than that with surface water, which has caused groundwater to be included more often in legislation on the subsurface than in a special groundwater section of water law.

At the heart of groundwater law* – as in water law in general – lies the question of ownership. Roman law, which is behind most modern legislation, distinguishes in this respect three types of legal status of water:

1   *Privately owned water.* Water ownership is in this case usually linked to land and only the landowner (either an individual or a group) is entitled to use it, without any restriction.
2   *Water as a common property* ('res communis'). In this case there is no owner, anybody may use the water without permission.
3   *State- or publicly-owned water.* Water is the property of the public embodied by the State and its use requires authorisation granted by the competent public authority.

In different regions of the world different types of legal status have been embraced. In cases where groundwater ownership was traditionally considered to be linked to land ownership – with the perception that the latter would include the subsurface –, groundwater was given the same legal status as land: water is privately owned under private lands and state-owned in areas belonging to the state. In practice, the ownership rights are limited to the right of access and ownership of the abstracted water.

Linking groundwater to land ownership and dealing with it in the same way as is done with an extractable solid raw material, however, does not correspond with the physical reality of the subsoil, as is better known nowadays. In fact, groundwater is mobile and extends across the borders of land parcels. The status of water as a common property or as public water is more consistent with the physical reality and is even compatible with the continuity between groundwater and surface water.

The current trend in groundwater law is that user rights are gaining importance at the expense of ownership rights, as the latter have proven to be ill-suited to fluids. The status of public property, with strong regulation of users rights, is generally becoming dominant, not only in countries where groundwater exploitation is intensive and captures a major share of the renewable resources or even exceeds them.

In summary, where groundwater property rights, originally intended to guarantee the rights of land owners, have become an obstacle to resources management, legislation has stepped in to allow regulatory management tools or financial instruments such as imposing taxes to be used.

* 'Law' is used here in a wider sense, including both formal law and customary law.

withdrawal. These regulations may in the first place require groundwater abstractions above a certain threshold abstraction rate to be declared and the creation or abandonment of groundwater abstraction works beyond a certain depth to be registered, for statistical purposes. Secondly, they may also include the obligation, usually above higher thresholds than those for declaring and registering, to obtain a permit before a new well is constructed or any volume of groundwater is abstracted. Sometimes, additional studies have to be carried out to assess the potential side effects of a proposed groundwater abstraction, after which – depending on the outcomes of the study and local circumstances – a permit is granted or refused. A permit may be granted for a finite period of time only and include additional obligations for the applicant, e.g. to pay a certain levy and to monitor changes in the groundwater state.

A specific aim of groundwater management regulations may be to protect existing groundwater abstraction sites by prohibiting new wells. Protection zones in areas where groundwater is traditionally withdrawn by gravity systems (qanats), such as in Oman, form a typical example. Regulating groundwater abstraction is also a priority in areas where stable phreatic levels are essential for the preservation of wetlands and agricultural production, such as in most of the territory of The Netherlands. Some regulations are intended to control actions that may reduce natural groundwater recharge or even to augment recharge. In India, for example, a restriction was imposed on the density of the built-up area in some cities (such as Delhi), while elsewhere (e.g. Jakarta, Indonesia) there are regulations requiring new houses to be provided with a 'roof catchment' in order to enhance local groundwater recharge by means of intercepted rain. Other regulations aim for groundwater pollution control, by the introduction of zones with different degrees of restrictions on land use or subsurface activities, or by establishing protection zones around well fields for public water supply. The implementation of all these regulations or the inspection of their compliance may come under the authority of 'water police' or a mandated water management institution.

In the particular case of transboundary aquifers, the management of shared resources has to follow rules agreed upon on a case-by-case basis, because generally accepted international law on groundwater is still lacking (Burchi and Mechlem, 2005). These rules may be based on the content of the 'Draft Articles of the Law on Transboundary Aquifers', jointly elaborated by the United Nations International Law Commission (UNILC) and UNESCO's International Hydrological Programme (UNESCO-IHP) during 2002–2008 and adopted by resolution A/RES/63/124 of the UN General Assembly in December 2008 (Stephan, 2009). Important principles included in these Draft Articles are the obligation to co-operate and to exchange information, the no-harm principle and the objective to protect, preserve and manage the aquifer resources. More information on transboundary aquifers will follow in the final part of Section 7.2.

• *Economic and financial instruments*
The behaviour of private or public economic actors, in their capacity to exploit or influence groundwater, can be attuned towards the direction desired by the groundwater management authorities by means of legally embedded financial instruments.

These can work in two directions:

- As *incentives* (financial support). Examples are the stimulation of exploitation by investments credits, by subsidies, by free insurance of exploration or drilling risks, by tax-free energy supply for pumping and by assistance with efforts for mitigating pollution. However, such support must be underpinned and be in line with the knowledge of the state of the resources, in order to avoid encouragement of pumping in cases where groundwater is wasted or overexploited.
- As *disincentives* (taxes or fines). A typical example is restricting or discouraging groundwater abstraction through taxes on withdrawals beyond a certain threshold. Such a threshold may vary according to area, season or type of use, as guided by a resource allocation policy. Taxes may also be applied to activities that cause pollution, following the 'polluter pays' principle adopted in several countries.

Under the influence of a more liberal ideology, water markets have appeared in several parts of the world, not only for surface water but also for groundwater . They often emerge and develop more or less spontaneously and tend to be characterised by informal behaviour and little or no regulation. On Gran Canaria and Tenerife, water markets have been functioning for decades, mostly for groundwater and small reservoirs. In Southern Asia water markets play an important role in improving access for small farmers to irrigation water (Shah, 1993). In the western states of the USA and in Australia they contribute in the first place to water being used for purposes that yield the highest economic returns, for example in California with its Water Bank. Water markets are also operational in Chile, after the country adopted in its Water Code of 1981 free-market mechanisms as the economic philosophy in water resources management. Even in individual cases, there are often pronounced differences in opinion as to whether water markets are beneficial or not. One of the lessons learned is that a certain degree of regulation of water markets is useful to combat negative side effects such as the development of monopolies, groundwater depletion and environmental externalities.

- *Educational, informational and awareness-raising instruments*

Efforts to inform all stakeholders on the groundwater conditions and issues in their area, to raise their understanding of the objectives of interventions for the common interest as well as of the rationale behind measures and their anticipated impacts, may contribute considerably to the effectiveness of the implemented measures. The reason is that stakeholders will then be more motivated to be co-operative and adapt their behaviour.

The institutional setting, available funds and the professional capacity of staff in charge of water resources management activities considerably influence the success of implementing any of the described categories of instruments. Most often these instruments are under the responsibility of different authorities whose areas do not always coincide, which does not facilitate smooth coordination. Therefore it is desirable that the mandates for technical interventions, regulating, using financial instruments and sharing information with stakeholders in a certain region are all concentrated in one single groundwater management authority with a jurisdiction over complete physical management units (aquifer systems), regardless of their size and their possible spread over several administrative or political districts. Effective coordination is also required with other institutions and authorities responsible for activities that may interact with groundwater (surface water management, land use planning, agriculture and livestock, tourism, mining, etc.). This should be regulated at a higher administrative level.

More decentralised and more community-oriented management is a present-day trend, entrusting the management responsibility to the community of those who exploit and use the groundwater resources. This is in line with the so-called *subsidiarity principle* stating that matters ought to be handled at the lowest competent administrative level. Examples are the *Comunidades de Usuarios de Aguas Subterráneas* or COTAS (Communities of Groundwater Users) in Mexico since 1993, or the groundwater user associations in Spain since 1975. Another example is the management of the Astien Sands aquifer in France (Hérault), intensively exploited, on the basis of an *aquifer contract* signed by its water users in 1997 in order to prevent the risk of saltwater intrusion and to maintain pressure and productivity.

7.1.4   Planning, decision-making and implementation of measures

The emergence of groundwater resources management takes place only after people have become aware of opportunities or problems related to their groundwater resources and after they have developed ideas on how to address these. Consequently, considerable time has often elapsed between the start of intensive groundwater exploitation in an area (or other stresses) and the onset of groundwater resources management. Initial interventions were and are usually to a large extent based on technical instruments, as these do not restrict local stakeholders in their activities and thus are relatively easily accepted or even supported by them. Examples are aquifer-wide well drilling programmes or the construction of artificial recharge facilities, usually carried out as *ad hoc* activities, not yet as part of a comprehensive groundwater resources management strategy or plan. Governments of many countries also encourage private initiatives for exploiting and using groundwater by providing incentives such as technical assistance, financial support or subsidies. Increasing awareness of the threats to groundwater or confrontation with groundwater-related problems catalyses the implementation of restrictive measures, such as the regulation of abstractions by licensing systems, taxation of the volumes of groundwater abstracted or imposing fines upon groundwater polluters. They also trigger the improvement of existing legislation related to water or the development of new laws addressing groundwater satisfactorily.

Different countries follow different paths. In quite a number of countries, groundwater resources management consists of isolated measures implemented in selected areas or even nationwide, in some cases with significant effects, in other cases without. More comprehensive groundwater resources management approaches have emerged in several other countries. Such approaches tend to follow an evolution towards a gradually increasing scope, in tune with the development of awareness, institutional capacity, socio-economic conditions and political setting. Typical initial steps are the integration of groundwater and surface water, water quantity and water quality, as well as all uses and functions of water into one single water resources management plan. This means that an integrated water resources management (IWRM) approach is adopted. Although the IWRM approach is widely advocated, it is not always easy to incorporate groundwater in a way that does justice to this component of the water cycle, as is illustrated in Box 7.2. Further widening of the scope is obtained by adding the socio-economic dimensions and the interaction with

Box 7.2: Integrated water resources management (IWRM): how does groundwater fit in?

IWRM is not an end in itself, but a means to make water-resource development more balanced – maximising socio-economic welfare in an equitable manner without compromising ecosystem sustainability. The axioms of IWRM are gaining increased acceptance by groundwater specialists and the approach is fully compatible with so-called 'adaptive management', in which provisional decisions and measures are taken based on best-available scientific evidence with full stakeholder consultation.

Nevertheless, there are a number of differences between groundwater and surface water systems that in some cases suggest or require a modified approach for the groundwater component. In the first place, while river systems are flow-dominated, most aquifers are characterised by large stocks and comparatively low fluxes. The consequences are, among others, that upstream-downstream considerations are less predominant in groundwater systems and that the groundwater storage buffer enables uncertainty to be accommodated more easily. Secondly, the river basin is the fundamental spatial unit for application of the IWRM process, but this has to be reconciled with the fact that groundwater bodies or aquifers are the appropriate spatial units for groundwater management and protection. Aquifer systems are in some cases perfectly embedded within one single river basin, but the interaction is more complex (if the boundaries are not related) or only very weak in many other cases. Extensive deep aquifer systems in arid regions form an example: a river-basin approach is not helpful to manage their groundwater resources, thus it is preferable to concentrate on the groundwater system. For aquifers that are small in size compared to the interconnected river basins it is usually advisable to develop and aggregate specific groundwater-body management plans as the hydrological realities dictate.

There are more factors that may complicate the integration of groundwater into IWRM. One of them is that many senior water managers have a limited grasp of the scales, dynamics and vulnerabilities of groundwater, while most hydrogeologists have little affinity with socio-economic drivers and institutional frameworks. Furthermore, one may question whether IWRM reaches sufficiently down through society, especially as much groundwater abstraction is concentrated in the 'informal sector' and many small-scale users are involved.

There is no blueprint for incorporating groundwater into IWRM. An interdisciplinary pragmatic framework as developed by GW-MATE may be helpful to structure the process of developing groundwater management plans.

(This text box summarises elements of a paper by Foster and Ait-Kadi (2012), published under the same title)

land use, and by including ecosystems as a component of equal importance as socio-economic systems ('*social-ecosystems approach*'). This evolution leads to improved coordination between government policy domains that used to be separate (water resources, environment, nature conservation, spatial planning, etc.) or even to their integration.

Whatever approach followed – integrated and comprehensive or focusing on single interventions–, common to all is that decisions have to be made and action taken. This requires firstly that sufficient and – above all – reliable information has to be made available, as mentioned earlier already: information on the groundwater systems considered and their setting, on the relevant opportunities, problems and threats, on promising management measures to be taken, on the expected effects of these measures and on the feasibility of their implementation. Secondly, it requires that views and preferences of the main categories of stakeholders are known and taken into account. As the next step, potential strategies and measures are identified and analysed. Numerical models are commonly used as a decision support tool to explore the merits of alternative strategies and thus underpin the finally adopted groundwater resources management strategies and measures. For gaining the public support required for successful action, it is important that the organisations with the mandate for managing the groundwater resources act transparently and can be held accountable by the groundwater users they serve. Involvement of stakeholders in the management process may serve the same purpose.

Compared to surface water management, groundwater resources management faces a number of additional difficulties, during the assessment and planning phase as well as during the subsequent implementation of measures:

- Delineating the boundaries of aquifers is much more difficult than those of river basins. Not only because they are underground and thus invisible, but also because of the three-dimensional geometry of aquifers, the complexity and high cost of exploring their boundaries (see e.g. Figure 7.1) and the fact that the boundaries may shift over time due to natural regime characteristics or in response to groundwater withdrawal. Shallow aquifers are usually embedded within a single river basin, which facilitates the integrated management of surface water and groundwater. However, this may be different for deeper aquifers, as many of these are located under two or more river basins.
- Groundwater management depends much less than surface water management on a physical hydraulic infrastructure that can be operated with reasonably predictable results. It relies primarily on changing people's behaviour, either by regulations or by means of incentives and disincentives. The effects of such non-technical interventions are much more uncertain.
- Decision support on the basis of numerical simulation models is common practice nowadays. Nevertheless, groundwater systems often suffer more than surface water systems from scarcity of data and knowledge required for these models. In addition, the simulation models used tend to focus on hydrodynamic or water quality processes and include sometimes also the economic evaluation of measures, but human behaviour – of utmost importance in groundwater resources management – is not a component of current modelling practice (see Box 7.3).

Box 7.3: Groundwater management models

Computer models that simulate the hydrodynamic behaviour of aquifers are first of all an instrument of study and synthesis, with the purpose of verifying the consistency between the observed data, hypotheses and simulation output, as well as helping to optimise the collection of additional information. They are subsequently used to simulate aquifer behaviour, which allows the consequences of proposed actions on the aquifer system to be predicted and – in particular – groundwater exploitation scenarios to be developed and analysed, in preparation for resources management. Their results permit the feasibility to be determined and the merits of different groundwater development alternatives to be compared, be it renewable or non-renewable groundwater.

The outputs provided by such models are estimates of how piezometric levels will change over time, or the natural groundwater outflows and, in some cases, salinity. These, in turn, can be translated by postprocessors into socio-economic impacts (e.g. change of groundwater production cost in response to changes in water level) or into impacts on ecosystems or the environment. These models are therefore important decision support tools and although they are currently commonplace, their use is still not sufficiently widespread.

It is essential that these models are sufficiently representative to simulate variations in the system's state reliably, even if these variations are outside the range of those in the data series used for calibration, especially in long-term non-equilibrium regimes. The validity of the conceptual model (especially in the case of multi-layer systems) and the reliability of the adopted hydraulic model parameters are of paramount importance. Calibration and verification by comparison with field data should be included in the modelling protocol (Anderson and Woessner, 1992). Assigning a margin of uncertainty or confidence interval to the numerical results is highly desirable.

A hydrodynamic groundwater simulation model can also form a component of an economic management model, especially if designed as an optimisation model, and then it is usually integrated into a more complex basin management modelling framework. Such models, periodically updated and corrected as knowledge about the effects of the withdrawals advances by feedback, can guide management by means of periodically revised projections.

Numerous intensively exploited aquifer systems have been modelled by different investigators, to serve the authorities responsible for groundwater resources management. Groundwater simulation models are not restricted to hydrodynamics, but may also address water quality, in particular the transport of polluting solutes. Groundwater quality simulation models, however, are more complex than hydrodynamic simulation models and are usually only applied to a limited part of an aquifer. Modelling results should always be used with care, even if the model is duly calibrated, validated and updated, because numerical models always present a highly simplified picture of reality.

- Monitoring to what extent goals are achieved is more difficult in the case of groundwater, among others because there are no locations where variables are spatially lumped, such as at a stream gauging station. In addition, change in groundwater systems usually takes more time. Observing change in groundwater is based on piezometric monitoring networks, water quality sampling campaigns and well inventories. Monitoring densities, frequencies of observation or sampling and other performance indicators vary considerably, but they often still leave much to be desired. Many countries have made good progress in groundwater monitoring, but the systems and practices are heterogeneous and often deficient, and in several cases in decline[1].

Recognition of the importance of local stakeholders is gradually increasing in many countries, which has resulted in different modalities of stakeholder involvement in groundwater resources management, varying from a rather passive role (being informed or interviewed) to participating in decision-making and in the implementation of measures. Figure 7.2 shows some examples. Stakeholder involvement may be decisive for developing a cooperative attitude among stakeholders and thus for success in groundwater resources management. In line with this, there is a recent trend to use the term 'groundwater governance' for approaches that incorporate the roles, linkages and accountabilities of multiple actors at several levels: government institutions as well as civil, professional and private stakeholders. Governance emphasises the processes by which decisions are made and put into effect, rather than the results of the decisions (Linton and Brooks, 2011).

Finally, apart from monitoring the relevant physical variables, also the groundwater resources management process and especially its impacts have to be monitored and periodically assessed. This requires appropriate indicators to be defined. The outcome of the periodic assessment provides feedback and allows updated and adjusted operational plans to be prepared for the next management period.

7.2 GROUNDWATER RESOURCES MANAGEMENT IN PRACTICE: EXAMPLES AND COMMENTS

Most of the examples presented below focus on a single groundwater resources management issue or intervention. However, it should be understood that this is usually only one component of a much more complex integrated water resources management programme for a basin or region.

7.2.1 Catalysts in regional groundwater development

Groundwater exploitation and use did not start booming exclusively as a result of private initiatives. In many countries, governments and non-governmental organisations – sometimes supported by donors in bilateral or multilateral co-operation projects – have catalysed groundwater development, usually with objectives such as stimulating regional development, achieving self-sufficiency in food produc-

---

1 The WMO estimated that some 146 000 groundwater observation wells would exist worldwide by 1994 and 189 national organisations would be in charge of collecting groundwater data (WMO, 1995).

(a) Villagers drilling jointly by hand a shallow well in weathered basement rock, for community drinking water supply in Tanzania.

(b) Rural representatives gathered in a meeting in Punata, Bolivia, to exchange ideas on the needs and options for managing and protecting the groundwater resources in their area.

(c) Planning groundwater management measures at village level in Gujarat, India

Figure 7.2 Some examples of the many forms of stakeholder participation in groundwater resources development and management. [*Photos*: Jac van der Gun].

tion, enhancing employment in agriculture, alleviating poverty or establishing rural drinking water supplies. In India, for instance, government policies that subsidise credit and rural energy supplies have encouraged rapid development of the groundwater resources since the 1960s (World Bank, 1998). Numerous bilateral and multilateral co-operation projects were carried out during the second part of the 20th century in Africa, Latin America and Asia to boost profitable exploitation of groundwater, often including sizeable well construction activities. More recently, Millennium Goal 7C ('*Halve by 2015 the proportion of people without sustainable access to safe drinking water and basic sanitation*') has become a new catalyst for regional groundwater development, especially in rural areas in developing countries. Progress between 1990 and 2008 related to this goal is shown in Figure 7.3, although without specifying which part of the observed progress can be attributed to groundwater.

7.2.2   Using the groundwater buffer and the buffer capacity of aquifers more intensely

Groundwater resources development can in most aquifers go beyond intercepting part or all of the natural flux, which is comparable to harvesting. It may also include a deliberate use of the groundwater buffer provided by the aquifers. The groundwater reserves may buffer not only between wet and dry seasons but also between wet and dry years, and compensate for the lack of surface water during dry periods, like surface water reservoirs do. Such an active and carefully planned utilisation of the reserves should not be confused with overabstraction, as long as temporary depletion of the reserves is followed by periods of recovery and a long-term hydrodynamic equilibrium is pursued. In practice it means that the exploitation of aquifers with significant reserves is no longer dependent and based on the use of springs and

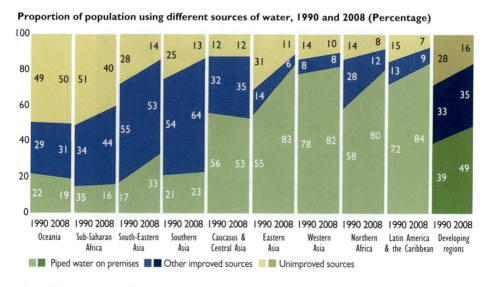

Figure 7.3 Progress in drinking water supply around the world. [*Source*: United Nations (2011)].

infiltration galleries at variable discharge rates, but on pumping at imposed discharge rates and under a regime of fluctuating groundwater levels.

Several examples taken from the Mediterranean basin demonstrate the feasibility of such management practices (all implemented a long time ago and still continuing today):

– Control of the Berzoug spring to supply Constantine in Algeria, since the 1950s.
– Management of the Yarkon-Tanninim aquifer in Israel, since 1950.
– Regulation of the Lèz spring to supply Montpellier in France, since 1982.
– Regulation of the El Algar spring for domestic and agricultural supply in the Valencia region in Spain, since the 1970s.

Artificial recharge is a related practice, making more intensive use of the buffer capacity of aquifers by augmenting the resources. Often it is based on diverting surface water peak flows and then it also contributes to a smoother stream flow regime, like surface water storage reservoirs do, but requiring less space and losing less water by evaporation. An overview of artificial recharge techniques and their application around the world has already been presented in Section 4.9.

In an attempt to assess potential future man-made modifications of the water cycle, L'Vovich estimated already in 1974 that more intense use of the groundwater buffers could raise the global groundwater flux by 5 000 cubic kilometres per year, from approximately 12 000 to around 17 000 cubic kilometres per year, thus augmenting the share of all regulated fluxes. This is considerable, given the fact that at that time 2 000 cubic kilometres per year were regulated by storage dams and that it was judged that additional dams could regulate another 3 500 (L'Vovich, 1974).

It is undoubtedly possible to make more intensive use of the buffer capacity of sufficiently large aquifers to control water resources with irregular regime, complementary to or instead of surface water reservoirs. The groundwater buffers will also prove to be invaluable in mitigating predicted water scarcity problems in the near future caused by climate change.

7.2.3   Combating undesired groundwater depletion

As shown in previous chapters (Sections 5.4, 6.2 and 6.3), progressive depletion of groundwater reserves is observed in many parts of the world's arid and semi-arid zone. The challenge is to stop or even reverse depletion, in order to conserve the groundwater reserves and thus enable groundwater to be exploited sustainably. However, this is often a very difficult task.

The *High Plains* in the USA may serve as the first example. Once barren and of marginal value for agriculture, these plains have been converted into an economically flourishing region since mechanical pumps – introduced in the 1940s – enabled large quantities of groundwater to be abstracted from the huge underlying phreatic Ogallala aquifer (covering around 450 000 km²). It is currently the most intensively exploited aquifer in the USA and has turned the region into one of the most productive agricultural areas in the world,. Nevertheless, groundwater exploitation has caused steady and significant declines in the groundwater level in a significant part of the plain, as shown in Figure 7.4. The spatial differences in decline are partly caused by variation in legal systems and management approaches adopted in the different states

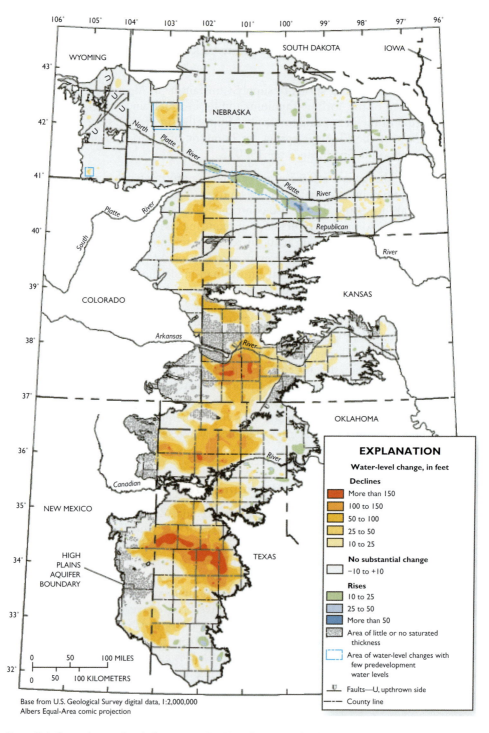

Figure 7.4 Groundwater level changes in the High Plains aquifer, from predevelopment until 2007. [*Source*: McGuire (2009)].

involved. Although there is a growing awareness of the need to stop depleting the resource, reducing groundwater abstraction to a sustainable level would have large repercussions for the local economy and is not readily accepted by many individual stakeholders who risk losing their business or part of their income if groundwater abstractions are curtailed. In spite of all efforts in the different water management districts, groundwater levels are still declining and controlled depletion seems the only feasible option in the central and southern part of the plains (McGuire, 2009; Sophocleous, 2010).

The situation on the High Plains is typical for numerous intensively exploited aquifers around the world, large and small. Among these are the much smaller *Sana'a basin* and other highland basins in Yemen, all of them in a severe state of overabstraction. Building effective partnerships between different categories of water users and reaching agreement on how to reduce and allocate groundwater abstraction seem the only way to prevent catastrophic water shortages in the near future.

In neighbouring *Oman*, however, the groundwater reserves are better under control – although not perfectly – due to a combination of thirty-two large recharge dams to augment the groundwater resources and licensing systems to regulate drilling and groundwater abstraction. The presence in Oman of more than 3 000 operational qanats (locally called 'aflaj'), sensitive to groundwater level declines, certainly has a positive influence on the willingness of local stakeholders to comply with government regulations.

India relies largely on indirect measures to control groundwater overabstraction. The main instruments used are energy supply and rural credits. The State Electricity Boards may restrict the number of hours of daily supply of the highly subsidised electricity in areas where groundwater is either currently overabstracted or is close to being. Likewise, the National Bank for Agriculture and Rural Development does not provide rural credits for new wells and pumps in areas where groundwater is under severe stress. The criterion for applying conservational measures is the degree of groundwater development stress observed in individual administrative spatial units (blocks, mandals or talukas), according to the classes shown in Figure 7.5: safe, semi-critical, critical and overexploited. The periodic update of this classification for the entire country is based on a 'Groundwater Estimation Methodology' developed for this purpose and contributes to keeping groundwater level monitoring networks operational (Chatterjee and Purohit, 2009). In addition, there are numerous small groundwater recharge schemes scattered over India, often developed and operated by local communities.

The *Great Artesian Basin* in Australia offers a completely different picture. Artesian wells have been tapping this aquifer since the end of the 19th century, but in spite of the steadily increasing number of wells drilled, their total yield has been declining since 1918, due to a gradual loss of artesian pressure (see Figure 7.6). However, the declines can be slowed down very significantly by technical measures that eliminate the massive waste of water caused by freely flowing artesian wells and by water conveyance losses. Although their implementation is expensive, these measures are not controversial because there is no explicit conflict of interest. A corresponding programme – the Great Artesian Basin Sustainability Initiative – has been implemented since the beginning of this century, with the feasible target of saving 211 million cubic metres annually by 2014 (Habermehl, 2006; Sinclair Knight Mertz, 2008; GABCC, 2009).

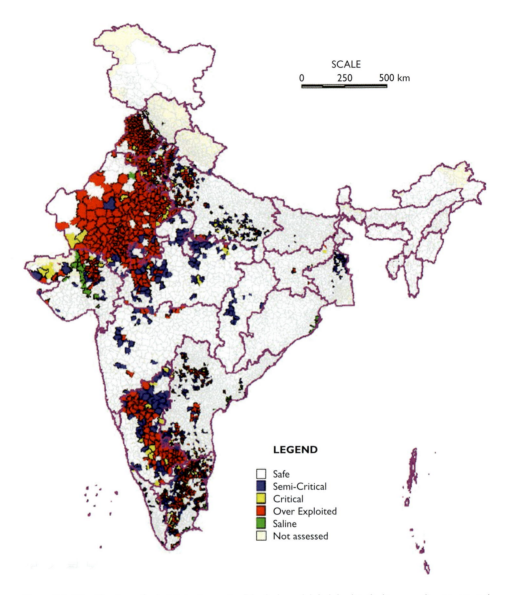

SCALE

0   250   500 km

LEGEND

☐ Safe
■ Semi-Critical
■ Critical
■ Over Exploited
■ Saline
☐ Not assessed

Figure 7.5 Classification of administrative units (blocks/mandals/talukas) in India according to ground-water development stress as of 2009. [*Source*: CGWB (2012)].

7.2.4   Controlling the impacts of groundwater withdrawal on ecosystems and the environment

The National Park *Tablas de Daimiel* in Spain, a wetland catalogued since 1981 as a Biosphere Reserve Area in the framework of UNESCO's Man and Biosphere programme, enjoyed until the early 1970s the natural outflow from the western La Mancha

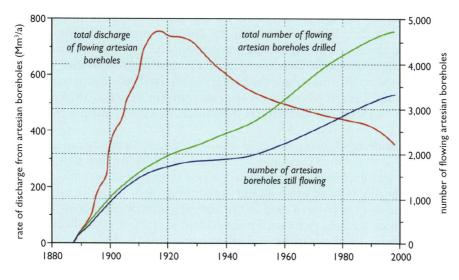

Figure 7.6 Trends in artesian groundwater outflow from the Great Artesian Basin, 1880–2000.
[*Source*: Habermehl (2006)].

aquifer. Between the mid-1970s and the late 1980s, however, new irrigation areas were established, abstracting huge amounts of groundwater and affecting the wetland. To stop depletion of the groundwater reserves and to protect the National Park, a Water Regeneration Plan was established in 1987, the aquifer was declared overexploited in 1994 and funds were raised to provide financial compensation to farmers who voluntarily restricted the volume of groundwater abstracted (López-Geta *et al.*, 2006). Current results of these interventions are encouraging.

Strict law enforcement is observed in *The Netherlands* in similar cases. Groundwater levels have long been controlled by the water boards, with the objective of maintaining optimal soil moisture conditions for agriculture and ecosystems, while interference by gradually increasing groundwater exploitation has triggered dedicated laws and regulations since the middle of the 20th century. Abstracting groundwater is currently subjected to obligatory reporting, registration and application for a license. In the case of large abstractions, the applicant is obliged to subcontract an independent study to predict the likely side effects of the abstraction. The license is granted only in cases where the predicted side effects are compatible with the targets of the provincial groundwater resources management plan and follow-up monitoring is required to validate the predictions. In the case of groundwater abstraction for nonconsumptive purposes, such as for temporary drainage at construction sites or for pump-and-treat remediation systems, it is increasingly made obligatory to re-inject the pumped groundwater.

Land subsidence caused by groundwater abstraction is occurring at many locations around the world, in particular in coastal areas. *Venice* and *Shanghai* are examples of successful implementation of measures to control land subsidence.

Intensive groundwater pumping in the industrial area of Porto Marghera near *Venice* during the period 1950–1970 caused the surface under the historical centre of Venice to subside by approximately 10 cm. This is comparatively little, but it constitutes a real threat to the city, located only between 0.7 and 1.0 m above mean sea level and thus susceptible to flooding. After the industrial groundwater withdrawal was shut down in 1970 to reduce the problems, the land subsidence ceased almost immediately and 2 cm of rebound was even measured in Venice in 1975 (Carbognin *et al.*, 1995). Sea level rise and natural subsidence, however, remain a threat to the city's future.

Land subsidence in *Shanghai* began around 1920, accelerated during the 1950s and reached a cumulative maximum of 263 cm in 1964. At that time, a number of measures was implemented, including a major reduction in groundwater pumping and the reallocation of groundwater withdrawals both from the urban area to elsewhere and from shallow to deep in the 300 metres thick multi-layer aquifer system; in addition, artificial recharge was introduced to stabilise groundwater levels in the upper aquifer. Land subsidence was effectively stopped by these measures, but in recent years it has restarted at a low rate due to draining the numerous construction pits needed for high-rise buildings under construction in this very rapidly expanding city (Chai *et al.*, 2004; Chai *et al.*, 2005; Wei, 2006).

In many other areas in the world, land subsidence due to groundwater withdrawal is still continuing at significant rates, as documented in Box 6.3 (Chapter 6). Documented cases of effective control are rare. It should be remembered that impacts of groundwater abstraction on ecosystems and the environment are not restricted to conditions of hydrological overabstraction (see Box 6.3).

7.2.5   Controlling seawater intrusion and saline groundwater upconing

Seawater intrusion and saline groundwater upconing are a threat to groundwater quality in many coastal aquifers where groundwater is intensively exploited. Since the hydraulic mechanism was understood at the end of the nineteenth century, a variety of technical measures have been devised and implemented to prevent or reduce the encroachment of saline water into the fresh groundwater domain. A first category of technical measures is based on simple adaptation to local conditions and includes measures such as control of abstractions (keeping them rather low), relocating wells to locations of maximum fresh groundwater thickness, minimising well depths, plugging the deeper parts of existing wells and using skimming techniques for groundwater abstraction. Proper use of these adaptations may be effectively supported by regulations and be guided by numerical model simulations and monitoring.

A technically more advanced category of measures consists of creating barriers to stop saline water from encroaching. The feasibility of physical barriers requires very specific conditions that are rarely present, and hence practical applications tend to focus on hydrodynamic barriers. Hydrodynamic barriers have been successfully formed by artificial recharge in the *Central and West Coast Basins of Los Angeles*, California. During the first half of the twentieth century, intensive groundwater abstraction for potable and agricultural purposes resulted there in groundwater levels sinking to more than 30 metres below mean sea level and in the intrusion of more than 700 million cubic metres of seawater by the late 1950s. Starting in 1953, 1969 and 1964, respectively, the Los Angeles Flood Control District has implemented three barrier projects, together stretching over 17 kilometres in length, including 290 injection wells and 773 observation wells (Figure 7.7). The configuration of the barriers corresponds to the pathways of historic saltwater intrusion as defined by the complex geology. Since 2008, some 35 million cubic metres of water has been injected annually into the aquifers, to depths down to 200 metres. These barrier projects have been adequately protecting the freshwater aquifers of the Central and West Coast Basins for over 50 years (Johnson, 2007; Barlow and Reichard, 2010). A similar hydraulic barrier, injecting treated wastewater, has been recently completed in Barcelona (Ortuño *et al.*, 2011).

Artificial recharge for combating coastal groundwater salinization is also practised in *The Netherlands*, but the leading concepts there are preventing salt water upconing and making use of storage capacity rather than creating barriers against lateral inflow of saline water. The belt of dunes along the North Sea coast, covering approximately 400 km², contains thin freshwater lenses overlying saline groundwater. Their exploitation for urban water supply started in 1854. In response to salinization problems observed since the beginning of the 20th century, artificial recharge has been applied since 1940, initially using exclusively open recharge systems such as basins, ponds

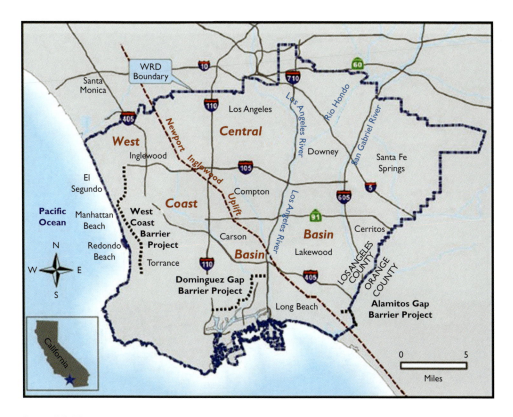

Figure 7.7 The three seawater intrusion barriers to protect groundwater in the Los Angeles coastal area, California, USA. [*Source:* Johnson (2007)].

and canals (see Figure 7.8), and since 1990 also injection wells. In 2008 ten artificial recharge sites spread over this dune area were artificially recharged by some 180 million cubic metres of pre-treated surface water, mainly originating from the Rhine and Meuse. The recharge works currently allow 190 million cubic metres to be abstracted from the dune aquifers without causing salinity problems, but only a minor part of it (10 million cubic metres) is derived from natural recharge. Because artificial recharge may affect the rich flora and fauna of the coastal dune area, the water supply companies have also been entrusted with the responsibility for nature conservation.

7.2.6 Groundwater pollution control

Groundwater pollution control is an enormously varied field. Apart from developing an enabling environment by awareness-raising, dedicated policies, laws and institutions responsible for pollution control, it includes more operational aspects such as assessing pollution risks and detecting the pollution actually present, studies on the behaviour and impacts of pollutants, devising potential interventions for preventing or mitigating pollution, prioritising potential interventions, plan development and approval, implementing measures and monitoring compliance and impacts. A wealth of theoretical and generic literature on many of these aspects is easily accessible, as well as reports and papers on identified pollution and its causes in specified areas,

Figure 7.8 Open recharge canal in the dune area of The Netherlands, supplied by a surface water transport pipeline. [*Photo*: PWN].

often with recommendations for action. Area- or country-specific descriptions of the implemented interventions and their impacts, however, are sparsely found in internationally accessible information sources. Below are some examples and comments on only a few of the abovementioned aspects.

Prevention of pollution starts with *assessing pollution risks*, which basically includes the periodic inventory of potential sources of pollution, mapping the aquifer's vulnerability to pollution and assessing the potential damage if groundwater were to be polluted (value of threatened current and future fresh groundwater functions). Such activities have become standard practice in several industrialised countries, but not yet in most developing countries.

One of the early applied measures to control groundwater pollution is the introduction of *groundwater protection zones* around well fields for public water supply, complementary to the more local well-head protection and sanitary completion of the wells concerned. The idea behind the concept is protecting water supplies rather than groundwater in the entire aquifer. Increasingly heavier restrictions on land use and waste disposal are being imposed in zones closer to the public water supply source. Figure 7.9 shows how this methodology is applied on the coral limestone island of Barbados, on the basis of a Development Control Zoning policy established in 1963 (Chilton and Alley, 2006). The overall aim of using protection zones is to reduce the potential for contaminants to reach the groundwater that is abstracted and to slow down the arrival of remaining contaminants. Approaches to delineating groundwater protection zones vary from very simple (adopting an arbitrary fixed radius) or simple (analytical methods to define mean travel time) to sophisticated (numerical flow and transport modelling). Table 7.1 presents examples of typical protection zone dimensions in a number of countries.

Figure 7.10 shows a simple example of how to involve the general public in groundwater pollution control in groundwater protection zones. A road sign informs people that they are entering a groundwater protection zone and requests them to report any pollution incident they may detect. This will enable quick action to minimise the damage caused by accidental pollution.

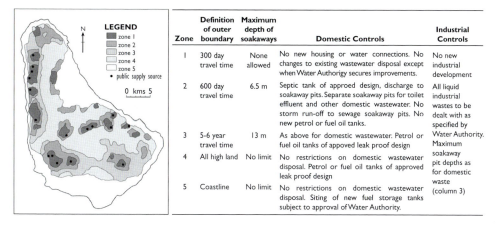

Zone	Definition of outer boundary	Maximum depth of soakaways	Domestic Controls	Industrial Controls
1	300 day travel time	None allowed	No new housing or water connections. No changes to existing wastewater disposal except when Water Authorigy secures improvements.	No new industrial development
2	600 day travel time	6.5 m	Septic tank of approed design, discharge to soakaway pits. Separate soakaway pits for toilet effluent and other domestic wastewater. No storm run-off to sewage soakaway pits. No new petrol or fuel oil tanks.	All liquid industrial wastes to be dealt with as specified by Water Authority. Maximum soakaway pit depths as for domestic waste (column 3)
3	5-6 year travel time	13 m	As above for domestic wastewater. Petrol or fuel oil tanks of appoved leak proof design	
4	All high land	No limit	No restrictions on domestic wastewater disposal. Petrol or fuel oil tanks of approved leak proof design	
5	Coastline	No limit	No restrictions on domestic wastewater disposal. Siting of new fuel storage tanks subject to approval of Water Authority.	

LEGEND
■ zone 1
■ zone 2
□ zone 3
□ zone 4
□ zone 5
• public supply source

0 kms 5

Figure 7.9 Groundwater protection zones in Barbados. [*Source:* Chilton, Schmoll and Appleyard (2006)].

Table 7.1 Examples of groundwater protection zone dimensions in a number of countries.

	Travel time and/or radius of zone		
Country	Wellhead protection zone or inner zone	Middle zone	Outer zone
Australia	50 m	10 years	Whole catchment
Austria	< 10 m	60 days	Whole catchment
Denmark	10 m	60 days or 300 m	10–20 years
Germany	10–30 m	50 days	Whole catchment
Ghana	10–20 m	50 days	Whole catchment
Indonesia	10–15 m	50 days	Whole catchment
Ireland	100 days or 300 m	50 days	Whole catchment or 1 000 m
Oman	365 days	10 years	Whole catchment
Switzerland	10 m	Individually defined	Double size of middle zone
United Kingdom	50 days and 50 m minimum	400 days	Whole catchment

Source: After Chave *et al.* (2006).

Figure 7.10 Road sign indicating the boundary of a groundwater protection area and showing a telephone number to be called to report any detected case of pollution. [*Photo*: Jac van der Gun].

In transforming preventive groundwater pollution control to an aquifer-wide or region-wide activity, three main lines of action can be observed, complementary to each other.

The first one is *land use planning and management*. Various countries have laws and regulations controlling land use. Among others, these can be used for preventing hazardous activities from taking place in the most vulnerable or sensitive areas from the groundwater point of view.

The second line is the *implementation of general regulations* on groundwater pollution control or environmental management, based on legislation, directives and protocols at sub-national, national and international levels. International exchange of ideas is reflected in the fact that many countries have adopted similar practices, such as the prohibition of certain harmful substances (e.g. DDT), compulsory environmental impact assessments and the adoption of the 'polluter pays' principle. In Europe, the European Union plays an important role. For example, by the obligation imposed by the EU Groundwater Directive of 1980 upon member states to prohibit the direct discharge of dangerous substances into controlled water, to require permits for their indirect discharge (i.e. through the unsaturated zone) and to limit the discharge of other pollutants. Similarly, the new Groundwater Directive of 2006 – a daughter directive of the European Water Framework Directive – obliges member states to stop or reverse pollution trends and achieve 'good chemical status' by 2015.

The third line of action consists of *sector-specific regulations and actions*. Examples in the agricultural sector are the prohibition of spreading slurry on farmland during the main recharge season, as well as the introduction of nutrient and pesticide management at farm level, with the objective to minimise contaminants being flushed down into the groundwater domain. Improved sanitary engineering practices, strict regulations on landfills and waste disposal, replacement of leaking sewerage systems, sewage treatment, the maintenance of septic tanks and the removal of buried domestic oil tanks are examples of interventions in the domestic sector. In the industrial and mining sector, many countries have issued laws and strict regulations on all phases of the production processes, including on handling liquid and solid waste, as well as on adequate contingency plans. In many countries, industries are encouraged to take initiatives for good environmental behaviour on a voluntary basis, e.g. by complying with ISO-14001 standards or even by completely recycling their wastewater after treatment.

Implementing these measures in all sectors requires institutions to have a legal mandate and sufficient operational capacity and funds. These requirements are more easily fulfilled in industrialised countries than in most developing countries, but even rich countries have to set priorities, because simultaneously implementing the full range of potentially useful interventions is not usually feasible.

Groundwater pollution control is not limited to prevention, but also includes different types of mitigation or remedial action after pollution has occurred. If there is an ongoing source of pollution, then *source control* is a priority (e.g. by improving landfills), otherwise groundwater remediation will not be very effective. '*Pump-and-treat*', implemented since the 1980s in North America and Europe, is probably the most widely used groundwater remediation technology. It basically consists of pumping the polluted groundwater to the surface, treating it and then discharging it again. Under favourable hydrogeological conditions and subject to the type of contaminants, pump-and-treat may be successfully used in cases of point pollution, if the geometry of the pollution plume is known and if removal of the polluted groundwater is considered feasible. It may take tens of years or even more to meet the clean-up goals, if they are met at all. By 2003, the US Environmental Protection Agency estimated over 700 pump-and-treat systems to be operating at National Priority List sites in the USA, requiring considerable funding (OIG, 2003). Pump-and-treat remediation of a gasoline spill site near Antwerp (Belgium), guided by a mass transport simulation model continuously updated on the basis of monitoring data, is described by Haest *et al.* (2010). The spill occurred in 1989 at an approximately 600 m upgradient from a drinking water production site (see Figure 7.11). Remediation started in 1998 and based on the progress observed by 2010 it is expected that the contaminant plume will be fully remediated by 2015, without affecting the drinking water production site.

Often, especially in cases of diffuse pollution over larger areas, remediation is not feasible. The preferred option then is preventing the body of polluted groundwater from migrating (*containment*). This can be done in principle by constructing *groundwater cut-off walls*, but in most cases it is more practical to apply *hydrodynamic isolation* techniques. These consist of creating 'zero hydraulic gradient' conditions around the polluted groundwater body, which makes it immobile. In some cases, such conditions are created coincidentally, e.g. below the Bolivian city of Santa Cruz (see Figure 7.12). Shallow groundwater there is severely polluted, but the pollution front does not appear to have penetrated below about 90 m depth, despite heavy pumping from the deep public supply boreholes. This is probably because the continuing abstraction from the shallower private wells intercepts and recycles a portion of the shallow polluted water. From the point of view of pollution control it would be favourable to continue the shallow groundwater abstractions (Chilton and Alley, 2006).

Finally, if available groundwater does not fully meet the quality requirements for the intended use, then treatment remains a technical solution.

7.2.7   Groundwater management and natural disasters

Groundwater resources users and managers may be confronted with sudden unexpected natural disasters such as floods, droughts, cyclones, tsunamis, earthquakes and landslides. These disasters may result not only in loss of life and property, but also in acute shortage of drinking water and degradation of aquifers and wells. Lessons have been learned from disasters in the past on how to establish emergency water supplies, how to rehabilitate affected aquifers and wells, but in particular how to anticipate by making the population prepared in disaster-prone areas. A few examples are given below.

The earthquake of January 1995 in Kobe (Japan) killed 6 400 people and cut off more than a million households from drinking water. Emergency water supplies were initially based on bottled water and water tankers, but these sources were quickly replaced by private wells scattered over the city, which contributed to rapid restoration of the municipal water supply. The disaster triggered the registration of suitable emergency wells in the city, which resulted in 517 suitable wells being registered by the end of 1998. Similar registration systems have also been launched in other Japanese cities such as Tokyo and Yokohama (Tanaka, 2011).

The super cyclone hitting the coast of the Indian state of Orissa in October 1999 caused many casualties and destruction of property, while drinking-water supplies were damaged by sea storm surges and additional flooding caused by heavy rains. Hydrochemical surveys immediately after the cyclone revealed that even in the worst affected areas only the shallow aquifers were salinized, while medium-deep and deep aquifers remained unaffected. The drinking water supply was fully restored within four months by drilling new medium-deep wells (Sukhija and Narasimha Rao, 2011).

The December 2004 tsunami flooded many coastal zones in Asia and even in East Africa, one result of which was that many wells became saline. An important question

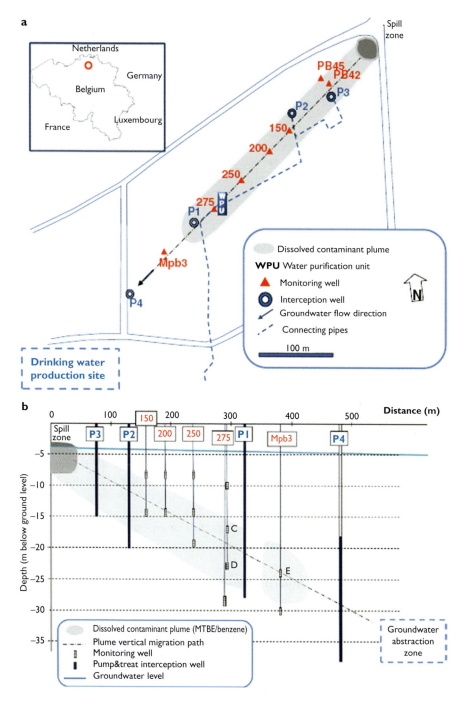

Figure 7.11 Pollution plume, interception wells and observation wells at the pump-and-treat remediation site near Antwerp *(a: Plan; b: Vertical cross section).* [*Source*: Haest et al. (2010)].

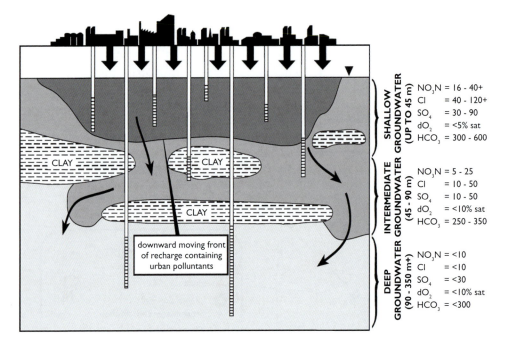

Figure 7.12 Stratification of diffuse urban pollution below the Bolivian city of Santa Cruz. [*Source:* Chilton (2006)].

was how to rehabilitate these wells: by pumping or by natural flushing and recovery processes? In the shallow unconfined sandy aquifers along the coast of eastern Sri Lanka the second option proved to be the most effective. Early pumping for a short period immediately after the tsunami contributed somewhat to salinity reduction inside the wells, but continued pumping seemed to disrupt natural recovery. Under local conditions, natural restoration required approximately one and a half years, as shown in Figure 7.13 (Villholth *et al.*, 2011).

Guidelines on how to deal with groundwater in emergency situations are presented by Vrba and Verhagen (2011).

7.2.8   Transboundary aquifer resources management

Like river basins, but in a less visible way, aquifers of all sizes may also be shared by different countries or by different jurisdictions within a country. Behaviour and functions of transboundary aquifers do not differ from those of other aquifers, but administrative borders that cross them render coordinated development and management of their groundwater resources more complex. Due to this complexity and due to the fact that attention to transboundary aquifers only began a few decades ago, most activities carried out so far in relation to transboundary management are of a preparatory or enabling nature; real interventions on the ground are still relatively rare.

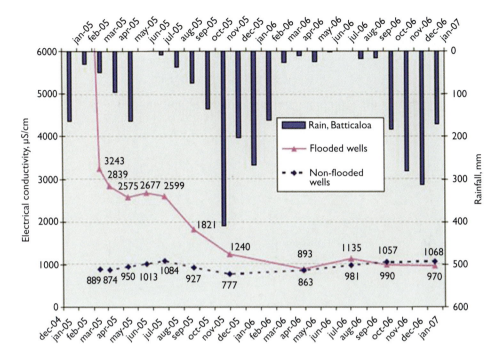

Figure 7.13 Average well water salinity and monthly rainfall after the December 2004 tsunami in the Batticaloa area, eastern Sri Lanka. [*Source*: Villholth *et al.* (2011)].

Physically, three phenomena have to be addressed in transboundary groundwater management: fluxes of groundwater, propagation of effects (in particular changes in pressure and water level) and transport of solutes across the boundaries. Groundwater fluxes across the boundary are in general small compared to transboundary surface water fluxes, although for some arid countries (like Egypt, Israel, Jordan and Syria) they may form a significant external contribution to the water resources. Therefore, not the allocation of groundwater fluxes, but rather the control of cross-boundary propagation of effects and transboundary migration of solutes, forms the main reason behind international consultations and agreements on transboundary groundwater.

Considerable achievements have been made since transboundary aquifers were put on the international agenda at the end of the 20th century. Regional inventories and characterisation of transboundary aquifer systems have been conducted under the umbrella of the United Nations Economic Commission for Europe (UNECE) in Europe (UNECE, 1999; UNECE, 2011) as well as in the Caucasus, Central Asia and South-East Europe (UNECE, 2007; INWEB, 2011; UNECE, 2011); and under the ISARM programme in the Americas (UNESCO, 2007 and 2008), Africa (UNESCO, 2010a and 2011) and Asia (UNESCO, 2010b). Largely based on the outcomes of the regional activities, a first global map of 'Transboundwater Aquifer Systems' was compiled by WHYMAP (2006) and later updated by IGRAC (2009 and 2012). Figure 7.14 contains a copy at reduced size of the most recent map 'Transboundary

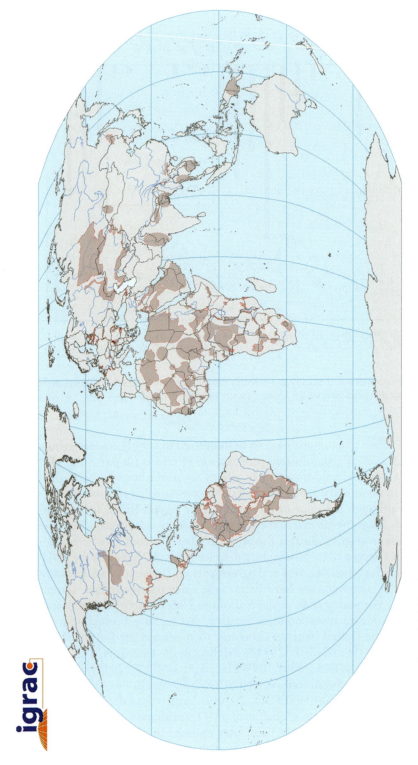

Figure 7.14 Transboundary aquifer systems around the world. The map shows the pattern of inventoried aquifer systems as inventoried before March 2012 (The original map should be consulted for the legend and more details). [Source: IGRAC (2012)].

Aquifers of the World' (IGRAC, 2012), showing the locations and selected properties of 448 identified transboundary aquifers and groundwater bodies around the world. ISARM's 'Atlas of Transboundary Aquifers' (Puri and Aureli, 2009) is based on similar information sources and presents some information for each of the aquifers in the form of indicators.

At the level of individual aquifers, important projects have been carried out or are still ongoing, especially in the framework of the International Waters focal area of the Global Environmental Facility (GEF). One of these, the Guaraní Aquifer project, has developed transboundary aquifer management instruments for the transboundary Guaraní Aquifer System in South America. In addition, in 2010 the presidents of the four countries involved signed an official agreement on this aquifer system, which marks an important step towards its coordinated or joint management. Another GEF-supported project, related to the North Western Sahara Aquifer System, has enhanced the knowledge of the aquifer system and explored how to vary the spatial pattern of groundwater abstractions from this non-renewable aquifer in order to produce minimal negative impacts at locations where natural groundwater discharge is still occurring (oases).

The *Carboniferous Limestone aquifer* shared between France and Belgium (regions of Flanders and Wallonia) provides a European example of cooperation at the level of a single transboundary aquifer. This aquifer, approximately 3 000 km^2 in extent and partly karstic, is important for drinking water and industrial water supply on both sides of the international boundary. However, intensive groundwater abstraction has long since produced a hydrological imbalance, resulting in dramatically falling groundwater levels, particularly in the confined part of the aquifer (see Figure 7.15). In Belgium, this development prompted an agreement between Flanders and Wallonia in 1997 to reduce groundwater abstractions. This and other factors, such as a major reduction in the industrial water demands in the French part of the area and recycling drainage water from quarries on the Belgian side, caused a stabilisation in the groundwater levels at the beginning of the 21st century and even some recovery afterwards.

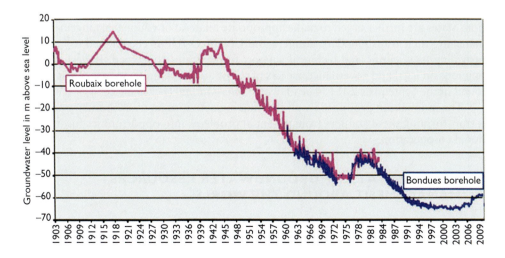

Figure 7.15 Evolution of groundwater levels in the French-Belgian Carboniferous Limestone aquifer, 1903–2011. [*Source:* After BRGM (2011) and Crastes de Paulet (2010)].

The challenges of managing the groundwater resources of this aquifer are not limited to ensuring sustainable water supplies, but also include other issues such as preventing the development of sinkholes (which has happened before) and controlling groundwater quality degradation problems (e.g. manganese and sulphates) triggered by rising groundwater levels. The aquifer system has not been exhaustively explored and relevant processes are not yet fully understood, which calls for additional assessment, analysis and monitoring efforts. On top of the obligations imposed by the European Water Framework Directory in this regard, in recent years various initiatives have been developed in transboundary cooperation between France and Belgium (Flanders and Wallonia) to address these needs. Among others, cooperation between BRGM and the University of Gembloux has been established on this subject, the International Scheldt Commission has declared the aquifer in 2006 to be a 'pilot transboundary aquifer', and the European project ScaldWIN (2008–2015) started with the objective to improve water quality in the Scheldt river basin. Increasing transboundary cooperation is expected to lead to an agreement that will formalise the French-Belgian co-operation on this aquifer and to sustainable joint exploitation of the Carboniferous Limestone aquifer (Crastes, 2010; De Smedt, 2010; BRGM, 2011, Anonymous, 2011).

Institutional arrangements and legal instruments are part and parcel of transboundary aquifer resources management. A study on groundwater in international law by Burchi and Mechlem (2005) shows that until recently international law rarely took account of groundwater and that by 2005 only a few legal instruments existed that were exclusively designed for groundwater: for the Geneva Aquifer, the Nubian Sandstone Aquifer and the NW Sahara Aquifer System. The development of such instruments is expected to be catalysed in the near future by the Draft Articles on the Law of Transboundary Aquifers. As mentioned before, these articles have been elaborated jointly by the United Nations International Law Commission (UNILC) and UNESCO's International Hydrological Programme (IHP) during 2002–2008 and they were adopted by resolution A/RES/63/124 of the UN General Assembly in December 2008 (Stephan, 2009). The UN General Assembly reaffirmed in its sixty-sixth session on 9 December 2011 the importance of transboundary aquifers and the related Draft Articles. In a new resolution, States are further urged to make appropriate bilateral or regional arrangements for managing their transboundary aquifers and UNESCO-IHP is encouraged to continue providing its scientific and technical support to the States concerned. Reflections and opinions on legal and governance aspects of transboundary aquifer management can be found in the September 2011 issue of *Water International* (Eckstein, 2011).

The momentum produced by all the ongoing activities on transboundary aquifers is transforming these aquifers from potential problems into opportunities for international co-operation.

7.3 ACTUAL STATE OF GROUNDWATER RESOURCES MANAGEMENT AROUND THE WORLD

After this brief outline of groundwater resources management and the presentation of some examples, one may wonder how widespread the implementation of groundwater management measures is in different parts of the world and to what extent their objectives are being met.

Because of the diversity of situations, opportunities and problems, as well as differences in availability of information on this subject, an overall judgement on the realities of groundwater resources management around the world does not make much sense and in any case would be too simplistic. This diversity even discourages classifying the different countries on the basis of indicators. Another complication in assessing the state of groundwater management is that it is easy to get confused between efforts made (measures taken, including regulations issued) and envisaged results achieved.

The number of identified problematic situations in the world (groundwater over-abstraction, overexploitation, pollution, resource degradation, conflicts, negative side effects of pumping, etc.) and the overviews presented of corresponding country-related data clearly indicate that effective groundwater resources management is virtually non-existent or still in its infancy in many countries or areas. Several factors that explain these shortcomings are known and have been repeatedly analysed and explained. They include the lack of awareness and understanding, negligence, selfish behaviour, conflicts of interests and poor governance. However, it needs to be recognised that there are often more fundamental underlying causes. Structural water scarcity, constraints inherent to the local groundwater resources, externalities related to groundwater as a resource, difficult socio-economic conditions and an adverse political setting create conditions in some areas that make any progress towards achieving adopted groundwater resources management goals very difficult. Despite administrative attempts to curb abstractions, intensive groundwater exploitation with harmful side effects is still increasing in many countries, while pollution often proliferates beyond control. Moreover, the slow pace of the physical processes in many aquifers often causes much time to elapse before the results of management efforts are observed, which does not encourage policy makers.

Rather than making an inventory of cases where groundwater resources management is evidently missing, deficient or failing, it is more meaningful to identify positive examples. In most of these cases, groundwater resources management is reactive (correcting undesired trends and conditions) rather than proactive (preventing them).

In various aquifers where overabstraction or overexploitation has become a problem at a certain point in time, effective groundwater management policies have resulted in restored equilibrium conditions – or at least in a reversed trend towards restoring equilibrium or in the reduction or elimination of negative side effects. These policies are usually based on a reduction in groundwater abstraction and the implementation of artificial recharge. Some examples:

- In Israel, hydrodynamic equilibrium has nearly been restored in the coastal aquifer by a combination of artificial recharge (90–135 million m³/year) and reduced groundwater withdrawal (by approximately 100 million m³/year, during the 1990s).
- Groundwater withdrawals have been reduced, or at least their increase was stopped, in several overexploited aquifers in Spain (Lower Llobregat, Plain of Valencia) and in Iran (where since the 1990s new tube wells for irrigation are prohibited in one-third of the plains).
- In California (USA), after a period characterised by overabstraction from the Central Valley aquifer system (1950s and 1960s), several artificial recharge schemes and the partial substitution of groundwater abstraction by surface water transfers,

complemented by the establishment of a market for water rights, have contributed to a partial improvement in the groundwater budget. Locally this has caused water levels to rise and land subsidence to stop. However, the overall situation is still far from stable, as has been very clearly demonstrated by recent assessments based on GRACE (see Table 5.5). According to H. Vaux from the University of California (cited by Shah, 2006), only 19 out of 431 individual groundwater basins in California are 'actively managed', implying some restrictions on pumping.

- In Oman, the hydrodynamic equilibrium of the groundwater systems in several areas have been completely or nearly restored with the help of large recharge dam schemes and a licensing system for drilling wells and abstracting groundwater.
- While on a large part of the High Plains in the USA a policy of planned groundwater depletion applies, a sustainable pumping policy was adopted in South Dakota and the eastern part. This may partially explain the differences in observed depletion.
- Land subsidence in Venice and Shanghai has been stopped by effective groundwater management interventions.
- Depletion of fresh groundwater lenses overlying saline water in the dune area along the west coast of The Netherlands has already been effectively prevented for many years by means of artificial recharge.
- Damage to agriculture and ecosystems that are dependent on very shallow water tables was virtually eliminated in The Netherlands several decades ago, by a combination of still ongoing meticulous surface water level control and strict groundwater abstraction licencing practices.
- Groundwater conservation measures and cutting down on water use have in recent years reversed the trend of total groundwater withdrawal in several European countries from increasing to slightly decreasing. Protection of ecosystems and of water-table fed agriculture is in some cases known to be behind the conservation policy.

Managing non-renewable groundwater resources, subject to mining, is different from managing renewable groundwater. Like in the case of mineral deposits or oil, one has to strike a balance between intensity and duration of the exploitation, taking into account the intended use of the abstracted water. Experience in managing non-renewable groundwater resources is still very limited and relevant concepts need still further development[2], but efforts in some countries have already produced noteworthy results:

- In Saudi Arabia, for example, the rate of mining of these non-renewable resources for irrigation purposes has been reduced by 738 million m³/year between 1990 and 2000 and an additional decrease by 38% was foreseen for the period 2000–2010 (Abdurrahman, 2002).
- As mentioned before, the Great Artesian Basin in Australia – with groundwater classified as a non-renewable resource – has been under intensive exploitation since the end of the 19th century, but freely flowing artesian wells and water conveyance losses have led to massive waste of water. The ambitious Great Artesian

---

2 Such as the concept of 'socially sustainable management' of non-renewable groundwater resources (Foster and Loucks, 2006).

Basin Sustainability Initiative, under implementation since the beginning of the 21st century, has already eliminated a substantial part of these losses and is expected to meet its target of saving 211 million cubic metres annually by 2014 (Habermehl, 2006; Sinclair Knight Mertz, 2008; GABCC, 2009).

With regard to groundwater pollution control, well-documented examples of significant reversals in groundwater degradation on a large spatial scale are rare, but this does not mean that groundwater pollution control has been ineffective so far. On the contrary, much of the effort spent is intended to prevent pollution from affecting valuable aquifers, areas or abstraction sites, and the impacts are impressive. There is little doubt that groundwater quality would have been much more problematic, at least in the industrialised countries, without the numerous interventions made in the form of protection zones, the enforcement of general environmental regulations and the introduction of environmentally safe practices in the domestic, agricultural and industrial sectors. Almost equally important are the effects of countless remediation operations, again mainly in industrial countries, that have stopped improperly constructed landfills or accidental spills from polluting groundwater. Groundwater pollution control is still at a low level in many developing countries, for a variety of reasons, including the lack of sufficient funds.

In many countries, a growing awareness can be witnessed of the need for improving groundwater resources management to serve the general interest. This is paving the way for corresponding projects and for putting in place appropriate regulations and specialised institutions. For example, in numerous countries it has triggered legislation that adequately addresses groundwater and the development of regulations that facilitate the implementation of related management measures. Many government institutions have understood that changing people's behaviour is crucial for achieving many of the groundwater management targets. Therefore, they have adopted corresponding enforcement tools (e.g. 'polluter pays' principle) or they have embarked upon participatory management approaches.

International initiatives, such as the new Framework Directive of the European Union on groundwater (see Box 7.4), the Millennium Development Goals (United Nations, 2011) and ISARM or Internationally Shared Aquifer Resources Management (Puri, 2001; Puri and Aureli, 2009) act as important catalysts for action. Nevertheless, increasing demographic pressures, climate change and other global change processes are expected to create huge groundwater resources management challenges in many countries, especially in the developing world. Unorthodox responses will be needed – from inside and outside the water sector – to complement the more conventional water management interventions. This will require the full commitment of the countries involved and sizeable international efforts.

To conclude, groundwater is at present widely used by humans and influenced by human activities, but from region to region in very different ways. Global estimates of the pressures on groundwater resources and their impact therefore have limited practical meaning and often the same is true for indicators of average conditions in a country. Such indicators may give information on trends and on the probability of occurrence of opportunities and problems, but real diagnostics should come from the local level. It is wise to refrain from certain over-simplified allegations hinting at catastrophes, as often favoured by the media.

Box 7.4: The European regulatory framework for the protection of groundwater

By Philippe Quevauviller[3]

Rules for the protection of groundwater against pollution were established within the European Community in the late 1970s by adopting Directive 80/68/EEC, that establishes a system to prevent discharges (direct or indirect) of certain hazardous pollutants (List I) and to make the discharge of less dangerous pollutants (List II) subject to a permit procedure that has to be preceded by a case-specific impact assessment. Growing concerns about the extent of diffuse pollution and the risk of over-exploitation of phreatic groundwater triggered studies and reports by the European Environment Agency in the early 1990s, and led the European Commission to develop an action program in 1996 in the form of a communication. The program aimed to strengthen the existing protection regime and its recommendations were taken into account in the drafting of the Framework Directive (2000/60/EC), that after being adopted on 22 December 2000, established an ambitious regulatory framework for all terrestrial and coastal waters (surface and groundwater).

This new legal instrument defines the general groundwater protection regime and provides for the expiration of Directive 80/68/EEC in 2013. The Water Framework Directive defines environmental objectives of 'good status' for groundwater quantity and quality in Europe to be reached before the end of 2015 and leaves the task of taking the necessary steps to achieve these goals to the Member States. The framework requires, however, well-defined binding steps to be implemented according to a fixed timetable. These steps include in particular an analysis of pressures and impacts and a characterisation of water bodies 'at risk', an economic analysis, a monitoring program and a program of measures largely based on compliance with related laws (requiring, for example, action programs for the protection of groundwater against diffuse sources of agricultural or industrial origin, or against point sources associated with landfills, as well as measures to balance groundwater abstraction and renewal). These steps lead to the development of management plans for river basin districts, of which the first one, to be published in 2009[4], is expected to achieve the objectives set by the Directive by 2015. These plans are made for a period of six years and, open to public consultation (participatory approach), they should involve possible revisions of the technical annexes on the basis of scientific progress.

While the provisions related to the quantitative status of groundwater bodies are clearly expressed in the Water Framework Directive, the directive is less specific in its regulations on criteria related to the chemical status ('environmental quality objectives'). Indeed, the different conceptual approaches have failed to reach agreement on defining detailed stipulations during the consultation process of the Framework Directive. This justified a request to be submitted to the European Commission for the development of a 'daughter' directive

---

3 European Commission, DG of the Environment, Brussels, Belgium.
4 This text box is a translation of an original text in French written before 2009.

and for establishing special stipulations regarding the chemical status of groundwater bodies and criteria related to the identification and reversal of pollution trends.

This 'daughter' directive, complying with the requirements of Article 17 of the Water Framework Directive, was developed in close consultation with a technical group composed of experts from environmental agencies and ministries, representatives of the industry, NGOs and scientists who have also worked with the European Commission to develop guidelines that should help implementing the regulatory framework of the various directives. It was adopted in December 2006 after a long period of negotiation between the European Parliament and the Council.

The new Directive on the protection of groundwater against pollution and deterioration establishes in this way criteria related to the assessment of good chemical status of groundwater bodies. These criteria are to be based especially on environmental quality standards established at the level of the European Union and on 'threshold values' to be established – and revised if necessary – by the Member States at the most appropriate level – national, regional or local, depending on the hydrogeological characteristics and human pressures that lead to characterising water bodies as 'at risk'. The Directive establishes additional criteria for identifying trends of increasing pollution and for reversing them (linked to the program of measures required by the Water Framework Directive), and includes an additional regulation for the prevention and limitation of the introduction of pollutants like this was stipulated in Directive 80/68/EEC (that will be withdrawn in 2013, as already mentioned), given their impact on the chemical status of water bodies.

This regulatory framework puts in place at the scale of the European Union a comprehensive program of inspection, monitoring, evaluation and, where appropriate, restoration of the quantity and quality of water masses. Its implementation will be complex because the challenge is immense, and protection entails a gamut of regulations that need to be coordinated (directives related to agriculture, industry and health). For the first time at this scale, it will be possible to assess the status of European water bodies and to monitor progress through data made publicly accessible by the development of the internet portal WISE (Water Information System for Europe) - a joint initiative of the European Commission, the European Environment Agency and Eurostat. Finally, the continuous augmentation of scientific and technical knowledge will feed into the revision of regulations at regular intervals (first revision in 2012, afterwards every six years).

This set of regulatory obligations, together with a strong commitment on the part of legislators to work closely with experts from various sectors for effective implementation, marks a turning point in the approach to establishing the protection of a valuable resource that is still too unknown to the general public.

Information on the development of regulations and on activities in support of their implementation are regularly published on the website of the European Commission[5].

---

5 http://ec.europa.eu/environment/water/water-framework/groundwater.html

In some areas, groundwater is relatively unknown and only used in limited quantities. The example of India shows that expanding groundwater abstraction for irrigation can drastically change a country's food production within a few decades and at the same time contribute enormously to poverty reduction (Moench, 2003). It is not unthinkable that similar developments could take place elsewhere in the world as well. Groundwater could in some areas also play a greater role in buffering variations in water resources availability, which is particularly important in zones where global change (demographic pressure, climate change) threatens to produce water scarcity in the near future. On the contrary, in other areas the exploitation of groundwater has already become excessive, which has sometimes led to its depletion, to economic damage or to harmful side effects on ecosystems and the environment. In the majority of such cases, returning to balanced and sustainable exploitation is highly desirable, necessary or even urgent. Various human activities that threaten the sustainability of groundwater and degrade its quality are still insufficiently under control, in different degrees.

Despite significant advances in knowledge and in monitoring conditions in different countries, short-term development of groundwater is often still prevailing over its long-term management. Proper long-term management should prevent overexploitation and ensure – wherever feasible – that the resources are sustainable with respect to quantity, quality and usability, also remaining available to future generations. The terms and conditions of such management activities are complex, but known. Nevertheless, in a large part of the world, groundwater resources management remains more an ambition to be pursued than a reality.

REFERENCES

Abdurrahman, W., 2002. *Development and management of groundwater in Saudi Arabia.* GW-MATE/UNESCO Expert Group Meeting, Socially sustainable management of groundwater mining from aquifer storage. Paris, 6 p.

Anderson, M.P., & W.W. Woessner, 1992. *Applied groundwater modeling: Simulation of flow and advective transport.* San Diego, California, Academic Press Inc., 381 p.

Anonymous, 2011. L'aquifère du Calcaire carbonifére: un enjeu transfrontalier franco-belge. *Géologues*, No. 171, December 2011.

Appelgren, B. (ed.), 2004. *ISARM Africa. Managing Shared Aquifer Resources in Africa.* Proceedings International Workshop, Tripoli, Libya, 2–4 June 2002. UNESCO-IHP, Paris, 238 p.

Barlow, P. & E. Reichard, 2010. Saltwater intrusion in coastal regions of North America. *Hydrogeology Journal*, Vol. 18, No. 1, pp. 247–260.

BRGM, 2011. *Vers une gestion concertée des Systèmes Aquifères Transfrontaliers. Partie 1: Constat Préliminaire – Analyse Génerale.*

Burchi, S., & K. Mechlem, 2005. *Groundwater in international law. Compilation of treaties and other legal instruments.* FAO Legislative study 86, FAO, Rome.

Carbognin, L., L. Tosi & P. Teatini, 1995. Analysis of actual land subsidence in Venice and its hinterland (Italy). In: *Proceedings of the Fifth International Symposium on Land Subsidence*, Balkema, Rotterdam, The Netherlands, pp. 129–138.

CGWB, 2012. *Ground Water Year Book – India, 2011–2012.* Faridabad, Central Groundwater Board, Ministry of Water Resources, Government of India, 63 p. Available from: http://cgwb.gov.in/documents/Ground%20Water%20Year%20Book%20-%202011-12.pdf.

Chai, J-C., S.L. Shen, H.H. Zhu & X.L. Zhang, 2004. Land subsidence due to groundwater drawdown in Shanghai. Technical note, *Géotechnique*, Vol. 54, No. 2, pp. 143–147.

Chai, J-C., S.L. Shen, H.H. Zhu & X.L. Zhang, 2005. 1D analysis of land subsidence in Shanghai. *Lowland Technology International*, Vol. 7, No. 1, pp. 33–41.

Chatterjee, R., & R.R. Purohit, 2009. Estimation of replenishable groundwater resources of India and their status of utilisation. *Current Science*, Vol. 96, No. 12.

Chave, P., G. Howard, J. Schijven, S. Appleyard, F. Fladerer & W. Schimon, 2006. Groundwater protection zones. In: O. Schmoll, G. Howard, J. Chilton and I. Chorus (editors), 2006. *Protecting Groundwater for Health. Managing the quality of Drinking-water sources.* World Health Organisation and IWA Publishing, London, UK, pp. 465–492.

Chilton, J., 2006. Assessment of aquifer pollution vulnerability and susceptibility to the impacts of abstraction. In: O. Schmoll, G. Howard, J. Chilton & I. Chorus (eds), 2006. *Protecting Groundwater for Health. Managing the quality of Drinking-water sources.* World Health Organisation and IWA Publishing, London, UK, pp. 199–242.

Chilton, J., O. Schmoll & S. Appleyard, 2006. Assessment of groundwater pollution potential. In: O. Schmoll, G. Howard, J. Chilton & I. Chorus (eds), 2006. *Protecting Groundwater for Health. Managing the quality of Drinking-water sources.* World Health Organisation and IWA Publishing, London, UK, pp. 375–409.

Chilton, J., & W. Alley, 2006. Hydrological management. In: O. Schmoll, G. Howard, J. Chilton and I. Chorus (eds), 2006. *Protecting Groundwater for Health. Managing the quality of Drinking-water sources.* World Health Organisation and IWA Publishing, London, UK, pp. 517–536.

Crastes de Paulet, F., 2010. *Scaldwin Project, Sustainable aquifer management.* ISARM 2010 Conference.

De Smedt, 2010. *Reviewing the criteria for the sustainable management of the Carboniferous Limestone Aquifer at the Belgium-France border.* UNESCO-IAH-UNEP Conference, Paris, 6–8 December 2010.

Eckstein, G. (ed.), 2011. *Strengthening cooperation on transboundary groundwater resources. Special issue of Water International*, Vol. 36, No. 5, 143 p.

FAO, 2003. *Groundwater management, the search for practical approaches.* Water Report 25, FAO, Rome, 39 p.

Foster, S.S.D., A.R. Lawrence & B.L. Morris, 1997. *Groundwater in urban development: assessing management needs and formulating policy strategies.* World Bank Technical Paper 390. Washington DC, World Bank.

Foster, S., M. Nanni, K. Kemper, H. Garduño & A. Tuinhof, 2003. *Utilization of non-renewable groundwater, a socially-sustainable approach to resource management.* World Bank, GW-MATE Briefing Note Series B, No 11, Washington DC.

Foster, S., & D. Loucks, 2006. *Non-renewable groundwater resources. A guidebook on socially-sustainable management for water policy-makers.* Paris, UNESCO-IHP, IHPVI, Series on Groundwater No. 10.

Foster, S., & K. Kemper (eds.), 2002–2009. *Sustainable Groundwater Management. Concepts & Tools.* GW-MATE, Briefing Note Series, Nos 0–16, The World Bank, Washington DC (versions in English, Spanish and Arabic).

Foster, S., & M. Ait-Kadi, 2012. Integrated Water Resources Management (IWRM): How does groundwater fit in? *Hydrogeology Journal*, Vol. 20, pp. 415–418.

GABCC, 2009. *Great Artesian Basin Strategic Management Plan: Progress and achievements to 2008.* GABCC Secretariat, Manuka, Australia.

Galloway, D.L., & T.J. Burbey, 2011. Review: Regional land subsidence accompanying groundwater extraction. *Hydrogeology Journal,* Vol. 19, No 8, pp. 1459–1486.

Habermehl, M., 2006. The Great Artesian Basin, Australia. In: Foster, S., & D. Loucks, 2006. *Non-renewable groundwater resources.* UNESCO-IHP, IHPVI, Series on Groundwater No. 10, pp. 82–88.

Haest, P.J., R. Lookman, I. Van Keer, J. Patyn, J. Bronders, M. Joris, J. Bellon & F. De Smedt, 2010. *Containment of groundwater pollution (methyl tertiary butyl ether and benzene) to protect a drinking-water production site in Belgium.* Hydrogeology Journal, 2010, Vol. 18, pp. 1917–1925.

IAH, 2006. Social and economic aspects of groundwater governance. Special issue of *Hydrogeology Journal*, Vol. 14, No 3 (13 articles). Springer, Heidelberg, pp. 269–432.

IAH, 2012. Economics of groundwater management. Special issue of *Hydrogeology Journal*, Vol. 20, No. 5 (16 articles). Springer, Heidelberg, pp. 817–1006.

IGRAC, 2009. *Transboundary Aquifers of the World, 1: 50 000 000, update 2009.* Special Edition for the Fifth World Water Forum in Istanbul.

IGRAC, 2012. *Transboundary Aquifers of the World, 1: 50 000 000, update 2012.* Special Edition for the Sixth World Water Forum in Marseille.

INWEB, 2011. *Inventory of internationally shared aquifers.* International Network of Water-Environment Centres for the Balkans. Available from: http:www.inweb.gr/index.php?option=com_aquifers_db.

Johnson, T., 2007. Battling Seawater Intrusion in the Central and West Coast Basins. Technical documents on Groundwater and Groundwater Monitoring. *WRD Technical Bulletin*, Vol. 13, Fall 2007.

Linton, J., & D.B. Brooks, 2011. Governance of transboundary aquifers: new challenges and new opportunities. In: Eckstein (ed.), 2011. *Strengthening cooperation on transboundary groundwater resources. Special issue of Water International*, Vol. 36, No 5, pp. 606–618.

Llamas, M.R., 2001. Considerations on ethical issues in relation to groundwater development and/or mining. In: *UNESCO IHP-V Technical Documents in Hydrology No 42*, UNESCO, Paris, pp. 467–480.

Lloyd, J. W., 1999. An overview of Groundwater Management in Arid Climates. In: Goosen, M., & W. Shayya: *Water Management, Purification & Conservation in Arid Climates. Volume 1: Water Management.* Technomic, Lancaster, USA, pp. 9–52.

Lopez-Geta, J.A., J.M. Fornés, G. Ramos & F. Villaroya, 2006. *Groundwater. A natural underground resource.* IGME, UNESCO and Fundación Marcelino Botín, 108 p.

L'Vovich, M.I., 1974. *World Water Resources and their future.* Mysl' P.H. Moscow. English translation A.G.U., Washington, 1979, 415 p.

Machard de Gramont, H., C. Noel, J.L. Olivier, D. Pennequin, M. Rama & R.M. Stephan, 2011. T*oward a joint management of transboundary aquifer systems. Methodological guidebook* (available in English and in French). Collection 'À Savoir' No 3, Agence Française de Développement, 123 p.

McGuire, V., 2009. *Water-level changes in the High Plains aquifer, predevelopment to 2007, 2005–06 and 2006–07.* Scientific Investigations Report 2009–5019, USGS, Reston, Virginia.

Moench, M., 1994. Approaches to groundwater management: To control or enable? In: *Economic and Political Weekly,* 24 Sept. 1994.

Moench, M., 2003. Groundwater and poverty: exploring the connection. In: Llamas and Custodio (eds), 2003, *Intensive Use of Groundwater. Challenges and Opportunities.* Balkema, Rotterdam, pp. 441–456.

Moench, M., 2004. Groundwater: the challenge of monitoring and management. In: P.H. Gleick et al., *The World's Water 2004–2005*, pp. 79–100.

Morishima, A., & S. Ohgaki, 2006. Groundwater management and policy: its future alternatives. Special issue of *International Review for Environmental Strategies*, Vol. 6, No 2, Institute for Global Environmental Strategies, Kanagawa, Japan, 274 p.

NGWA, 2005. Transboundary Ground Water. In: *Groundwater*, Vol. 43, No 5, pp. 645–770.

OIG, 2003. *Improving nationwide effectiveness of pump-and-treat remedies requires sustained and focused action to realize benefits.* Memorandum Report, Report no. 2003-P-000006, Office of Inspector General, US, EPA.

Ortuño, F., J. Molinera, I. Juarez, D. Arcos, J. Fraile, J. Niñerola & T. Garrido, 2011. *Barrera hidráulica contra la intrusion marina en el acuífero principal del delta del Llobregat (Barcelona, España).* I Congreso Internacional de Medio Ambiente Subterráneo, 15–18 Sept. 2009, Sao Paolo.

Planning Commission of India, 2007. *Groundwater Management and Ownership.* Report of the Expert Group, Government of India, 2007.

Puri, S. (ed.), 2001. *Internationally Shared Aquifer Resources Management.* Framework document. IHP-VI, Non Serial Documents in Hydrology. Paris, UNESCO, 72 p.

Puri, S., & A. Aureli, 2009. *Atlas of transboundary aquifers. Global maps, regional co-operation and local inventories.* UNESCO-IHP, ISARM Programme, Paris, 328 p.

Shah, T., 1993. *Groundwater markets and irrigation development: Political economy and practical policy.* Oxford University Press, Oxford, UK, 241 p.

Shah, T., 2006. *Institution groundwater management in the United States: Lessons for South Asia and North China.* An Agricultural Law Research Article. The National Agricultural Law Center, University of Arkansas School of Law. Originally published in Kansas Journal of Law and Public Policy, 2006, pp. 567–571. Available from http://nationalaglawcenter. org/assets/bibarticles/shah_institutional.pdf.

Sinclair Knight Merz, 2008. *Great Artesian Basin sustainability initiative. Mid-term review of Phase 2.* Report prepared for Australian Government, Dep. of the Environment and Water Resources.

Sophocleus, M., 2010. Review: groundwater management practices, challenges and innovations in the High Plains aquifer, USA – lessons and recommended actions. *Hydrogeology Journal,* Vol. 18, pp. 559–575.

Stephan, R., 2009. *Transboundary aquifers: Managing a vital resource. The UNILC Draft Articles on the Law of Transboundary Aquifers.* UNESCO-IHP, Paris.

Sukhija, B., & B.S.R. Narasimha Rao, 2011. Impact of the October 1999 super cyclone on the groundwater system and identification of groundwater resources for providing safe drinking water in coastal Orissa, India. In: GWES, 2011: *Groundwater for Emergency Situations: a Methodological Guide.* Edited by Jaroslav Vrba and Balthazar Verhagen. IHP-VII Series on Groundwater No 3, UNESCO-IHP, Paris.

Tanaka, T., 2011. Groundwater use in an emergency: impact of, and experience gained in the huge Hanshin-Awaji, Japan earthquake. In: GWES, 2011: *Groundwater for Emergency Situations: a Methodological Guide.* Edited by Jaroslav Vrba and Balthazar Verhagen. IHP-VII Series on Groundwater No. 3. UNESCO-IHP, Paris.

UNECE, 1999. *Inventory of Transboundary Groundwater, Volume 1.* UN-ECE Task Force on Monitoring and Assessment, RIZA, Lelystad, The Netherlands.

UNECE, 2007. *Our waters: joining hands across borders. First assessment of transboundary rivers lakes and groundwaters.* United Nations, New York & Geneva, 372 p.

UNECE, 2011. *Second assessment of transboundary rivers lakes and groundwaters.* United Nations Economic Commission for Europe. New York & Geneva, 430 p. Available from: http://www.unece.org/env/water/publications/pub/second_assessment.html

UNESCO, 2007. *Sistemas acuíferos transfronterizos en las Américas.* UNESCO-PHI & OEA, Series ISARM Americas No 1, Montevideo/Washington, 178 p.

UNESCO, 2008. *Marco legal e institucional en la gestión de los sistemas acuíferos transfronterizos en las Américas.* UNESCO-PHI & OEA, Series ISARM Americas No 2, Montevideo/Washington, 110 p.

UNESCO, 2010. *Managing shared aquifer resources in Africa.* Proceedings of the Third International ISARM Conference, Tripoli, 25–27 May 2008. IHP-VII Series on Groundwater No. 1, UNESCO, Paris.

UNESCO, 2010. *Transboundary aquifer in Asia.* IHP-VII Technical Document in Hydrology, UNESCO, Beijing & Jakarta.

UNESCO & World Bank, 2002. *Socially-sustainable management of Groundwater Mining for Aquifer Storage.* UNESCO Expert Group Meeting/GWMATE, World Bank Group, Global Water Partnership, September 2002, Paris.

United Nations, 2011. *Millennium Development Goals Report 2011.* Sales No. E.11.I.10, United Nations, New York.

Villholth, K., P. Jeyakumar, P. Amerasinghe, A. Manamperi, M. Vithanage, R. Goswami & C. Panabokke, 2011. Tsunami impacts and rehabilitation of groundwater supply: lessons learned from Eastern Sri Lanka. In: GWES, 2011. *Groundwater for Emergency Situations: a Methodological Guide.* Edited by Jaroslav Vrba and Balthazar Verhagen. IHP-VII Series on Groundwater No 3, UNESCO-IHP, Paris.

Vrba, J., & B. Verhagen, 2011. *Groundwater for Emergency Situations: a Methodological Guide.* IHP-VII Series on Groundwater No 3, GWES project, UNESCO-IHP, Paris.

Wei, Q., 2006. *Land subsidence and water management in Shanghai.* MSc thesis, Technical University Delft, The Netherlands, 79 p.

WHYMAP, 2006. *Groundwater Resources of the World, Transboundary Aquifer Systems, 1: 50 000 000.* Special Edition for the Fourth World Water Forum in Mexico City. BGR, Hannover, and UNESCO, Paris.

WMO, 1995. *INFOHYDRO Manual, 2nd Edition.* Operational Hydrology Report No 28, World Meteorological Organization, Geneva, Switzerland.

World Bank, 1998. *India Water Resources Management Sector Review: Groundwater Regulation and Management Support.* Report prepared in collaboration with the Central Ground Water Board, Ministry of Water Resources, Government of India.

Chapter 8

Final comments

- *What can be concluded about the information and knowledge regarding the world's groundwater. How to improve?*
- *What picture of the role, importance and key features of groundwater emerges from this attempt to present a global synopsis?*
- *Some final comments on groundwater resources management*

8.1 ABOUT THE INFORMATION AND KNOWLEDGE ON THE WORLD'S GROUNDWATER

8.1.1 A wealth of information – but with limitations

The preceding chapters show that a wealth of information is available on the world's groundwater. Compiling this information would have been impossible without the enormous and continuous efforts of many generations of dedicated individuals around the world, with support of the agencies they belong to, who have been participating in exploring, observing, assessing, monitoring and analysing groundwater and aquifer systems in many parts of the world, who have been processing and storing the collected data, who have been studying groundwater, its use and its management, and who have written reports and papers on it and who have shared all this information with the international community.

In addition to these actors at the level of producing new area-specific information, various international institutions and programmes are also of crucial importance for enhancing and disseminating knowledge on groundwater around the world. Most prominent in this respect are UNESCO with its International Hydrological Programme (UNESCO-IHP) and the International Association of Hydrogeologists (IAH) with its conferences and publications. But significant contributions are also being produced by several other entities, among others: the UN Food and Agricultural Organization (FAO), UNESCO's International Groundwater Resources Assessment Centre (IGRAC), the UN Environmental Programme (UNEP), the International Atomic Energy Agency (IAEA), the World Bank (IBRD), the Global Environmental Facility (GEF), the World Meteorological Organization (WMO), the International Water Management Institute (IWMI), the UN Department of Economic and Social Affairs (UNDESA) and regional organisations such as the European Union (EU), the Observatoire du Sahara et du Sahel (OSS) and the Organization of American States (OAS). These entities commonly establish regional or worldwide co-operative networks for carrying out their projects and programmes.

Only a minor fraction of all the globally available information could be accessed for the compilation of this book and fitted into its scope. But even this modest selection – without much spatial detail but lumped over large spatial units – contains many gaps, uncertainties and inconsistencies, including possible flaws which encroached during the processing and interpretation by the compilers. Part of the data inconsistencies simply result from the lack of uniformly adopted accurate definitions of certain variables (e.g. groundwater abstraction, groundwater recharge) and concepts (e.g. non-renewable groundwater, overexploitation). Most of the presented information is related to the physical aspects of groundwater rather than to socio-economic and institutional aspects. Another limitation is that many features and processes relevant at a local scale have been suppressed by the chosen macroscopic perspective, and thus are not or hardly visible.

8.1.2   Towards more and better information

The level and quality of information on groundwater varies enormously from one country to another. Dedicated efforts are being made by several agencies, especially by UNESCO in the framework of its International Hydrological Programme (IHP) and by FAO with its AQUASTAT, to reduce gaps and inconsistencies in the global information on groundwater fluxes, stored volumes and quality, as well as to update information on groundwater withdrawal and its socio-economic and environmental impacts. These efforts are significant and include surveys, studies, enquiries and statistical summaries, with priority for countries that are least advanced in this matter, not necessarily coinciding with least developed countries. Improving the consistency of approaches for assessing groundwater resources and for estimating and classifying their withdrawal and use is equally important for meaningful comparisons between data collected in different countries and for different aquifers. This is particularly important for enhancing the validity and homogeneity of macroscopic comparisons between resources and their use in an international context. In addition, it is necessary to broaden the scope and include all the information relevant for obtaining a good picture of governance conditions.

8.1.3   Towards improved understanding

Information is only useful if it leads to better understanding of the relevant systems and processes considered. The first priority is to understand the behaviour of groundwater and its role in nature, both in the water cycle and, more broadly, in the environment and the biosphere. It must be understood that reducing the concept 'groundwater' to an available volume or flux of water to be tapped is too simplistic, because the strengths, limitations and vulnerabilities of groundwater, as well as the usual interactions with surface water, ecosystems and the environment are characteristics that should not be ignored in groundwater resources management.

It should be understood that these resources are accessible to individuals and communities alike and that they should not only be exploited, but also managed in a way that strikes a balance between private and public interests, and between the interests of current and future generations. Understanding human behaviour is extremely crucial in this context, in order to design interventions that are really feasible and effective.

Finally, the efforts to be agreed upon for assessing and conserving these resources must be in proportion to their economic value, to the services rendered to society by their development and use, and to the value of their *in situ* functions.

8.1.4 Paying attention to the diversity of situations, opportunities and problems

When deepening knowledge and developing a vision on groundwater, one should beware of views that are too general, uniform or simplified regarding the occurrence and state of groundwater. On the other hand, relying on only very local descriptions and analysis is not advisable either, because that approach would insufficiently incorporate all relevant spatial interdependencies. It is wise not to refrain from identifying major features and dominant trends.

Groundwater forms a world of endless variation – whether one considers size and complexity of aquifer systems, volumes of groundwater stored and renewed, groundwater quality and vulnerability, accessibility of groundwater and exploitation methods, present exploitation levels or trends over time, etcetera.

The world of today is faced with fundamental and very diverse problems, but also offers a wide variety of opportunities. Together they prompt a range of goals to be achieved and activities to be carried out, each in its own setting and geographical context. Formulating these goals and activities correctly is only possible if the specific local conditions have been assessed and are known in sufficient detail. Inter alia, this would imply the exploration of the groundwater systems, the assessment of locally exploitable groundwater resources in terms of quantity and quality, the identification of pollution risks, the inventory of groundwater use and non-extractive functions of groundwater, and the detection of relevant trends (withdrawal, water levels, water quality, environmental impacts), including their underlying factors. The resulting area-specific information will allow a reliable diagnostic analysis to be carried out, revealing realistic opportunities and relevant problems that may be addressed in a water resources management strategy.

8.2 ABOUT THE KEY FEATURES OF GROUNDWATER, ITS ROLE AND OBSERVED TRENDS

8.2.1 Key features of groundwater

As documented in previous chapters, by far the largest share of all unfrozen freshwater on Earth is groundwater and groundwater is also a major component of the terrestrial water cycle. Its exceptionally high stock-to-flow ratio, compared to that of other terrestrial components of the water cycle, produces a number of very special groundwater characteristics: a high buffer capacity, relatively low vulnerability to pollution of a large part of the reserves, and low resilience to recover from disturbances in its state. Another special feature of groundwater is the large spatial extent of its occurrence, not only laterally but also vertically, often to hundreds or thousands of metres deep.

The *buffer capacity* of groundwater systems keeps water available during periods without recharge. It varies enormously from one system to another. At the low end of the scale, in small and shallow aquifer systems, groundwater reserves may bridge a dry season of several months or even a single dry year, already sufficient for survival in arid and semi-arid zones. Many aquifers, however, may easily buffer over much longer periods, usually accompanied after some time by negative side effects but without exhausting the reserves in the short or medium term. The most extreme cases of buffer capacity are found in mega aquifers with non-renewable groundwater, as presented in Chapter 3 and discussed elsewhere. These aquifers still constitute huge sources of water even after thousands of years without significant replenishment.

The *vulnerability* of groundwater to pollution varies with depth, but mainly as a function of the local hydraulic conditions and the spatial flow pattern. Shallow groundwater may be polluted rather easily, especially if a protecting aquitard or similar cover is missing. Deep groundwater is often very well protected against pollution entering from the surface, because it takes very long (hundreds to thousands of years) to arrive there and overlying poorly permeable strata could even isolate it almost completely from the active water cycle.

The other side of the coin is the *low resilience* of groundwater systems, also related to the inertia inherent to the high stock-to-flow ratio. Once an aquifer is seriously depleted, it requires many years under strict control to recover the reserves to their original level. With respect to pollution, the resilience of groundwater systems is even less. In many cases, observed pollution of an aquifer is virtually irreversible, at a human time scale.

8.2.2 Role and importance of groundwater

In addition to the role of groundwater in various natural processes and cycles, it may be clear from the presented information that groundwater is very important to mankind, in a direct sense by providing a source of water to be tapped and used, and indirectly by supporting ecosystems and the environment. As an exploitable resource, groundwater has a number of properties that offer comparative advantages above other sources of water, from the points of view of survival, reliability and potential for added value. The buffer capacity of groundwater systems plays a prominent role in this respect. It enables human life in several zones on Earth where survival without groundwater would be impossible. It also contributes to the high reliability of groundwater as a source of supply, in terms of quantity and quality, which – in turn – often causes groundwater to be more valuable per unit of volume than surface water. The economic importance of groundwater use should therefore be assessed on the basis of added value rather than on the basis of its volumetric share in supplies. Given the role played by groundwater in domestic, agricultural and industrial water supplies, as described in Chapter 5, groundwater is of enormous importance in nearly all the countries of the world.

8.2.3 Groundwater is in transition worldwide

Groundwater conditions are changing rapidly around the world. An unprecedented increase in groundwater withdrawal during the past century, observed in almost all countries (see Chapter 5) and known as the *'silent revolution'* (Llamas & Martínez-Santos, 2005), has produced and is still producing huge benefits but also numerous negative side effects. To mention a few: groundwater reserves are being depleted in numerous aquifers around the world, groundwater levels are falling, artesian wells and springs are becoming rare and may in some areas disappear completely within only a few decades, baseflows in streams are decreasing, wetlands degrading and land subsidence affecting sensitive urban zones.

In addition, modern lifestyles and production processes have introduced pollution that is progressively affecting groundwater quality – especially in near-surface aquifer domains.

The drivers behind these changes – especially demographic and economic ones – are powerful. Therefore, it is not realistic to expect that stopping or even reversing the observed trends would be easy. Even more so, because predicted climate change is likely to result in additional problems in many water-scarce areas of the world.

8.3 ABOUT GROUNDWATER RESOURCES MANAGEMENT

8.3.1 Major issues and a few specific ones

As outlined in the Chapters 6 and 7, major issues in groundwater resources management deal with promoting or enhancing profitable use of groundwater, while at the same time conserving the quantity and quality of the groundwater resources, and avoiding detrimental environmental impacts of groundwater withdrawal. Examples of how to address these issues have been presented. Although already included in some way or another, it is worthwhile highlighting a few specific issues that deserve particular attention and effort:

- *Exploitation and management of non-renewable groundwater resources*
Experience in developing strategies for the exploitation and management of non-renewable groundwater resources is still lacking. Nevertheless, this is a major issue for the water economy of some countries in the arid zone. Hence, the subject deserves to be studied in more detail, in order to develop scientifically well-founded approaches. Duration and intensity of mining will have to be defined, sufficiently profitable uses of the pumped water have be chosen and special attention is needed for using the profits enjoyed in the short term for a smooth transition in the near future to other sources of water or to a significant reduction in water use.

- *Strategies in cases where sustainable groundwater resources management is not feasible*
Although countries and water management institutions generally aspire to ensure the sustainability of their groundwater resources, in some situations it is not realistic to expect that gradual degradation of the groundwater resources can be stopped or even

reversed. Shallow fresh groundwater resources of small and topographically very flat islands (such as atolls in the Pacific and Indian Ocean) form one category of such situations. If sea-level rise causes permanent inundation of part of such islands by the sea, then the fresh groundwater lenses will irrevocably shrink. Non-renewable groundwater (already mentioned) is a second category: its exploitation necessarily depletes the resource. A third category consists of intensively exploited renewable aquifers in arid and semi-arid regions. In theory they can be exploited at sustainable rates, but in practice this is often an illusion due to the huge socio-economic importance of current groundwater withdrawal and the absence of alternative sources of water. A fourth category is shallow groundwater in urbanised and industrial areas exposed to severe pollution. All these cases of degrading fresh groundwater resources constitute 'creeping' problems that remain dormant for some time, but may produce hazardous conditions in the longer run. It is extremely important to identify them and prepare for a future independent of these dwindling groundwater resources. It is better to plan and control the reduction of water use in such cases rather than to undergo it as a catastrophe. This may require a complete transformation to less groundwater-dependent economies in the regions concerned.

- *Anticipating climate change*

In spite of much uncertainty in the currently available predictions, a progressively clearer underpinned picture of how the climate is likely to change in the near future is emerging for many regions around the world. Instead of waiting for more accurate predictions or even for what is going to happen on the ground, it is preferable to be pro-active and to analyse what repercussions the expected climate change may have for water availability and water demands in each country. One of the priorities is to identify those groundwater systems that are likely to be affected most severely by climate change, and to devise and analyse potential management interventions for such cases. Another priority is to identify groundwater systems that may offer solutions for adaptation in areas where the overall water availability is likely to become critical. Suitable groundwater buffers there might be used to bridge prolonged dry periods or to buy time for a smooth transition to adjusted levels of sustainable water use.

8.3.2   Improving groundwater resources management

Groundwater resources management may be improved in many countries by rooting it more strongly in reality, by being more ambitious and by integrating it better into overall land use and water resources management.

- *Management rooted in reality*

Groundwater resources management has in many countries hardly passed the stage of intentions and more or less formalised institutional arrangements. Rather than on the implementation of measures, the attention focuses in many cases more on the development of regulatory instruments. In addition, the applicability and implementation of these instruments often require more means and willingness than are available. As long as a proposal or plan for groundwater resources management does not go beyond describing these instruments, indicating who is responsible for their implementation and defining the different administrative tasks, then this is no more than 'potential' management. It is desirable that groundwater institutions in the future make more room for 'real' management that should, in one form or another, be community oriented and participatory. This implies that groundwater managers should become more aware of and better informed about common goals and priorities, and that management mandates are linked to units defined by relevant natural features. Groundwater governance is an important aspect in this context and that is the reason that many of the international agencies mentioned at the beginning of this chapter have joined forces to explore the governance dimension of groundwater management.

- *More ambitious and more active management*

Groundwater resources management is in numerous cases more than simply intercepting part of the natural groundwater fluxes and this will possibly be true in many more cases in the future. It includes taking more advantage of the groundwater reserves and influencing aquifer recharge, by the following actions:

- Expanding groundwater exploitation strategies by utilising the interannual buffer capacity of certain groundwater reservoirs. This is an option to compensate for shortages of surface water during dry years, a phenomenon that in a significant part of the world will probably become more widespread and more serious due to climate change, while the capacities of surface water storage reservoirs are declining due to siltation. Groundwater in such cases either substitutes surface water or is used as a support during dry periods.
- Developing artificial recharge, often in combination with the preceding strategy.
- Using, maintaining or even improving the self-purification capacity of certain aquifers, in particular for the re-use of wastewater.
- Setting certain areas aside for the production of water of excellent quality for public water supply, especially where appropriate land management and agricultural policies allow this.

- *More integrated and sustainable management*

The objectives of the groundwater resources management should not be limited to the interests of the groundwater exploiters or the direct users of abstracted groundwater. They need to be geared to the objectives of water resources management in general – both as a resource that can be withdrawn and as a component of the environment. These objectives should also be tuned to those of land use and use of the subsurface, as well as to those of environmental management. This may result in variable priorities for different objectives in the related policy fields, depending on cases and areas. Linking up these policy fields will also lead to a broader dialogue between the different parties involved, with sometimes diverging goals and interests, and to the incorporation of political preferences, in particular to account for general long-term interests – the interests of future generations.

The heritage value of groundwater is especially manifest in the case of aquifers that form strategic reserves of water that can be mobilised in emergency situations when water supplies have been disrupted by natural or man-made disasters. This is addressed by the project 'Groundwater for Emergency Situations (GWES)', initiated by UNESCO in 2006 (Vrba and Verhagen, 2011).

8.3.3 Linking groundwater resources management to sustainable development: 'getting out of the water box'

Groundwater resources management is not a purpose in itself, but should serve society's general objections and ambitions. Although there has been a clear evolution in groundwater resources management towards more integration and more coordination with other policy fields, it is still – like water resources management in general – a typical water sector activity with limited influence at the higher political levels. As is formulated in the first chapter of the Third World Water Development Report (Bullock et al., 2009): '*The sector is beginning to recognize that decisions by people outside the water sector determine how water will be used, but the other sectors are seen as cross-cutting. The approach within the sector has been to invite those working in other socioeconomic sectors to join in integrated water resources management. But the societal and political questions that determine the real allocation and management of water resources also need to take into account the technological aspects of integrated water resources management.*' Many problems, such as the structural reduction of available resources in the future in groundwater overabstraction areas or the linked crises of water, climate change, energy, food supply and financial markets cannot be solved within the narrow limits of a single sector (such as the 'water-box'), but need a fully integrated approach to decision making. The authors of the cited chapter present sustainable development as the framework for water resources management and state that '*leaders in government, the private sector and civil society (.....) must learn to recognize water's role in obtaining their objectives.*' It is a challenge to groundwater professionals and other water sector specialists to be open-minded and recognise situations and problems that would benefit from being addressed on a wider platform than offered by their own sector. The opportunities to turn this into reality vary from country to country. At the international level, various sector linkages that did not receive much attention in the past are currently being addressed in projects and conferences. Examples are the integration of groundwater in integrated subsurface management (including development of oil, gas and geothermal energy, mining activities, subsurface storage and other uses of the subsurface space) and the water-food-energy nexus (Martin-Nagle et al., 2012), underlying the pursued transition towards a green economy (UNEP, 2011). Initiatives such as the Millennium Development Goals, the Earth Summits on Sustainable Development and the World Water Fora with their Ministers' Conferences may provide effective mechanisms to enhance political commitment and support.

REFERENCES

Bullock, A., W. Cosgrove, W. Van der Hoek & J. Winpenny, 2009. Getting out of the box – linking water to decisions for sustainable development. In: *WWDR3*, Chapter 1, pp. 3–23.

Llamas, M.R. & P., Martínez-Santos, 2005. Intensive groundwater use: a silent revolution that cannot be ignored. *Water Science and Technology Series*, 2005, Vol. 51, No 8, pp. 167–174.

Martin-Nagle, R., E. Howard, A. Wiltse & D. Duncan, 2012. *The water, energy and food security nexus – Solutions for the Green Economy*. Bonn 2011 Conference Synopsis, Federal

Ministry for the Environment, Nature Conservation and Nuclear Safety (BMU) and Federal Ministry for Economic Cooperation and Development (BMZ), Germany, 28 p.

UNEP, 2011. Towards a Green Economy: Pathways to sustainable development and poverty eradication, 44 p. Available from: www.unep.org/greeneconomy.

Vrba, J., & B. Verhagen, 2011. *Groundwater for Emergency Situations: a Methodological Guide*. IHP-VII Series on Groundwater No. 3, GWES project, UNESCO-IHP, Paris.

Appendix 1

Glossary

Abstraction
1. The process of taking water from a source, either temporarily or permanently.
2. The volume of water taken during this process per unit of time.
Synonyms: withdrawal, extraction.

Alluvial Geological term, indicating an origin related to terrestrial flowing surface water. Example: alluvial sediments or deposits (usually composed of gravel, sand, silt and/or clay), as found within reach of present-day or former stream systems.

Aquifer A hydraulically continuous body of relatively permeable unconsolidated porous sediments or porous or fissured rocks containing groundwater. It is capable of yielding exploitable quantities of groundwater.
Synonym: water-bearing formation.

Aquitard Groundwater-filled body of poorly permeable formations, through which still significant volumes of groundwater may move, although at low flow rates.

Aquiclude Groundwater-filled bodies of poorly permeable formations, through which no or almost no flow of groundwater passes.

Arid region A region of low precipitation, characterized by a severe lack of available water, to the extent of hindering or even preventing the growth and development of plant life. In agriculture this term is used to indicate extremely dry areas where without irrigation no crops can be grown.

Artesian Groundwater pressure condition that allows groundwater to rise to above the local ground surface, if a well is drilled to the zone where such pressure is present ('artesian aquifer'). In other words: the piezometric level of groundwater in an artesian aquifer is located above ground level. A well tapping an artesian aquifer is called a 'flowing well' or 'artesian well'.

Baseflow The steady flow component of stream flow that continues for a long time after rainfall has stopped. It is mainly fed by the discharge of groundwater systems and is sometimes called 'dry-weather flow'.

Basement The oldest rocks recognized in a given area, a complex of metamorphic and igneous rocks – usually Precambrian or Palaeozoic in age – that underlies all sedimentary formations.

Bedrock Solid rock that is relatively unaltered and has not become soft (thus it is rather dense, and usually of low porosity).

Black water Wastewater containing faeces or other remnants of sanitary use.

Blue water Natural surface water and groundwater.

Borehole Hole made in the soil or the earth's upper crust by drilling. Its diameter is usually small, and its purpose is commonly either geological exploration or the construction of a well for abstracting or injecting liquids (e.g. water).

Brackish water   Water containing dissolved solids in a concentration between 1 000 and 10 000 milligram per litre.

Cainozoic (also: Cenozoic)   Era in geological history, running from roughly 66 million years ago until the present. Often also used as an adjective, to indicate the age of a geological formation.

Capillary fringe   Subsurface zone immediately above the water table, in which the interstices are completely filled with water, but under pressure less than atmospheric pressure (due to suction forces).

Carbon cycle   Biogeochemical cycle by which carbon is exchanged among the biosphere, pedosphere, geosphere, hydrosphere and atmosphere of the Earth.

Carbonate rock   A rock composed of carbonate minerals, especially limestone and dolomite.

Carboniferous   Period in geological history, belonging to the Palaeozoic Era. Often also used as an adjective, to indicate the age of a geological formation.

Catchment, catchment area   Total area having a common outlet for its surface water discharge. *Synonyms: watershed, river basin.*

Climate   Synthesis of local weather conditions, presented as statistics (mean values, variances, probabilities of extreme values, etc.) over a long period of time (usually taken as 30 years).

Climate change   Long-term modification of the climate (manifested by a change in the long-term statistical properties of climatic variables, in particular in averages and/or variability).

Collector well   Well provided with horizontal tubular drains arranged in several radial directions and which increase its effective radius.

Confined aquifer   Fully saturated aquifer (i.e. pressure everywhere greater than atmospheric pressure) directly overlain by an impermeable or almost impermeable formation (confining bed). The confining bed prevents the aquifer from interacting directly with the atmosphere and with surface water bodies (except for surface water bodies that intersect the aquifer).

Confining bed   Impermeable or poorly permeable formation overlying a fully saturated aquifer (see also: 'Confined aquifer').

Connate water   Water entrapped in the interstices of a sedimentary rock at the time the rock was formed.

Consolidated
1. *Geological term*: indicates that in principle a rock forms a solid mass, not an accumulation of uncemented loose materials such as gravels, sands, silts and clays *(the latter are unconsolidated sediments).*
2. *Soil mechanics term*: indicates that the volume of an earth layer has decreased (mainly due to reduced porosity) in response to an external load or to the reduction of hydrostatic pressure.

Consumptive use   Part of the water used that is not returned to aquifers, streams or seas, but is either incorporated in products or organisms, or discharged in the form of vapour into the atmosphere.

Cretaceous   Period in geological history, belonging to the Mesozoic Era. Often also used as an adjective, to indicate the age of a geological formation.

Darcy's law   One of the main physical laws in groundwater hydraulics, stating that the volume of flow passing through a section in a porous medium during a given time interval is proportional to the hydraulic gradient and the hydraulic conductivity of the saturated medium.

Delta   A land form at the mouth of a river, where this flows into an ocean, sea, estuary, lake, reservoir, another river or a flat arid area, characterized by the deposition of a large share of the sediments carried by that river (alluvial sediments).

Depletion   Reduction of the stored volume of groundwater in an aquifer (also: reduction of the stored water volume of any other component of the local hydrological cycle).

Deposit   Geological synonym of sediment.

Domestic water use   Use of water for drinking water and other household purposes, in offices and for public water services outside the agricultural and industrial spheres. The water may be supplied either by municipal or public water supply systems or withdrawn by the users themselves (self-supply).

Drainage   Natural discharge or artificial removal of excess water.

Drawdown   Lowering of the groundwater level or piezometric surface caused by abstraction of groundwater (including not only pumping, but also outflow from an artesian well or discharge from a spring).

Ecosystem   A dynamic complex of plant, animal and micro-organism communities and their non-living environment interacting as a functional unit.

Endorheic   Adjective indicating that water from the area, aquifer or stream concerned is not discharged into the open sea and oceans, but the area's discharge returns directly to the atmosphere. Endorheic aquifers are either drained by streams in endorheic river basins or by evaporation in closed depressions. Endorheic river or lake basins are usually called 'closed basins'.

Evaporite   A chemical sedimentary rock consisting of minerals precipitated by evaporating waters. Mainly salt (rock salt) and gypsum.

Evaporation
1. The physical process of emission of water vapour to the atmosphere.
2. The amount of water evaporated. *(The maximum rate, which is the rate of evaporation from a water surface, is called potential evaporation)*

Evapotranspiration
1. Release of water to the atmosphere through evaporation from the ground and from water surfaces, and also escaping from the leaf surfaces of plants to the atmosphere (transpiration).
2. The amount of water lost by evapotranspiration. *(The maximum rate, which is the rate of evapotranspiration from a fully developed vegetation cover, well supplied with water, is called potential evapotranspiration)*

Exhaustion   Depletion that has reached the stage of no remaining exploitable resources.

Facies   The set of characteristics of a rock that indicates its particular conditions when it was formed and that distinguishes it from other facies in the same rock formation. In the case of sedimentary rocks it indicates its particular environment of deposition, in the case of metamorphic rock the particular range of pressure and temperature under which metamorphism occurred.

Fault   *See Fracture.*

Fissures   Secondary interstices, which have originated after the rock was formed.

Flux   Numerical value of the rate of flow across a given surface.

Flux density   Flux per unit of area.

Folded mountains   Mountains created by an originally planar earth structure (usually a sequence of sedimentary beds) being bent. The deformation may have been produced by horizontal or vertical forces in the crust.

Formation   A set of rocks or unconsolidated sediments that are or once were horizontally continuous and that share some distinctive features of lithology.

Fossil groundwater   Stored groundwater that has entered a rock formation during a remote period of time (under climatic and/or geological conditions different from the present day ones) and is not renewed under present-day conditions.
Synonym: non-renewable groundwater.

Fractures   Cracks in rock masses that develop if forces in the crust exceed a certain critical point (below this critical point folding may occur). There are two categories of fractures: joints and faults. A joint is a crack along which no displacement has occurred. If there is relative displacement of rocks on both sides of the fracture and parallel to it, then the fracture is called a fault.

Free aquifer   Unconfined aquifer, phreatic aquifer.

Freshwater   Water of relatively low content of dissolved solids (usually less than 1 000 mg per litre).

Geyser   A hot spring that forcibly ejects hot water and steam into the air. The heat is thought to result from the contact of groundwater with magma bodies.

Geothermal   Related to the heat energy of the earth's crust.

Graben   Geological term indicating a downthrown block of the earth's crust, limited by parallel boundaries (known as 'normal faults').

Green water   Water originating from precipitation that does not run off or recharge groundwater, but is stored in the soil or temporarily stays on top of the soil or vegetation. Eventually, this part of precipitation evaporates or transpires through plants.

Grey water   Polluted water that results from uses of water other than for sanitary purposes.

Groundwater   Subsurface water in which the pressure is equal to or higher than the local atmospheric pressure. In other words: water below the water table or phreatic level.

Groundwater development, groundwater exploitation   Human activity consisting of abstracting groundwater and making it available for beneficial use.

Groundwater discharge   Outflow of water from a groundwater system (1. Phenomenon; 2. Flux).

Groundwater level   Elevation to which groundwater will or does rise in a piezometer connected to a point in the groundwater domain. It is a time-dependent variable, varies from point to point within the groundwater domain, and indicates the potential energy of groundwater in any point considered (in metres of water column relative to a selected topographic reference level).
Synonyms: piezometric level, piezometric head, hydraulic head, groundwater hydraulic potential.

Groundwater mining   Abstracting groundwater that will not be renewed anymore (fossil or non-renewable groundwater).

Groundwater recharge, groundwater renewal, groundwater replenishment
1. Process of inflow of water into a groundwater system or aquifer.
2. Flux related to this inflow process.

Groundwater runoff   Part of stream flow contributed by groundwater (by springs or by diffuse seepage into the stream bed).
See also: baseflow (almost a synonym).

Groundwater table   Surface defined by the phreatic levels in an aquifer (*i.e. surface of atmospheric pressure within an unconfined aquifer*).

Humid region   Region characterized by a surplus of precipitation over potential evapotranspiration. In other words: the available water in the area is sufficient to sustain the growth and development of plant life without irrigation.

Hydraulics, hydraulic   *Hydraulics* is the applied science dealing with the mechanics of flow of water in pipes, open watercourses and porous media. *Hydraulic* is the adjective.

Hydraulic conductivity   Capacity of a porous medium to transmit water.
Synonym: permeability.

Hydraulic gradient   Change in water level (piezometric level) per unit of horizontal distance.

Hydrograph   Graph of water level or flow rate versus time.

Hydrological cycle   The circulatory flow of water at or near the Earth's surface (in which water is moving between the subsystems atmosphere, surface water, unsaturated zone, groundwater and oceans).
Synonyms: Hydrologic cycle, water cycle, H_2O cycle.

Hydrological overabstraction   Abstraction from a renewable groundwater resource that is excessive with respect to the hydrological budget of the corresponding aquifer. The abstraction intensity is too high to allow a new dynamic equilibrium to be established in the intermediate or long term. *(Some groundwater specialists rank this under 'ground-water overexploitation', which may be confusing).*

Hydrology   The science of the terrestrial part of the water cycle.

Hydrosphere   The waters of the Earth, as distinct from the rocks (lithosphere), living things (biosphere) and the air (atmosphere)

Igneous rock   A rock formed by the solidification of a magma.

Impermeable, impervious   Incapable of transmitting water *(property of a porous medium).*

Infiltration

1. The process of water entering a porous or fissured formation.
2. The quantity or rate of infiltrating water.

Infiltration gallery   Man-made construction to tap groundwater and to conduct it to the surface without the need for external energy (the system is based on flow by gravity).
Synonyms: Foggara, qanat, karez, falaj, rhettara, ghail.

Intensive development of groundwater   Groundwater development to such a degree that the natural flow in the aquifer or aquifer system has been significantly changed (significant change of flow regime).

Interface   Contact surface between two distinct bodies, e.g. between saline and fresh groundwater. Depending on local conditions, the interface between saline and fresh groundwater may be rather sharp or consist of a thick transition zone (zone where fresh and groundwater are mixed).

Interstices (voids)   Open spaces within subsurface unconsolidated sediments and rock formations (i.e. the portion not occupied by solid matter), allowing fluids (air, water, oil, gas, etc.) to flow or be stored underground. Interstices may be subdivided into original or primary interstices (mainly pores) and secondary interstices that have originated after the rock was formed (mainly fissures, sometimes widened by dissolution). The corresponding sediment/rock categories are indicated as porous or fissured, respectively. The term porosity is used to indicate the percentage of the bulk volume of the rock that is occupied by interstices (pores and fissures combined).

Irrigation   Artificial application of surface water or groundwater to the land or soil, usually with the objective of promoting crop growth.

Joint   See Fracture.

Karst   Phenomenon that causes irregularities in land surface and drainage features (both internal and external) as a result of dissolution of solid rock, mainly limestone.

Land subsidence   Sinking elevation of the ground surface, which may result from compression of a sediment due to reduced water pressures, which may occur when water is drained or abstracted from subsurface strata, e.g. from an aquifer.

Lithology   1. Type of rock. 2. Branch of earth science dealing with the distinction and description of different types of rock and their physical properties.

Marine   Geological term, indicating an origin related to seas or oceans.

Marine regression   Sea-ward shift of the boundary between sea and land, lasting for a geologically significant period of time.

Marine transgression   Land-ward shift of the boundary between sea and land, lasting for a geologically significant period of time.

Mesozoic   Era in geological history, running from roughly 250 to 66 million years before present. Often also used as an adjective, to indicate the age of a geological formation.

Metamorphic rock   A rock whose original mineralogy, texture or composition has been changed due to effects of pressure, temperature, or the gain or loss of chemical components.

Mineral water   Natural water with a content of dissolved salts greater than a defined threshold and often considered to have therapeutic properties.

Mining Depleting the stock of a non-renewable natural resource.

Monitoring Systematically repeated observation of a certain variable or phenomenon.

Natural resource A natural resource is commonly defined as something (usually a substance or 'raw material') found in nature and necessary or useful to humans. Typical examples are solar energy (perpetual resource), forests and water (renewable resources), or fossil fuels and minerals (non-renewable resources).

Non-renewable groundwater A groundwater body or unit that under current climatic conditions receives no or negligible quantities of recharge.
Synonym: fossil groundwater

Overexploitation Intensive exploitation considered to be excessive. Among groundwater specialists there is no generally shared interpretation of groundwater overexploitation, and in particular it is often not explicitly specified with respect to what criterion the exploitation is considered excessive. Here, overexploitation is interpreted as intensive exploitation characterized by a less favourable balance between benefits and negative side effects than would have been the case at a lower rate of exploitation. *(See also: Hydrological overabstraction).*

Palaeozoic (also: Paleozoic) Era in geological history, running from roughly 590 to 250 million years before present. Often also used as an adjective, to indicate the age of a geological formation.

Permeable, pervious Capable of transmitting significant quantities of water *(property of a porous medium).*

Permeability Capacity of a porous medium to transmit water.
Synonym: hydraulic conductivity.

Permafrost Soil or rock layer in which the temperature has been continuously below 0°C for some years at least.

Phreatic aquifer ('water-table aquifer') Aquifer in which the upper boundary of the groundwater mass forms a surface (water table) that is in direct contact with the atmosphere. This condition favours the aquifer being actively involved in the water cycle.
Synonyms: water-table aquifer, unconfined aquifer, free aquifer.

Phreatic level Level in a phreatic aquifer where water pressure is exactly equal to the local atmospheric pressure. It is the level observed in a shallow well in such an aquifer.

Phreatophytes Water-loving plants supplied permanently or intermittently by groundwater through their roots reaching the capillary fringe.

Piedmont An area lying at the foot of a mountain or mountain range and formed by its erosion products.

Piezometer Device or well that is designed to measure the potential energy level in a point of a water body (stream, lake, aquifer, aquitard, etc.).

Piezometric level Elevation to which water will rise in a piezometer connected to a point in an aquifer.
Synonyms: potentiometric level, piezometric head, hydraulic head.

Piezometric surface Surface defined by the piezometric levels corresponding to a predefined layer or topographic level inside an aquifer.
Synonym: potentiometric surface.

Plateau Extensive relatively flat and horizontal upland area, usually higher than the surrounding area and bounded at least at one side by steep slopes.
Synonyms: meseta, altiplano (the latter usually restricted to a highland intermontane plateau).

Platform Extensive relatively flat and horizontal area, usually higher than the surrounding area.

Pores, porous, porosity See Interstices.

Potential evaporation, potential evapotranspiration See Evaporation and Evapotranspiration.

Precambrian Eon in geological history, running from roughly 4 000 to 590 million years before present. Often also used as an adjective, to indicate the age of a geological formation.

Precipitation
1. Rain, hail, snow, dew and other liquid or solid condensation products of water vapour falling from clouds or deposited from the air on the ground.
2. Amount of water produced by rain, hail, snow, etc. falling on a unit of horizontal surface per unit of time.

Quaternary   Most recent period in geological history, belonging to the Cenozoic Era. Often also used as an adjective, to indicate the age of a geological formation.
Synonym: Anthropogen.

Recharge area   Area where most or all of the recharge of an aquifer occurs.

Regolith   Any solid material lying on top of bedrock, which includes soil, alluvium, and rock fragments weathered from the bedrock.

Renewable groundwater   Body of groundwater that is replenished under current climate and geological conditions.

Reserves   Volume stored.

Resilience   Ability of a system to recover from an unsatisfactory state (in particular, the ability to remain in or return to a state of dynamic equilibrium).

Return flow   Any flow of abstracted water that returns after use to a stream channel or to the groundwater reserves.

River basin   Catchment area of a river.

Runoff   The part of precipitation that appears as stream flow. Often three components are distinguished: (a) *surface runoff*, resulting from overland flow during and after a storm or precipitation event, and caused by that event; (b) interflow, which is a delayed flow passing at least partly through the upper soil layers; and (c) *groundwater runoff*, which is the outflow of groundwater to streams (forming the essential part of baseflow).

Runoff coefficient   Ratio of runoff volume and the volume of rainfall during the event that caused runoff.

Safe yield   Flux of groundwater that can be withdrawn from an aquifer without causing undesirable side effects (*outdated term*).

Saline water   Water containing dissolved solids in concentrations of more than 10 000 milligram per litre.

Salinity   Concentration of dissolved salts in water.

Salt water intrusion, seawater intrusion   Invasion of salt water (usually seawater) into a body of freshwater (either a surface water or a groundwater body).

Saprolite   Section of the decomposed/weathered zone (regolith) of crystalline basement aquifers, rich in clay minerals. It has low hydraulic conductivity but significantly higher storage capacity than the more permeable fissured zone below it (saprock).

Saturated zone   Part of the subsurface water-bearing formation in which all interstices (voids), large and small, are filled with water.

Scale   Ratio of distance on a map compared to distance in the real world (e.g. 1: 50 000 or 1:1 000 000). For a given map size, a small-scale map may present a relatively large area, but with limited detail.

Secondary groundwater resources   Groundwater resources due to recharge in addition to the natural recharge, such as irrigation water losses or return flows (by-product) and artificial recharge.

Sediment   Any of a number of materials (usually erosion products) deposited by natural processes on the Earth's surface.

Seepage
1. Slow movement of water in a porous medium.
2. Loss of water as a result of water entering a porous medium (infiltration) or emerging from such a medium along a line or surface (exfiltration).

Semi-arid region   A region of low precipitation during a significant part of the year, characterized at least seasonally by a lack of available water, to the extent of hindering the growth and development of plant life, at least during the dry season. In agriculture this term is used to describe dry areas where land cultivation is impracticable or much less profitable without irrigation.

Semi-confined aquifer   Confined aquifer, where either the confining bed on top of the aquifer or the poorly permeable basal formation underneath the aquifer have sufficient permeability (aquitards) to allow the exchange of water between the aquifer and the domains above or below it.

Semi-permeable bed   Earth layer with limited capacity to transmit groundwater.
Synonym: aquitard.

Sinkhole   Depression in the ground surface in karst terrain, acting as point of disappearance of surface water flow. It usually develops into a funnel-shaped cavity at the ground surface by dissolution of karstified limestone or dolomite (doline).
Synonyms: ponor, swallow hole.

Soil moisture   Water present in soil pores above the water table.

Speleology   Science dealing with caves.

Spring   Location of concentrated natural groundwater outflow.

Storage, stored volume, stock   Volume of water present in a component of the local hydrological system (e.g. aquifer, stream, lake, soil).

Stream   Body of water flowing in a natural surface channel. Generic term that includes rivers, rivulets, brooks and other natural water courses.

Subsidence, land subsidence   Sinking elevation of the ground surface, due to natural processes or caused by human activities. It results from compression of a sediment due to reduced water pressures, which may occur – inter alia – when water is drained or abstracted from subsurface strata, e.g. from an aquifer or an aquitard.

Surface water   Water located on the surface of the Earth, such as in streams, rivers and lakes.

Sustainable yield   Flux of groundwater that can be withdrawn from an aquifer without causing undesirable side effects, in particular without causing a permanent state of non-equilibrium of the hydrological budget of an aquifer.
(see also: Safe yield, Hydrological overabstraction).

Terrestrial   Related to land masses (as opposed to marine and atmospheric).

Tertiary   Period in geological history, belonging to the Cainozoic Era. Often also used as an adjective, to indicate the age of a geological formation.

Texture, rock texture   Physical properties of a rock related to the size, shape and arrangement of the mineral constituents it is composed of.

Transboundary aquifer   An aquifer that spans two or more political entities, separated by political boundaries.

Unconfined aquifer   Phreatic aquifer.

Unconsolidated sediments   Sediments consisting of a matrix of loose, uncemented particles.

Unsaturated zone   Zone below ground surface where the interstices are partly filled with water and partly with air. This zone is located above the water table or above the top of a confined aquifer. The water pressure in the unsaturated zone is lower than atmospheric pressure.

Virtual water   Water used in the production of goods or services.
Synonym: embedded water.

Water budget   A specification of inflows, outflows and change of stored volume of water for a defined water system (river basin, aquifer, soil system, area) during a specified period. In principle, the change of stored volume equals the difference between inflows and outflows.
Synonym: water balance.

Water cycle   The circulatory flow of water at or near the Earth's surface (in which water moves between the subsystems atmosphere, surface water, unsaturated zone, groundwater and oceans).
Synonyms: hydrologic cycle, hydrological cycle, H_2O cycle.

Water footprint   The total volume of freshwater used to produce the goods and services consumed by an individual or community or produced by a business.

Well   A man-made construction – usually of vertical cylindrical form – used to get access to groundwater and abstract it by lifting it to the surface.

Wetland   An area where the ground is permanently or seasonally saturated with water (swamp, marsh, peatland, shallow lake).

Withdrawal

1. The process of taking water from a source, either temporarily or permanently.
2. The volume of water taken during this process (usually expressed as a volume per unit of time).

Synonym: abstraction.

Statistics on renewable water resources – by country

#	Country or territory	Area	Internal water resources renewal (AQUASTAT, 2009)				Comparison in mm/yr (averaged over total area)				Indicators	
			Total internal renewable water resources	Surface water produced internally	Groundwater produced internally	Overlap between groundwater and surface water	Precipitation (IGRAC, 2008 update)	Total internal renewable water resources	Groundwater produced internally	Diffuse natural groundwater recharge (Döll & Fiedler, 2008)	Total internal renewable water as a % of precipitation	Per capita renewable water resources (reference year 2010)
		km²	km³/yr	km³/yr	km³/yr	km³/yr	mm/yr	mm/yr	mm/yr	mm/yr	%	m³/yr
1	Afghanistan	647 500	55				327	85		10	26	1 751
2	Albania	28 748	26.9	23.05	6.2	2.35	1 485	936	216	191	63	8 396
3	Algeria	2 381 740	11.25	9.763	1.487	0	89	5	1	3	5	317
4	American Samoa	199										
5	Andorra	468					800					
6	Angola	1 246 700	148	145	58	55	1 010	119	47	105	12	7 756
7	Anguilla	102					1 105					
8	Antarctica											
9	Antigua and Barbuda	443	0.052				1 030	117			11	584
10	Argentina	2 766 890	276	276	128	128	591	100	46	53	17	6 830
11	Armenia	29 800	6.859	3.948	4.311	1.4	562	230	145	26	41	2 218
12	Aruba	193					500					
13	Australia	7 686 850	492	440	72	20	534	64	9	34	12	22 094
14	Austria	83 858	55	55	6	6	1 110	656	72	163	59	6 552
15	Azerbaijan	86 600	8.115	5.955	6.51	4.35	447	94	75	33	21	883
16	Bahamas	13 940	0.02				1 292	1		130	0	58

	Country or territory	Area	Internal water resources renewal (AQUASTAT, 2009)				Comparison in mm/yr (averaged over total area)				Indicators	
			Total internal renewable water resources	Surface water produced internally	Groundwater produced internally	Overlap between groundwater and surface water	Precipitation (IGRAC, 2008 update)	Total internal renewable water resources	Groundwater produced internally	Diffuse natural groundwater recharge (Döll & Fiedler, 2008)	Total internal renewable water as a % of precipitation	Per capita renewable water resources (reference year 2010)
		km²	km³/yr	km³/yr	km³/yr	km³/yr	mm/yr	mm/yr	mm/yr	mm/yr	%	m³/yr
17	Bahrain	665	0.004	0.004	0	0	83	6	0		7	3
18	Bangladesh	144 000	105	83.91	21.09	0	2 666	729	146	245	27	706
19	Barbados	431					1 422					
20	Belarus	207 600	37.2	37.2	18	18	618	179	87	95	29	3 877
21	Belgium	30 510	12	12	0.9	0.9	847	393	29	275	46	1 120
22	Belize	22 966	16				1 705	697		327	41	51 282
23	Benin	112 620	10.3	10	1.8	1.5	1 039	91	16	86	9	1 164
24	Bermuda	53.3					1 507					
25	Bhutan	47 000	78				2 200	1 660		77	75	107 438
26	Bolivia	1 098 580	303.5	303.5	130.9	130.9	1 146	276	119	146	24	30 564
27	Bosnia and Herzegovina	51 129	35.5				1 028	694		226	68	9 441
28	Botswana	600 370	2.4	0.8	1.7	0.1	416	4	3	19	1	1 196
29	Brazil	8 511 965	5 418	5 418	1 874	1 874	1 783	637	220	326	36	27 792
30	British Virgin Islands	153					1 346					
31	Brunei Darussalam	5 770	8.5	8.5	0.1	0.1	2 722	1 473	17		54	21 303
32	Bulgaria	110 910	2	20.1	6.4	5.5	608	189	58	77	31	2 802
33	Burkina Faso	274 200	12.5	8	9.5	5	748	46	35	39	6	759
34	Burundi	27 830	10.06	10.06	7.47	7.47	1 274	361	268	104	28	1 200
35	Cambodia	181 040	120.6	116	17.6	13	1 904	666	97	268	35	8 530
36	Cameroon	475 440	273	268	100	95	1 604	574	210	234	36	13 958
37	Canada	9 976 140	2 850	2 840	370	360	537	286	37	83	53	83 782
38	Cape Verde	4 033	0.3	0.181	0.124	0.005	228	74	31		33	605
39	Cayman Islands	262					1 260					
40	Central African Republic	622 984	141	141	56	56	1 343	226	90	148	17	32 038
41	Chad	1 284 000	15	13.5	11.5	10	322	12	9	30	4	1 336
42	Channel Islands	194					800					
43	Chile	756 950	884	884	140	140	1 522	1 168	185	81	77	51 654
44	China	9 596 960	2 813	2 712.1	828.8	727.9	627	293	86	69	47	2 097

(Continued)

	Country or territory	Area	Internal water resources renewal (AQUASTAT, 2009)				Comparison in mm/yr (averaged over total area)				Indicators	
			Total internal renewable water resources	Surface water produced internally	Groundwater produced internally	Overlap between groundwater and surface water	Precipitation (IGRAC, 2008 update)	Total internal renewable water resources	Groundwater produced internally	Diffuse natural groundwater recharge (Döll & Fiedler, 2008)	Total internal renewable water as a % of precipitation	Per capita renewable water resources (reference year 2010)
		km²	km³/yr	km³/yr	km³/yr	km³/yr	mm/yr	mm/yr	mm/yr	mm/yr	%	m³/yr
45	Colombia	1 138 910	2 112	2 112	510	510	2 612	1 854	448	446	71	45 620
46	Comoros	2 170	1.2	0.2	1	0	900	553	461		61	1 633
47	Congo	342 000	222	222	122	122	1 646	649	357	349	39	54 910
48	Cook Islands	240					1 750					
49	Costa Rica	51 100	112.4	75.1	37.3	0	2 926	2 200	730	353	75	24 125
50	Côte d'Ivoire	322 460	76.84	74	37.84	35	1 348	238	117	136	18	3 893
51	Croatia	56 542	37.7	27.2	11	0.5	1 113	667	195	240	60	8 562
52	Cuba	110 860	38.12	31.64	6.48	0	1 335	344	58	111	26	3 386
53	Cyprus	9 250	0.78	0.56	0.41	0.19	498	84	44		17	707
54	Czech Republic	78 866	13.15	13.15	1.43	1.43	677	167	18	92	25	1 253
55	Dem. People's Rep. of Korea	120 540	67	66	13	12	1 054	556	108	99	53	2 752
56	Democratic Republic of the Congo	2 345 410	900	899	421	420	1 543	384	179	247	25	13 643
57	Denmark	43 094	6	3.7	4.3	2	703	139	100	363	20	1 081
58	Djibouti	23 000	0.3	0.3	0.015	0.015	221	13	1	3	6	337
59	Dominica	754					2 083					
60	Dominican Republic	48 730	21	21	11.7	11.7	1 410	431	240	120	31	2 115
61	Ecuador	283 560	432	432	134	134	2 087	1 523	473	297	73	29 865
62	Egypt	1 001 450	1.8	0.5	1.3	0	51	2	1	1	4	22
63	El Salvador	21 040	17.75	17.6	6.15	6	1 724	844	292	230	49	2 866
64	Equatorial Guinea	28 051	26	25	10	9	2 156	927	356	394	43	37 143
65	Eritrea	121 320	2.8	2.7	0.5	0.4	384	23	4	6	6	533
66	Estonia	45 226	12.71	11.71	4	3	626	281	88	172	45	9 478
67	Ethiopia	1 127 127	122	120	20	18	848	108	18	39	13	1 471
68	Faeroe Islands	1 399					1 250					
69	Falkland Islands (Malvinas)	12 173								226		
70	Fiji	18 270	28.55				2 592	1 563		289	60	33 159

(Continued)

Country or territory	Area	Internal water resources renewal (AQUASTAT, 2009)				Comparison in mm/yr (averaged over total area)				Indicators	
		Total internal renewable water resources	Surface water produced internally	Groundwater produced internally	Overlap between groundwater and surface water	Precipitation (IGRAC, 2008 update)	Total internal renewable water resources	Groundwater produced internally	Diffuse natural groundwater recharge (Döll & Fiedler, 2008)	Total internal renewable water as a % of precipitation	Per capita renewable water resources (reference year 2010)
	km²	km³/yr	km³/yr	km³/yr	km³/yr	mm/yr	mm/yr	mm/yr	mm/yr	%	m³/yr
71 Finland	337 030	107	106.8	2.2	2	537	317	7	128	59	19 944
72 France	547 030	200	198	120	118	867	366	219	200	42	3 185
73 French Guiana	91 000	134				2 895	1 473		270	51	580 087
74 French Polynesia	4167										
75 Gabon	267 667	164	162	62	60	1 831	613	232	323	33	108 970
76 Gambia	11 300	3	3	0.5	0.5	836	265	44	101	32	1 736
77 Georgia	69 700	58.23	57	17.23	16	1 026	835	247	137	81	13 380
78 Germany	357 021	107	106.3	45.7	45	700	300	128	201	43	1 300
79 Ghana	239 460	30.3	29	26.3	25	1 187	127	110	105	11	1 242
80 Gibraltar	6.5					800					
81 Greece	131 940	58	55.5	10.3	7.8	652	440	78	91	67	5 106
82 Greenland	2 166 086					350			7		
83 Grenada	344					2 350					
84 Guadeloupe	1 780					2 470					
85 Guam	549										
86 Guatemala	108 890	109.2	100.7	33.7	25.2	1 996	1 003	309	296	50	7 589
87 Guinea	245 857	226	226	38	38	1 651	919	155	212	56	22 641
88 Guinea-Bissau	36 120	16	12	14	10	1 577	443	388	205	28	10 561
89 Guyana	214 970	241	241	103	103	2 387	1 121	479	365	47	319 629
90 Haiti	27 750	13.01	10.853	2.157	0	1 440	469	78	100	33	1 302
91 Holy See	0.44					800					
92 Honduras	112 090	95.93	86.92	39	29.99	1 976	856	348	232	43	12 621
93 Hong Kong	1 092					1 750					
94 Hungary	93 030	6	6	6	6	589	64	64	73	11	601
95 Iceland	103 000	170	166	24	20	1 940	1 650	233	306	85	531 250
96 India	3 287 590	1 446	1 404	432	390	1 083	440	131	93	41	1 181
97 Indonesia	1 919 440	2 019	1 973.3	457.4	411.7	2 702	1 052	238	442	39	8 417
98 Iran (Islamic Republic of)	1 648 000	128.5	97.3	49.3	18.1	228	78	30	15	34	1737
99 Iraq	437 072	35.2	34	3.2	2	216	81	7	18	37	1111
100 Ireland	70 280	49	48.2	10.8	10	1 118	697	154	373	62	10 962
101 Isle of Man	572					2 631					
102 Israel	20 770	0.75	0.25	0.5	0	435	36	24	42	8	101
103 Italy	301 230	182.5	170.5	43	31	832	606	143	147	73	3 014

(Continued)

Country or territory	Area	Internal water resources renewal (AQUASTAT, 2009)				Comparison in mm/yr (averaged over total area)				Indicators	
		Total internal renewable water resources	Surface water produced internally	Groundwater produced internally	Overlap between groundwater and surface water	Precipitation (IGRAC, 2008 update)	Total internal renewable water resources	Groundwater produced internally	Diffuse natural groundwater recharge (Döll & Fiedler, 2008)	Total internal renewable water as a % of precipitation	Per capita renewable water resources (reference year 2010)
	km²	km³/yr	km³/yr	km³/yr	km³/yr	mm/yr	mm/yr	mm/yr	mm/yr	%	m³/yr
104 Jamaica	10 991	9.404	5.512	3.892	0	2 051	856	354	137	42	3 431
105 Japan	377 835	430	420	27	17	1 668	1 138	71	279	68	3 398
106 Jordan	92 300	0.682	0.485	0.45	0.253	111	7	5	12	7	110
107 Kazakhstan	2 717 300	75.42	69.32	6.1	0	250	28	2	10	11	4 706
108 Kenya	582 650	20.7	20.2	3.5	3	630	36	6	46	6	511
109 Kiribati	811										
110 Kuwait	17 820	0	0	0	0	121	0	0	1	0	0
111 Kyrgyzstan	198 500	48.95	46.46	13.69	11.2	533	247	69	11	46	9 177
112 Lao People's Democratic Republic	236 800	190.4	190.4	37.9	37.9	1 834	804	160	236	44	30 705
113 Latvia	64 589	16.74	16.54	2.2	2	641	259	34	159	40	7 433
114 Lebanon	10 400	4.8	4.1	3.2	2.5	661	462	308	100	70	1 135
115 Lesotho	30 355	5.23	5.23	0.5	0.5	788	172	16	16	22	2 409
116 Liberia	111 370	200	200	45	45	2 391	1 796	404	419	75	50 075
117 Libyan Arab Jamahiriya	1 759 540	0.6	0.2	0.5	0.1	56	0	0	2	1	94
118 Liechtenstein	160					1 250					
119 Lithuania	65 200	15.56	15.36	1.2	1	656	239	18	153	36	4 681
120 Luxembourg	2 586	1	1	0.08	0.08	934	387	31		41	1 972
121 Macao Special Admin. Region of China	25.4					1 750					
122 Madagascar	587 040	337	332	55	50	1 513	574	94	216	38	16 269
123 Malawi	118 480	16.14	16.14	2.5	2.5	1 181	136	21	164	12	1 083
124 Malaysia	329 750	580	566	64	50	2 875	1 759	194	481	61	20 422
125 Maldives	300	0.03	0	0.03	0	1 972	100	100		5	95
126 Mali	1 240 000	60	50	20	10	282	48	16	22	17	3 904
127 Malta	316	0.0505	0.0005	0.05	0	560	160	158		29	121
128 Marshall Islands	181.3					92					
129 Martinique	1 100					2 041					
130 Mauritania	1 030 700	0.4	0.1	0.3	0	92	0	0	4	0	116
131 Mauritius	2 040	2.751	2.358	0.893	0.5	2 041	1 349	438		66	2 118

(Continued)

Country or territory	Area	Internal water resources renewal (AQUASTAT, 2009)				Comparison in mm/yr (averaged over total area)				Indicators	
		Total internal renewable water resources	Surface water produced internally	Groundwater produced internally	Overlap between groundwater and surface water	Precipitation (IGRAC, 2008 update)	Total internal renewable water resources	Groundwater produced internally	Diffuse natural groundwater recharge (Döll & Fiedler, 2008)	Total internal renewable water as a % of precipitation	Per capita renewable water resources (reference year 2010)
	km²	km³/yr	km³/yr	km³/yr	km³/yr	mm/yr	mm/yr	mm/yr	mm/yr	%	m³/yr
132 Mayotte	374					1 750					
133 Mexico	1 972 550	409	361	139	91	752	207	70	51	28	3 606
134 Micronesia (Federated States of)	702										
135 Monaco	1.95					800					
136 Mongolia	1 565 000	34.8	32.7	6.1	4	241	22	4	2	9	12 627
137 Montenegro	14 026								118		
138 Montserrat	102										
139 Morocco	446 550	29	22	10	3	346	65	22	17	19	908
140 Mozambique	801 590	100.3	97.3	17	14	1 032	125	21	104	12	4 288
141 Myanmar	678 500	100 3	992.3	453.7	443	2 092	1 478	669	226	71	20 912
142 Namibia	825 418	6.16	4.1	2.1	0.04	284	7	3	10	3	2 698
143 Nauru	21					2 126					
144 Nepal	140 800	198.2	198.2	20	20	1 500	1 408	142	135	94	6 616
145 Netherlands	41 526	11	11	4.5	4.5	778	265	108	355	34	662
146 Netherlands Antilles	960										
147 New Caledonia	190 60								146		
148 New Zealand	268 680	327				1 732	1 217		334	70	74 863
149 Nicaragua	129 494	189.7	185.7	59	55	2 391	1 465	456	337	61	32 775
150 Niger	1 267 000	3.5	1	2.5	0	151	3	2	12	2	226
151 Nigeria	923 768	221	214	87	80	1 150	239	94	163	21	1 395
152 Niue	260					2 041					
153 Norfolk Island	34.6										
154 Northern Mariana Islands	477										
155 Norway	324 220	382	376	96	90	1 414	1 178	296	215	83	78 231
156 Occupied Palestinian Territory	6 220	0.812	0.072	0.74	0		131	119			201
157 Oman	212 460	1.4	1.05	1.3	0.95	125	7	6	3	5	503
158 Pakistan	803 940	55	47.4	55	47.4	494	68	68	12	14	317
159 Palau	458										

(Continued)

Country or territory	Area	Internal water resources renewal (AQUASTAT, 2009)				Comparison in mm/yr (averaged over total area)				Indicators	
		Total internal renewable water resources	Surface water produced internally	Groundwater produced internally	Overlap between groundwater and surface water	Precipitation (IGRAC, 2008 update)	Total internal renewable water resources	Groundwater produced internally	Diffuse natural groundwater recharge (Döll & Fiedler, 2008)	Total internal renewable water as a % of precipitation	Per capita renewable water resources (reference year 2010)
	km²	km³/yr	km³/yr	km³/yr	km³/yr	mm/yr	mm/yr	mm/yr	mm/yr	%	m³/yr
160 Panama	78 200	147.4	144.09	21	17.69	2 692	1 885	269	325	70	41 911
161 Papua New Guinea	462 840	801				3 142	1 731		458	55	116 798
162 Paraguay	406 750	94	94	41	41	1 130	231	101	113	20	14 562
163 Peru	1 285 220	1 616	1 616	303	303	1 738	1 257	236	280	72	55 577
164 Philippines	300 000	479	444	180	145	2 348	1 597	600	263	68	5 136
165 Pitcairn	47										
166 Poland	312 685	53.6	53.1	12.5	12	600	171	40	144	29	1 400
167 Portugal	92 391	38	38	4	4	855	411	43	142	48	3 559
168 Puerto Rico	9 104	7.1				2 054	780			38	1 894
169 Qatar	11 437	0.056	0	0.056	0	75	5	5	1	7	32
170 Republic of Korea	98 480	64.85	62.25	13.3	10.7	1 274	659	135		52	1 346
171 Republic of Moldova	33 843	1	1	0.4	0.4	450	30	12	42	7	280
172 Réunion	2 517					3 000					
173 Romania	237 500	42.3	42	8.3	8	637	178	35	93	28	1 969
174 Russian Federation	17 075 200	4 313	4 037	788	512	460	253	46	54	55	30 164
175 Rwanda	26 338	9.5	9.5	7	7	1 212	361	266	68	30	894
176 Saint Helena	410					763					
177 Saint Kitts and Nevis	261	0.024	0.004	0.02	0	1 427	92	77		6	462
178 Saint Lucia	616					2 301					
179 Saint Pierre and Miquelon	242										
180 Saint Vincent and the Grenadines	389					1 583					
181 Samoa	2 944					3 072					
182 San Marino	61.2					800					
183 Sao Tome and Principe	1 001	2.18				3 200	2 178			68	13 212
184 Saudi Arabia	1 960 582	2.4	2.2	2.2	2	59	1	1	1	2	87
185 Senegal	196 190	25.8	23.8	3.5	1.5	687	132	18	63	19	2 075
186 Serbia	77 474					795			118		

(Continued)

Country or territory	Area km²	Internal water resources renewal (AQUASTAT, 2009) — Total internal renewable water resources km³/yr	Surface water produced internally km³/yr	Groundwater produced internally km³/yr	Overlap between groundwater and surface water km³/yr	Comparison in mm/yr (averaged over total area) — Precipitation (IGRAC, 2008 update) mm/yr	Total internal renewable water resources mm/yr	Groundwater produced internally mm/yr	Diffuse natural groundwater recharge (Döll & Fiedler, 2008) mm/yr	Indicators — Total internal renewable water as a % of precipitation %	Per capita renewable water resources (reference year 2010) m³/yr
187 Seychelles	455					2 330					
188 Sierra Leone	71 740	160	150	25	15	2 526	2 230	348	394	88	27 267
189 Singapore	692.7	0.6				2 497	866			35	118
190 Slovakia	48 845	12.6	12.6	1.73	1.73	824	258	35	116	31	2 307
191 Slovenia	20 273	18.67	18.52	13.5	13.35	1 162	921	666	272	79	9 197
192 Solomon Islands	28 450	44.7				3 028	1 571		419	52	83 086
193 Somalia	637 657	6	5.7	3.3	3	282	9	5	10	3	643
194 South Africa	1 219 912	44.8	43	4.8	3	495	37	4	14	7	894
195 Spain	505 988	111.2	109.5	29.9	28.2		220	59	70	35	2 413
196 Sri Lanka	65 610	52.8	52	7.8	7	1 712	805	119	165	47	2 531
197 Sudan and South Sudan	2 505 810	30	28	7	5	416	12	3	22	3	689
198 Suriname	163 270	88	88	80	80	2 331	539	490	290	23	167 619
199 Svalbard and Jan Mayen	62 422								7		
200 Swaziland	17 363	2.64	2.64	0.66	0.66	788	152	38	38	19	2 226
201 Sweden	449 964	171	170	20	19	624	380	44	142	61	18 230
202 Switzerland	41 290	40.4	40.4	2.5	2.5	1 537	978	61	228	64	5 271
203 Syrian Arab Republic	185 180	7.132	4.288	4.844	2	252	39	26	31	15	349
204 Taiwan Province of China	35 980					2 515					
205 Tajikistan	143 100	66.3	63.3	6	3	691	463	42	35	67	9 638
206 Thailand	514 000	224.5	213.29	41.9	30.69	1 622	437	82	178	27	3 248
207 The former Yugoslav Republic of Macedonia	25 333	5.4				619	213		62	34	2 620
208 Timor-Leste	15 007	8.215	8.129	0.886	0.8		547	59	93		7 309
209 Togo	56 785	11.5	10.8	5.7	5	1 168	203	100	131	17	1 908
210 Tokelau	10										
211 Tonga	748					1 966					
212 Trinidad and Tobago	5 128	3.84				2 200	749			34	2 864
213 Tunisia	163 610	4.195				207	26		18	12	400

(Continued)

Country or territory	Area	Internal water resources renewal (AQUASTAT, 2009)				Comparison in mm/yr (averaged over total area)				Indicators	
		Total internal renewable water resources	Surface water produced internally	Groundwater produced internally	Overlap between groundwater and surface water	Precipitation (IGRAC, 2008 update)	Total internal renewable water resources	Groundwater produced internally	Diffuse natural groundwater recharge (Döll & Fiedler, 2008)	Total internal renewable water as a % of precipitation	Per capita renewable water resources (reference year 2010)
	km²	km³/yr	km³/yr	km³/yr	km³/yr	mm/yr	mm/yr	mm/yr	mm/yr	%	m³/yr
214 Turkey	780 580	227	186	69	28	593	291	88	56	49	3 120
215 Turkmenistan	488 100	1.36	1	0.36	0	161	3	1	2	2	270
216 Turks and Caicos Islands	430										
217 Tuvalu	26					3 000					
218 Uganda	236 040	39	39	29	29	1 180	165	123	95	14	1 167
219 Ukraine	603 700	53.1	50.1	20	17	565	88	33	50	16	1 168
220 United Arab Emirates	82 880	0.15	0.15	0.12	0.12	78	2	1	3	2	20
221 United Kingdom	244 820	145	144.2	9.8	9	1 220	592	40	339	49	2 337
222 United Republic of Tanzania	945 087	84	80	30	26	1 071	89	32	93	8	1 873
223 United States of America	9 629 091	2 818	2 662	1 383	1 227	715	293	144	88	41	9 079
224 United States Virgin Islands	352										
225 Uruguay	176 220	59	59	23	23	1 265	335	131	175	26	17 513
226 Uzbekistan	447 400	16.34	9.54	8.8	2	206	37	20	8	18	595
227 Vanuatu	12 200								359		
228 Venezuela	912 050	722.4	700.1	227	204.7	1 875	792	249	282	42	24 928
229 Viet Nam	329 560	359.4	322.98	71.42	35	1 821	1 091	217	186	60	4 091
230 Wallis and Futuna Islands	274					2 500					
231 Western Sahara	266 000								0		
232 Yemen	527 970	2.1	2	1.5	1.4	167	4	3	3	2	87
233 Zambia	752 614	80.2	80.2	47	47	1 020	107	62	108	10	6 127
234 Zimbabwe	390 580	12.26	11.26	6	5	657	31	15	32	5	975

Specification of information sources:

Column	Variable	Source
3	Area	GGIS Database IGRAC, update 2008.
4–7	Internal water resources renewal	FAO's AQUASTAT database, update 2009.
8	Precipitation	GGIS Database IGRAC, update 2008.
9	TRWR in mm	Calculated by dividing values in column 4 by those in column 3 and multiplying the result by 1 000 000.
10	Renewable groundwater in mm	Calculated by dividing values in column 6 by those in column 3 and multiplying the result by 1 000 000.
11	Diffuse recharge	Taken from: Döll, P., & K. Fiedler, 2008. Global-scale modelling of groundwater recharge. *Hydrol. Earth Syst. Sci.*, Vol.12, pp. 863–885.
12	TRWR as a% of precipitation	Calculated by dividing values in column 9 by those in column 8.
13	Renewable water resources per capita	Calculated by dividing values in column 4 by total population as specified for 2010 by UNDESA (2011) and shown in Appendix 5 (http://esa.un.org/unpd/wpp/Excel-Data/population.htm).

All estimated fluxes are long-term mean values.

Appendix 3

Some data on the world's mega aquifer systems

(for locations, see the map of Figure 3.5 in Chapter 3)

N°	Name	Countries	Extent (x 1 000 km²)	Hydrogeological setting (P = maximum thickness)	Theoretical reserves (x 1 000 km³)	Mean rate of recharge (km³/year)	References (see list of references at the end of Chapter 3)
1	Nubian Aquifer System (NAS) (including Nubian Sandstone Aquifer System (NSAS) and Post-Nubian Aquifer System (PNAS))	Egypt Libya Sudan Chad	2 199 (of which 1 800 freshwater)	Cambro-Ordovician to Oligocene multi-layer sequence dipping north, mainly consisting of continental sandstones; P = 3 500 m	542 freshwater	~1	Bakhbakhi 2002 UNESCO/ OSS 2005 CEDARE/ IFAD 2002 IME 2008
2	North Western Sahara Aquifer System (NWSAS)	Algeria Libya Tunisia	1 019	Cambro-Ordovician to Miocene multi-layer sequence: superposition of the *Continental Intercalaire* (CI – the deeper series; mainly sandstones) and the *Complexe Terminal* (CT; carbonates and detritic series); P = 1 600 m	60	~1	OSS 2003 UNESCO/ OSS 2005 IME 2008
3	Murzuk–Djado Basin	Libya Algeria Niger	450	Cambro-Ordovician to Cretaceous multi-layer sequence, mainly sandstones (two superimposed aquifer units); P = 2 500 m	4.8 in Libya	~0.15	UNESCO/ OSS 2005 IME 2008
4	Taoudeni–Tanezrouft Basin	Algeria Mauritania Mali	2 000	Infra-Cambrian to Tertiary multi-layer series, incl. *Continental Intercalaire* (CI) and *Complexe Terminal* (CT); P= 4 000 m	0.018 exploitable in Mali and in Mauritania	0.3 in Mali	UNESCO/ OSS 2005 IME 2008
5	Senegalo–Mauritanian Basin	Mauritania Senegal Gambia Guinea-Bissau	300	Multilayer aquifer system: main units are Maestrichtian sands (SEN), CI (MRT) and Oligocene-Miocene deposits (SEN, MRT, GNB); P = 500 m	1.5	~9	Diagana 1997 UNESCO/ OSS 2005 IME 2008

N°	Name	Countries	Extent (× 1 000 km²)	Hydrogeological setting (P = maximum thickness)	Theoretical reserves (× 1 000 km³)	Mean rate of recharge (km³/year)	References (see list of references at the end of Chapter 3)
6	Iullemeden–Irhazer Aquifer System (IAS)	Niger Algeria, Mali Nigeria	635	Multilayer system of Cambro-Ordovician to Eocene sandstones and sands (CI and CT), in three sub-basins; P = 1 500 m	10 to 15		Dodo 1992 UNESCO/OSS 2005
7	Lake Chad Basin	Niger Nigeria, Chad Cameroun C.A.R.	1 917	Multi-layer system, including Continental Intercalaire, Complex Terminal and Plio – Quaternary sediments in a closed basin; P = 7 000 m	0.6 in Niger (~ 0.4 exploitable in Chad)	3.6 in Niger	UNESCO/OSS 2005 Schneider 2001
8	Sudd Basin–Umm Ruwaba Aquifer	Sudan Ethiopia	365	Multi-layer system of Neogene - Quaternary sediments in basin structure; P = 3 000 m	0.11	0.34	ICID 1983 UN 1987 Safar-Zitoun 1993a
9	Ogaden–Juba Basin	Ethiopia Somalia Kenya	~ 1 000	Multi-layer complex of confined and unconfined aquifers; P = 12 000 m		~ 10	UN 1987
10	Congo Basin	D.R.Congo Congo Angola C.A.R. Gabon	750	Multi-layer system of Mesozoic sandstones ('Karoo'), a Tertiary-Quaternary sand/gravel/sandstone aquifer and overlying alluvial sediments. P = 3 500 m		~ 100	Zektser & Everett 2004 Safar-Zitoun 1993b
11	Cuvelai–Upper Zambezi Basin (Upper Kalahari)	Angola Botswana Namibia Zambia Zimbabwe	~ 700	Multi-layer Cretaceous to Neogene carboniferous aquifer ('Karoo')		~ 30 to 60	Ndengu 2004 Zektser 2004
12	Stampriet–Kalahari Basin (Lower Kalahari)	South Africa Botswana Namibia	~ 350	Multi-layer 'Karoo' system (sandstones, basalts) and overlying Neogene Kalahari sands		~ 1 to 2	Ndengu 2004 Zektser 2004

No	Name	Country		Description			Reference
13	Karoo Basin	South Africa	600	Multi-layer system composed of Cambrian to Lower Carboniferous pre-Karoo aquifers and Upper Carboniferous to Jurassic indurated Karoo formations; P = 7 000 m	3 to 5 (dolomites)	16 to 37	Safar-Zitoun 1993b Zektser 2004
14	Northern Great Plains Aquifer	Canada USA	~ 2 000	Multi-layer aquifer in synclinal basin, with mainly Palaeozoic carbonate rocks and sandstones of Cretaceous to Eocene age			UN 1976 USGS 2003, 2011
15	Cambrian–Ordovician Aquifer System (Central Region)	USA	250	Multi-layer sequence of Cambrian–Ordovician rocks, mainly marine sandstones and carbonates			USGS 2003, 2011
16	Central Valley, California	USA	80	Quaternary alluvial multilayer aquifer system, unconfined and confined; P = 600 m	1.13	~ 7	USGS 2003, 2011
17	High Plains Aquifer (Ogallala Aquifer)	USA	450	Unconsolidated Tertiary alluvial deposits (Ogallala) form the main aquifer; P = 150 m	~ 15	6 to 8	USGS 2003, 2011
18	Atlantic and Gulf Coastal Plains Aquifers	USA Mexico	1 150	Multi-layer complex of Palaeozoic to Tertiary adjoining aquifers; P = 12 000 m			USGS 2003, 2011
19	Amazon Basin	Brazil Colombia Peru, Bolivia	1 500	Multi-layer Palaeozoic to Cainozoic sediment series. Cainozoic strata form the most important aquifer units. P = 7 000 m	32.5		Rebouças 1988
20	Maranhão Basin	Brazil	700	Multi-layer system of Ordovician to Cretaceous sediments, dipping towards the centre; P = 8 000 m	17.5	4	Rebouças 1976, 1988

(Continued)

N°	Name	Countries	Extent (x 1 000 km²)	Hydrogeological setting (P = maximum thickness)	Theoretical reserves (x 1 000 km³)	Mean rate of recharge (km³/year)	References (see list of references at the end of Chapter 3)
21	Guaraní Aquifer System (also called Mercosur Aquifer System or Paraná sedimentary basin)	Brazil Argentina Paraguay Uruguay	1 195	Multi-layer system of Silurian to Cretaceous sediments dipping gently towards the centre of the intercratonic basin. Most important is the Triassic-Jurassic confined Botacu or Guaraní sandstone; P = 800 m (total basin: 5 000 m)	57	234	Rebouças 1976, 1988 Araújo et al. 1999
22	Arabian Platform Aquifer System	Saudi Arabia Jordan, Kuwait Bahrain, Qatar	1 485 in Saudi Arabia	Multi-layer Cambrian to Neogene sediment complex, dipping North-Eastward; P = 6 500 m	2 185 in Saudi Arabia	1.2	UN 1982 Abderrahman 2002
23	Indus Basin	Pakistan	~ 320	Unconfined alluvial aquifer; P = 300 m		66 (potentially usable)	ICID 2000
24	Indus – Ganges– Brahmaputra Basin (Bhabhar and Tarai Aquifers)	India, Nepal Bangladesh	~ 600	Unconfined and confined alluvial aquifer; P = 600 m			
25	West Siberian Basin	Russia	3 200	Multi-layer series of Palaeozoic to Cretaceous sediments, partly sub-permafrost; P = 6 000 m. Fresh and saline groundwater	1 000		UN 1986 Zektser & Everett 2004
26	Tunguss Basin	Russia	1 000	Multi-layer series of Cambrian to Triassic sediments, sub-permafrost; P = 4 000 m. Fresh and saline groundwater			UN 1986 Zektser & Everett 2004
27	Angara–Lena Basin	Russia	600	Multi-layer complex of Cambrian to Jurassic sediments; P = 3 000 m Fresh and saline groundwater		12.6	UN 1986 Zektser & Everett 2004

No.	Name	Country	Description			Reference
28	Yakut Basin	Russia	Multilayer complex of Upper-Cambrian to Cretaceous sediments, sub-permafrost; P = 12 000 m. Fresh and saline groundwater	720	18	UN 1986 Zektser & Everett 2004
29	North China Plain Aquifer System (Huang Huai Hai Plain)	China	Quaternary alluvial multi-layer aquifer system, unconfined and confined; contains artesian zones; P = 1 000 m	320	0.018	49.2 (of which 48.1 exploitable)	Wang et al. 2000 Zektser & Everett 2004 Foster & Garduño 2004
30	Song Liao Plain	China	Quaternary alluvial multilayer aquifer, unconfined and confined; contains artesian zones; P = 300 m	311	33.4 (23.8 exploitable)	Zektser & Everett 2004
31	Tarim Basin	China	Quaternary alluvial multilayer aquifer, unconfined and confined, in endorheic basin; P = 1 200 m	520		
32	Paris Basin	France	Multi-layer system of Triassic to Neogene sediments, dipping towards the centre of the basin; P = 3 200 m	190	0.5 to 1	20 to 30	UN 1990
33	Russian Platform Basins	Russia	Multi-layer complex of adjacent Infra-Cambrian to Quaternary aquifers; several artesian sub-basins (e.g. Moscow basin and Baltic basin); P = 20 000 m sub-permafrost in the North.	~ 3 100		UN 1990
34	North-Caucasus Basin	Russia	Multi-layer complex of Carboniferous to Neogene sediments; P = 10 000 m	230		UN 1990

(Continued)

N°	Name	Countries	Extent (x 1 000 km²)	Hydrogeological setting (P = maximum thickness)	Theoretical reserves (x 1 000 km³)	Mean rate of recharge (km³/year)	References (see list of references at the end of Chapter 3)
35	Pechora Basin	Russia	350	Multilayer complex of Ordovician to Tertiary sediments; P = 3 000 m; partly sub-permafrost			UN 1990
36	Great Artesian Basin	Australia	1 700	Multi-layer complex of Triassic to Cretaceous sediments; P = 3 000 m	20 8.7	1.1	Habermehl 2002 Anonymous 2003
37	Canning Basin	Australia	430	Multi-layer complex of Devonian to Cretaceous sediments; P = 1 000 m			Lau et al. 1987 Zektser & Everett 2004

Additional references on these mega aquifers:
See Appendix 6.3

Brief description of the global groundwater regions

Global groundwater regions were delineated by IGRAC for the first time in 2004. The main purpose is to help the understanding and memorising of major features of the hydrogeological setting at a global scale. After a few adjustments in 2005, each of the 36 global groundwater regions was subdivided into a number of groundwater provinces. The resulting 217 groundwater provinces are shown in the maps presented below. The principal criterion for delineating both the global groundwater regions and the groundwater provinces is the predominance of a certain overall hydrogeological setting that contrasts with neighbouring units. In the case of groundwater provinces, already existing delineations and names (e.g. in the USA, Australia and South America) have been respected, in order to be consistent with national and local practices. In general, the global groundwater regions can be broadly subdivided into four categories: basement regions (B), sedimentary basin regions (S), high-relief folded mountain regions (M) and volcanic regions (V). The description below builds upon earlier summaries by IGRAC.

4.1 NORTH AND CENTRAL AMERICA AND THE CARIBBEAN

4.1.1 Global Groundwater Region 1: Western mountain belt of North and Central America (M)

This region, the westernmost zone of the continent, includes folded areas at high elevation of North and Central America, belonging to the Pacific and Cordilleran Belts. Sedimentary and metamorphic rocks underlie the region. After these belts were formed, extensive less deformed basins dominated by volcanic rocks superimposed them. There are large variations in climate: above-average rainfall in mountain zones, low rainfall in arid zones, permafrost conditions in the North, and tropical temperature regimes in the South.

The region is subdivided into the following groundwater provinces:

1.01 Alaska
1.02 Cordilleran Orogen of Canada
1.03 Pacific Mountain System
1.04 Columbia Plateau
1.05 Basin and Range
1.06 Colorado Plateau
1.07 Rocky Mountains System
1.08 Central American ranges (including Mexican Sierras)

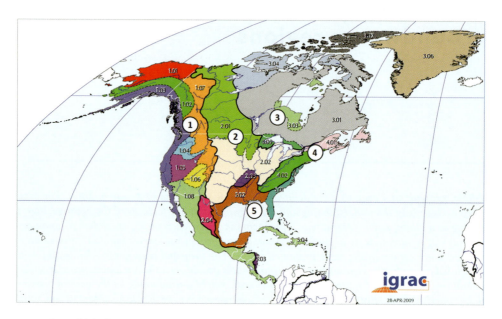

Figure A4.1 Groundwater provinces of North and Central America and the Caribbean.

Groundwater resources are variable. In the mountains they are associated with glacial and fluvial aquifers in down faulted troughs (e.g. Central Valley aquifer in California) or with intermontane basins. The volcanic regions in Central America contain some of the most productive aquifers. Coastal aquifers are present in structural basins filled with marine and alluvial sediments.

4.1.2 Global Groundwater Region 2: Central plains of North and Central America (S)

This region is located in the heart of North America and includes topographically low areas with gently rolling to flat topography. Except at its Northern edge, there is no boundary with the sea. Very thick sequences of sediments deposited in marine, alluvial, glacial and aeolian environments cover the Precambrian Basement and form huge reservoirs. The climate is predominantly dry. Permafrost is present in the North.

The region is subdivided into the following groundwater provinces:

2.01 Interior Platform of Canada
2.02 Interior Plains of USA
2.03 Interior Highlands
2.04 Sierra Madre Oriental

Groundwater resources are abundant and the region is characterised by large volumes of stored groundwater. Large regional aquifers occur in porous or fractured consolidated sediments and in large alluvial areas, with the High Plains and the Northern Great Plains aquifer systems as prominent examples. There are numerous aquifers in glacial or fluvio-glacial sediments in the Northern parts of the region.

4.1.3   Global Groundwater Region 3: Canadian shield (B)

This region includes the low to moderately elevated areas of the Canadian Shield. The region's topography results mainly from the effects of glacial action. Precambrian crystalline rocks cover most of the region. Locally, remains of sedimentary covers (mainly limestone) are found. The region receives moderate precipitation, often in the form of snow. Continuous permafrost is present in the Northern half of the region.

The region is subdivided into the following groundwater provinces:

3.01   Seven geological provinces of the Canadian shield
3.02   Innuitian Orogen
3.03   Hudson Bay Lowlands
3.04   Arctic Platform
3.05   St. Lawrence Platform (including Laurentian Platform of USA)
3.06   Greenland

Groundwater resources are limited. Presence of exploitable groundwater is restricted to local pockets of weathered or fractured consolidated rocks or to shallow aquifers consisting of fluvial and glacial sediments.

4.1.4   Global Groundwater Region 4: Appalachian highlands (M)

This region includes the high-elevation Eastern zone of North America that belongs to the Appalachian Orogenic belt. Metamorphic and igneous rocks are present. After the belt was deformed, extensive less deformed basins dominated by sedimentary rocks were superimposed upon it. Climate is predominantly humid.

The region is subdivided into the following groundwater provinces:

4.01   Appalachian Orogen in Canada
4.02   Appalachian Highlands in USA

Groundwater resources are variable. The principal aquifers are found in carbonate rocks and sandstones (e.g. Valley and Ridge aquifers). The high-elevation areas have limited ground-water resources, with local shallow aquifer systems in sand and gravel deposits of glacial and alluvial origin.

4.1.5   Global Groundwater Region 5: Caribbean islands and coastal plains of North and Central America (S)

This region includes the low-lying, flat areas along the Atlantic coast and the Gulf of Mexico, associated with the Ouachita tectonic depression. The coastal plains are huge wedges of uncon-solidated sedimentary rocks that thicken towards the sea and overly consolidated limestone and sandstones. The sediments are of fluvial, deltaic and shallow marine origin. The Caribbean island arc (several thousand islands) is a partially submerged cordillera, having a nucleus of igneous rocks overlain by sediments and volcanics. The region receives abundant precipitation.

The region is subdivided into the following groundwater provinces:

5.01   Atlantic Plains (incl. Florida Peninsula)
5.02   Mexican Gulf Plains (incl. Yucatan Peninsula)
5.03   Caribbean Plains
5.04   Caribbean Islands Arc

Groundwater resources are abundant. Large regional groundwater systems are developed in unconsolidated and (semi-) consolidated sediments (e.g. the Coastal lowlands aquifer system and Mississippi River Valley aquifer system). Karst aquifers are frequently found, e.g. the Floridan and Yucatan aquifer systems. Large islands have combined unconsolidated (alluvial), carbonate and volcanic aquifers. Seawater intrusion is a major problem, e.g. in the Biscayne aquifer (Florida) and on islands with shallow freshwater lenses.

4.2 SOUTH AMERICA

4.2.1 Global Groundwater Region 6: Andean belt (M)

This region includes the high-elevation areas of South America belonging to the Andean Mobile Belt. It includes many wide valleys of tectonic origin and valleys that have resulted from fluvial erosion.

The region is composed of heterogeneous rocks. Frequently, a core of granitic and metamorphic rocks is surrounded or partially covered by folded and fractured sedimentary rocks (marine limestone and continental conglomeratic sandstones). At many locations, volcanic rocks (pyroclastic material and lavas) have been ejected or have flowed out. A large variation in climatic conditions is reflected by striking contrasts between above-average rainfall high in the mountains and arid conditions in the southern Altiplano and on a significant part of the coastal plains.

The region is subdivided into the following groundwater provinces:

6.01 Andes
6.02 Altiplano
6.03 Coastal province

Groundwater resources are variable. Groundwater is often associated with colluvial aquifers in faulted troughs or basins in the mountains. The volcanic zones contain important groundwater resources. High fluoride concentrations are common in these zones. Coastal aquifers occur in structural basins filled with marine and alluvial sediments. They are prone to seawater intrusion.

4.2.2 Global Groundwater Region 7: Lowlands of South America (S)

This region includes the low to moderately elevated, flat areas of the South American sedimentary basins, relatively unaffected by tectonic events. It contains thick sedimentary sequences composed of consolidated (mainly conglomerates and sandstone) and unconsolidated sediments of alluvial, lacustrine and aeolian origin transported from neighbouring topographically high areas. Precipitation in the region varies from low to moderate in its Southern part (Patagonia) to very high in the Amazonian Selvas.

The region is subdivided into the following groundwater provinces:

7.01 Orinoco basin (Northern Ilanos Basins)
7.02 Amazon basin
7.03 Pantanal and Gran Chaco
7.04 Pampas with Rio de la Plata estuary
7.05 Paraná Basin
7.06 Patagonia plains

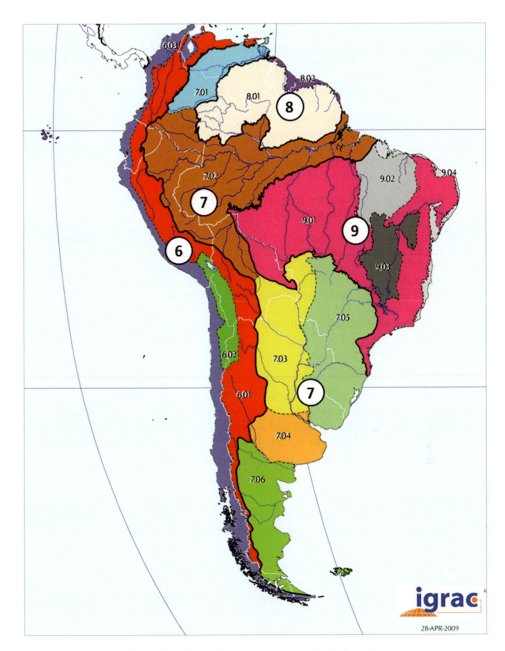

Figure A4.2 Groundwater provinces of South America.

Groundwater resources are abundant. Unconsolidated alluvial sediments, deposited by the major rivers (e.g. aquifers of the Amazon basin and the Argentinean Puelches Aquifer), sandstones (e.g. the Guaraní aquifer system) and volcanic rocks (Serra Geral aquifer) form very important aquifers. Saline groundwater and high concentrations of arsenic are widespread in the aquifers of the Argentinean Chaco-Pampanean plain.

4.2.3   Global Groundwater Region 8: Guyana Shield (B)

This region includes the moderately elevated, flat-topped Guyana Shield in the North-Eastern part of the continent. Crystalline rocks cover almost the whole area. The region has a warm and humid climate with annual rainfall in the range of 1 000–3 000 mm. Groundwater resources are limited due to geological conditions, except for relatively narrow sedimentary zones along the coast. Based on these different hydrogeological conditions, the region is subdivided into two groundwater provinces:

8.01   Guyana Shield province: mainly Precambrian igneous and metamorphic rocks with low potential for storing and transmitting groundwater.

8.02   Guyana Coastal province: deltaic multilayer sandy aquifer systems in the coastal lowlands. They are the main aquifers of the region.

4.2.4   Global Groundwater Region 9: Brazilian Shield and Associated Basins (B)

This region includes the low to moderately elevated, predominantly flat Brazilian Shield with the associated younger sedimentary basins. The Brazilian Shield is mainly composed of crystalline rocks. The associated basins contain a sequence of sandstones, volcanic rocks and unconsolidated sediments. Precipitation in the region is high, except for its North-eastern part.

The region is subdivided into the following groundwater provinces:

9.01   Brazilian Shield (North, Central, East and South)
9.02   Parnaiba Basin
9.03   Sao Francisco Basin
9.04   Brazilian Coastal Province

Groundwater resources are variable. On the shield, groundwater is restricted to local pockets of weathered or fractured zones in the crystalline rocks, or to shallow layers of sediments. In the sedimentary basins, both sandstones and unconsolidated sediments can form large aquifers, such as the Maranhão basin aquifer system. The volcanic areas (predominantly basalts) have limited groundwater potential.

4.3   EUROPE

4.3.1   Global Groundwater Region 10: Baltic and Celtic Shields (B)

This region includes the low to moderately elevated, predominantly flat areas of the Baltic Shield and the Ireland-Scotland Platform. Included are also topographically higher areas of the Norwegian Caledonides, Iceland and the Armorican Massif, all with a more pronounced relief. The Baltic Shield, the Scotland Platform and Armorican Massif are mainly composed of crystalline rocks. The Ireland Platform is composed of sedimentary rocks, while Iceland is built almost exclusively of young volcanic rocks, predominantly basalts. Annual precipitation in the region varies from low-medium in the North-east to high in the Western part.

The region is subdivided into the following groundwater provinces:

10.01   Baltic Shield
10.02   Norwegian Caledonides
10.03   Island of Iceland
10.04   Ireland-Scotland Platform
10.05   Armorican Massif

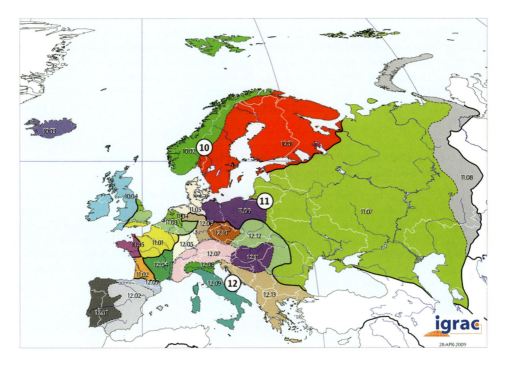

Figure A4.3 Groundwater provinces of Europe.

Groundwater resources are limited. Groundwater in crystalline rocks is restricted to local pockets of weathered or fractured hard rock. The only widespread aquifers with inter-granular permeability are found in the Quaternary deposits (glacio-fluvial deposits). Local karst aquifers occur in Ireland (e.g. the Waulsortian aquifer). The recent volcanoclastics are highly permeable and can form local aquifers in Iceland.

4.3.2   Global Groundwater Region 11: Lowlands of Europe (S)

This region includes the low-lying, flat areas of European sedimentary basins. A zone of higher elevation, associated with the Uralian Orogenic Belt, is included in this region as an administrative boundary between Europe and Asia. The region contains thick sedimentary sequences composed of consolidated and unconsolidated sediments of marine, aeolian and alluvial origin. Precipitation in the region varies from low in the South-east medium to medium in the Western part of the region.

The region is subdivided into the following groundwater provinces:

11.01   Anglo & Paris Basin
11.02   Aquitaine Basin
11.03   London & Brabant Platform
11.04   Dutch Basin
11.05   Northwest German Basin
11.06   German-Polish Basin
11.07   Russian Platform
11.08   Ural Mountains

Groundwater resources are abundant. Unconsolidated sediments form the most important aquifers in the deltas of the main rivers, e.g. in The Netherlands. The large Paris basin aquifer system consists of a multilayer sequence of sedimentary formations with unconfined outcrops on the periphery connected to deeper confined zones towards the centre. Glacial and aeolian aquifers are of local importance in the northern half of the region. In south-western and northern basins, limestone aquifers are found, e.g. the Chalk Aquifer in England. The central and eastern parts of the region also have extensive sandstone aquifers. The Ural Mountains, composed of crystalline rocks, have limited groundwater resources. All coastal aquifers are prone to saline intrusion.

4.3.3   Global Groundwater Region 12: Mountains of Central and Southern Europe (M)

This region includes the topographically high zones in Europe belonging to the Hercynian and Alpine Orogenic Belt. Also included are the sedimentary basins of low to medium elevation associated with these tectonic structures. The folded areas have complex lithology with alternations of crystalline, volcanic and sedimentary rocks. Thick sequences of predominantly carboniferous rocks or thick layers of unconsolidated sediments can be found in the sedimentary basins. Large variations in climate are reflected by an above average rainfall in the mountain zones and dryer conditions in the lowlands.

The region is subdivided into the following groundwater provinces:

12.01   Iberian Massifs (for example, the Hesperian Massif) and adjoining coastal plains
12.02   Iberian Basins
12.03   Pyrenees
12.04   Massif Central
12.05   Jura, Vosges and Ardennes
12.06   Southern German Basins
12.07   Alps
12.08   Po Basin
12.09   Apennines
12.10   Bohemian massifs
12.11   Pannonian Basin
12.12   Carpathian Mountains
12.13   Dinaric Alps

Groundwater resources are variable. Groundwater in crystalline rocks is restricted to local pockets of weathered or fractured hard rocks. Alluvial and colluvial fills in relatively flat areas form local aquifers in the mountain zones. In the sedimentary basins, limestone, sandstones, and unconsolidated sediments can form large interconnected aquifer systems, such as the Po Valley aquifers and the Hungarian Plain aquifers. The Dinaric Alps on the Balkan contain important karst aquifer systems.

4.4   AFRICA

4.4.1   Global Groundwater Region 13: Atlas Mountains (M)

This region includes the elevated areas of the Atlas Mountains, created during the Alpine orogenesis in North-western Africa. The Northern part is composed of folded sedimentary rocks (mainly limestone). In the Southern part, crystalline basement rocks are covered by shallow

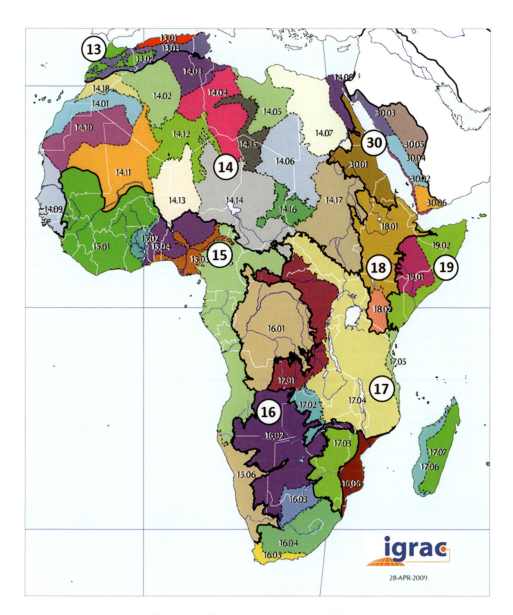

Figure A4.4 Groundwater provinces of Africa.

marine and alluvial sediments. Precipitation shows large spatial and temporal variation. The Southern areas are subject to desert influences.

The region is subdivided into the following groundwater provinces:

13.01 Northern Atlas mountain range (Anti, High, Middle and Tell Atlas)
13.02 El-Shatout depression
13.03 Saharan Atlas mountains in the South

Groundwater resources are limited to alluvial sediments in the mountain zones, karstic aquifers in the Northern part and shallow aquifers at the North and North-West coast.

4.4.2 Global Groundwater Region 14: Saharan Basins (S)

This region includes the North African Craton. It comprises a Precambrian basement unconformably overlain by a thick sequence of continental and marine sediments (clastic sediments covered by carbonates), structured into a number of flat basins at low to medium elevation, separated by zones of higher elevation. The flat areas are covered by aeolian sands and locally by alluvial deposits. The region has an arid climate and is subdivided into the following groundwater provinces:

14.01 Tindoef Basin
14.02 Grand Erg/Ahnet Basin
14.03 Trias/Ghadamedes Basin
14.04 Hamra Basin
14.05 Sirte Basin
14.06 Erdis/Kufra Basin (Nubian sandstone)
14.07 Dakhla Basin (Nubian sandstone)
14.08 Nile valley and delta
14.09 Senegal-Mauritanian Basin
14.10 Regubiat High
14.11 Taoudeni Basin
14.12 Hoggar High
14.12 Iullemeden Basin
14.13 Chad Basin
14.14 Tibesti (Quadai) Mountains
14.16 Ennedi-Darfour Uplift
14.17 Sudan interior basins (Nubian sandstone)
14.18 Ougarta Uplift

Groundwater resources are variable. Groundwater in crystalline rocks is restricted to local pockets of weathered or fractured zones or to shallow layers of fluvial sediments. Thick accumulations of sandstone and limestone in the sedimentary basins constitute eight very large regional aquifer systems that currently receive insignificant quantities of contemporary recharge: the Nubian aquifer system, the North-western Sahara aquifer system, the Murzuk basin, the Taoudeni-Tanezrouft basin, the Iullemeden basin, the lake Chad basin and the Sudd basin. Large alluvial aquifers with renewable groundwater have developed along the river Nile and in its delta.

4.4.3 Global Groundwater Region 15: West African Basements (B)

This region includes low to moderately elevated flat areas of the West African Shield and sedimentary basins associated with large rivers. A narrow coastal strip containing unconsolidated sediments is also included. The shield areas are dominated by outcropping crystalline basement rocks. The region has a dry climate in the Northern zone along the Sahara and in the South, while the climate is humid elsewhere.

The region is subdivided into the following groundwater provinces:

15.01 Eburneen Massif
15.02 Volta Basin

15.03   Niger Delta
15.04   Nigerian Massif
15.05   West Congo Precambrian Belt
15.06   Damer Belt

In general, groundwater conditions are relatively poor, with the exception of the deltas of the large rivers, such as the Volta and Niger rivers.

4.4.4   Global Groundwater Region 16: Sub-Saharan basins (S)

This region includes large inland depressions in the basement rocks of Central and Southern Africa that have been filled by sediments of various origins. The sedimentary areas are flat and of moderate elevation. A topographical high of crystalline rocks, separating two Southern basins, is included in this region. The region's climate is humid in the North and dry in the South. Groundwater resources are abundant.
The region is subdivided into the following groundwater provinces:

16.01   Congo Basin
16.02   Kalahari-Ethosha Basin
16.03   Kalahari Precambrian Belt
16.04   Karoo Basin
16.05   Cape Fold Belt
16.06   Coastal Basins of Mozambique

Large regional aquifers are found in unconsolidated sediments (e.g. in the Congo basin) and in fractured sandstones (e.g. Karoo aquifer system). Limestone and dolomite layers form local aquifers, but shales and crystalline rocks are generally poor aquifers. Some of the aquifers receive limited quantities of contemporary recharge.

4.4.5   Global Groundwater Region 17: East African basement and Madagascar (B)

This region includes moderately elevated flat areas of the East African Shield, affected in the Eastern parts by rifting. The region is dominated by outcropping crystalline basement rocks, with local occurrence of volcanic rocks and sediments. The climate varies from humid in the North to dry in the South. Groundwater resources are limited.
The region is subdivided into the following groundwater provinces:

17.01   East Congo Precambrian Belt
17.02   Luffilian Arch (Katanga system)
17.03   East Kalahari Precambrian Belt
17.04   East Africa Basement (including rifted zones)
17.05   Tanzania Coastal Basin
17.06   Sediments of Madagascar
17.07   Basement of Madagascar

Groundwater resources are limited. Groundwater in the basement areas is restricted to local pockets of weathered or fractured zones in the crystalline rock or to shallow layers of overlying alluvial sediments. Coastal sediments (e.g. Karoo Sandstone in Tanzania) have favourable groundwater conditions. High fluoride concentrations occur locally.

4.4.6 Global Groundwater Region 18: Volcanics of East Africa (V)

This region includes the moderately to high-elevation part of the East African Craton that has been affected by rifting and volcanism. In the rifted zone, large fault escarpments and steep slopes of volcanoes dominate the relief. An arid to semiarid climate prevails, with humid zones in higher elevated areas. Groundwater resources are variable.

The region is subdivided into the following groundwater provinces:

19.01 Amhara Plateau
19.02 Eastern Branch of East African Rift Valley

In the volcanic areas, groundwater occurs in fractured zones and in the sediments interbedded between successive lava flows ('old land surfaces'). Groundwater may contain high concentrations of fluoride and is often hot and brackish in the Rift Valley.

4.4.7 Global Groundwater Region 19: Horn of Africa basins (S)

This region includes large depressions in basement rocks that have been filled by sediments of various origin. Locally isolated uplifted basement complexes occur. The southern and central parts are flat and of moderate elevation, while the northern parts have a more pronounced relief. An arid to semiarid climate prevails, with humid zones at higher altitudes. The region is subdivided into the following groundwater provinces:

19.01 Ogaden Basin
19.02 Somali Coastal Basin

Groundwater resources are variable, but most of the area is considered to be underlain by the large and complex Ogaden–Juba basin aquifer system. Sandstones and fractured limestones are permeable and have good yields, although water levels are locally deep. Interbedded silt and clay horizons form barriers to groundwater flow. The most productive aquifers are found in coarse Quaternary alluvial sediments in the floodplains of major rivers. Dissolution of evaporates occurring in certain sediment layers causes increased salinity of groundwater.

4.5 ASIA

4.5.1 Global Groundwater Region 20: West Siberian platform (S)

This region east of the Ural includes the low to moderately elevated flat areas of the West Siberian craton that has been rifted and filled by thick sedimentary sequences. The upper part of these sequences consists predominantly of alluvial-lacustrine sediments. The climate is cold and dry, with a belt of permafrost in the Northern parts of the region.

The region is subdivided into the following groundwater provinces:

20.01 Yenisey Basin
20.02 West Siberian Basin
20.03 Turgay Depression (basin)

Groundwater resources are abundant. Unconsolidated sediments show large variations in grain size. The main aquifers are associated with layers of coarse material in alluvial sediments.

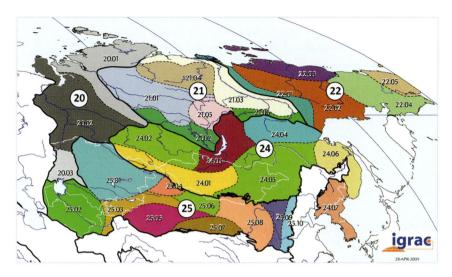

Figure A4.5 Groundwater provinces of Northern and Central Asia.

Fractured sandstones form local aquifers. The West Siberian basin can be considered as one huge and complex aquifer system.

4.5.2   Global Groundwater Region 21: Central Siberian plateau (B)

This region includes the moderately elevated areas of Central Siberian craton, including large basins separated by uplifted highs and arches of crystalline rocks. Numerous rivers have further modified the relief of the region. The climate is cold and dry, with a belt of permafrost in the Northern parts of the region.

The region is subdivided into the following groundwater provinces:

21.01 Tunguska Basin
21.02 Cis-Sayan Basin
21.03 Lena-Vilyuy Basin
21.04 Anabar-Olenek High
21.05 Nepa-Botuoaba Arch
21.06 Aldan uplift

Groundwater resources are moderate. The alluvial deposits associated with large rivers (e.g. Lena) and fissured limestones and sandstones form potential aquifers. The crystalline rocks have a low groundwater potential restricted to weathered zones. Groundwater availability is significantly influenced by permafrost.

4.5.3   Global Groundwater Region 22: East Siberian highlands (M)

This region includes the areas at moderate to high elevation of the East Siberian craton. This craton is largely covered by thick sequences of marine and continental sediments. The relief of the region is related to anticlinal structures in the sedimentary rocks. In the Eastern parts of the region, there are also crystalline rock crop outcrops in the folded structures. Climate is cold and dry. Almost the entire region belongs to the permafrost zone.

The region is subdivided into the following groundwater provinces:

22.01   Verkhoiansk Range
22.02   Cherskii Range
22.03   Kolyma Plain
22.04   Yukagir Plateau
22.05   Anadyr Range

Groundwater resources are limited. Groundwater occurrence is restricted to fractured or weathered zones in crystalline rocks and consolidated sediments. Locally aquifers in unconsolidated alluvial sediments may occur. Groundwater distribution is greatly influenced by permafrost.

4.5.4   Global Groundwater Region 23: North-western Pacific margin (V)

This region includes the areas at moderate to high elevation of North-eastern Asia associated with the unstable island arch of the West Pacific (part of the Circum-Pacific Belt). These areas have a pronounced relief, related to the uplift of sedimentary rocks and volcanic activity. The sedimentary formations consist mainly of marine sandstones and mudstones, intercalated by limestone and with granitic intrusions. The climate varies from cold and dry to hot and humid.

The region is subdivided into the following groundwater provinces:

23.01   Kamchatka Peninsula
23.02   Kuril Islands
23.03   Japan
24.04   Philippines

Groundwater occurs in fractured and fissured sandstones and limestone. Porous volcanic rocks and locally thick unconsolidated sediments also form productive aquifers (e.g. Tokyo Group Aquifer System). Thermal zones, associated with volcanic activity, affect the groundwater composition.

4.5.5   Global Groundwater Region 24: Mountain belt of Central and Eastern Asia (M)

This region includes the elevated areas of Central and East Asia associated with the Paleozoic Mobile Belt. As the result of intensive folding, the region has a very pronounced relief. Crystalline rocks dominate the surface in the Northern half of the region. In the Southern half, crystalline and volcanic rocks alternate with consolidated and unconsolidated sediments.

The region is subdivided into the following groundwater provinces:

24.01   The Altay-Sayan Folded Region (Central Siberia-Mongolia Border)
24.02   Mongol-Okhotsk Folded Region
24.03   Baikal-Paton Folded Region (surroundings of Lake Baikal)
24.04   Aldan Shield in Eastern Siberia
24.05   Yinshah Da and Xia Hinggannling Uplift (Yablonovy and Khingan ranges)
24.06   Sikhote-Alin Folded Region (South-East Siberia)
24.07   Korean Peninsula

Groundwater resources are limited to moderate. Local aquifers occur in intermontane alluvial systems, fractured volcanic rocks, and karstified carbonates. Crystalline rocks have a low groundwater potential, restricted to the thickness of the weathered zone. Highlands have low precipitation and high evaporation, while the coastal areas have a humid climate.

4.5.6   Global Groundwater Region 25: Basins of Central Asia (S)

This region includes the low to moderately elevated, relatively flat areas of West and Central Asia. The topographically low Western part of the region is associated with a huge geo-syncline containing a thick sequence of sedimentary rocks and forming an endorheic system with the Aral Sea as its drainage basis. In the Eastern parts, the sedimentary basins are separated by more elevated areas containing crystalline rocks. Unconsolidated alluvial and aeolian deposits cover large areas. Climate is arid to semi-arid.

The region is subdivided into the following groundwater provinces:

25.01   Central Kazakhstan Folded Region
25.02   Syr-Darya Basin
25.03   Tian Shan Fold Belt
25.04   Junggar Basin
25.05   Tarim Basin
25.06   Altushan Fold Belt
25.07   Jinguan Minle Wuwei Basin
25.08   Ordos Basin
25.09   Shauxi Plateau
25.10   Taihang Shan Yanshan Fold Belt

Groundwater resources are variable. The largest aquifer system is the Tarim basin, a huge multilayer aquifer system consisting of Quaternary deposits. Regional aquifers occur in fractured sandstones, karstified limestone (e.g. Erdos Basin aquifer) and alluvial sediments. Groundwater in extensive loess deposits is associated with the presence of permeable paleo-soils and forms an important water resource. Contemporary recharge is very limited.

4.5.7   Global Groundwater Region 26: Mountain belt of West Asia (M)

This region includes high-elevated areas of West Asia, belonging to the Alpine-Himalayan Mobile Belt (Taurus Mountains, Anatolian Plateau, Caucasus, Central Iranian Basins, Elburz Mountains and Zagros Fold belt and Trust zone). Also included are medium-elevated sedimentary basins associated with tectonic structures. The folded areas have complex lithology with altering crystalline, volcanic and sedimentary rocks. Sedimentary depressions containing predominately marine sediments (carboniferous rocks and sandstones). Basins are locally covered with thick layers of unconsolidated sediments. Climate is predominantly dry, with some moist zones in the higher altitudes.

The region is subdivided into the following groundwater provinces:

26.01   Taurus Mountains
26.02   Anatolian Plateau
26.03   Caucasus
26.04   Central Iranian Basins

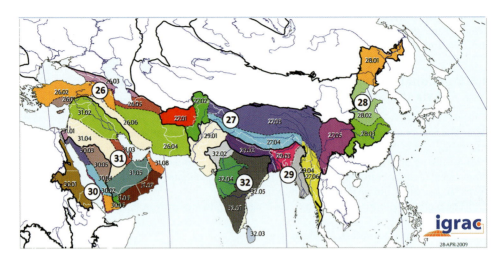

Figure A4.6 Groundwater provinces of Western and South Asia.

26.05 Elburz Mountains
26.06 Zagros Fold belt and Trust zone (Zagros Mountains)

Groundwater resources are limited to moderate. Alluvial and colluvial fills in relative flat zones in the mountains form local aquifers. In the sedimentary basins, karstified limestones (e.g. Midyat Aquifer in Turkey) form major aquifers. Fractured sandstones and unconsolidated sediments can form interconnected aquifer systems. Groundwater availability is good in unconsolidated alluvial deposits directly connected to the riverbeds.

4.5.8 Global Groundwater Region 27: Himalayas and associated highlands (M)

This region includes areas of high elevation in Central Asia, belonging to the Himalayan Mobile Belt. The region is an eastward continuation of the West Asian part of this belt (Region 24). The folded areas have complex lithology with alternations of crystalline, volcanic and sedimentary rocks. Climate varies from warm and humid to cold and arid. Large areas are covered by glaciers or seasonal snow.

The region is subdivided into the following groundwater provinces:

27.01 Hindu Kush
27.02 Pamir High
27.03 Tibetan Plateau
27.04 Himalayas
27.05 Shan Plateau
27.06 Tenasserim Mountains

Groundwater resources are limited. Alluvial and colluvial fills in relative flat zones in the mountain area sometimes form extensive aquifers (e.g. Kathmandu Valley). Local aquifers also occur in karstic limestone and fractured sandstones.

4.5.9   Global Groundwater Region 28: Plains of Eastern China (S)

This region includes the low to medium elevated areas of Great Plains of Eastern China. Thick sequences of alluvial and aeolian sediments were deposited here in sedimentary basins. Annual precipitation in the region is moderate, with a trend to increase from North to South.

The region is subdivided into the following groundwater provinces:

28.01   Manchurian Plain
28.02   North China Plain
28.03   Middle and Lower Chang Jiang (Yangtze) River Basin

Groundwater resources are abundant. Very large alluvial aquifers have large volumes of groundwater in storage, e.g. the Huang-Hai-Hai Plain (North China Plain) and the Song-Liao Plain aquifers systems. This region also has a very high population density: the majority of East Asia's population lives in this region.

4.5.10   Global Groundwater Region 29: Indo-Gangetic-Brahmaputra plain (S)

This region includes the low elevated and flat reaches of large Asian rivers draining the Himalayas. Thick layers of sediments accumulated in the foredeep, which underlie the Ganges Plain and neighbouring plains. Climate varies from arid to humid as the mean annual rainfall increases from west to east.

The region is subdivided into the following groundwater provinces:

28.01   Indus Basin
28.02   Ganges Basin
28.03   Brahmaputra Basin
28.04   Irrawaddy Basin

Groundwater resources are abundant. The extensive alluvial aquifer system, associated with major rivers draining the Himalayas, is one of the largest groundwater reservoirs in the world.

4.5.11   Global Groundwater Region 30: Nubian and Arabian shields (B)

This region includes the moderately high areas belonging to the Nubian and Arabian Shields, together with the low-lying coastal plains along the Red Sea and the rift-related volcanic areas at high elevation.

The region is subdivided into the following groundwater provinces:

30.01   Red Sea Hills in Africa
30.02   Red-Sea coastal plains (e.g. Tihama Plain)
30.03   North Western Escarpment Mountains (Midian & Hiraz)
30.04   Asir Mountains
30.05   Arabian Shield (e.g. Najd Plateau)
30.06   Yemen Highlands

Groundwater resources are variable. The crystalline rock areas have limited groundwater potential, restricted to local weathered zones. Larger aquifer systems are associated with unconsolidated sediments of the coastal plains and river deltas inside the Red Sea graben (e.g. Tihama aquifer system). Sandstones, fractured limestone and volcanic rocks form small but locally important aquifers in tectonic depressions in the Yemen Highlands (e.g. in the Sana'a basin).

4.5.12   Global Groundwater Region 31: Levant and Arabian platform (S)

This region includes the low to moderately elevated, predominantly flat areas of the Levant region and the Eastern part of the Arabian Peninsula. Some more elevated areas, such as the Oman Mountains, are also included. Large rift basins have been filled successively with sediments of different origin. Climate is arid in most of the region, but annual rainfall amounts are moderate along the Mediterranean.

The region is subdivided into the following groundwater provinces:

31.01   Sinai
31.02   Euphrates-Tigris Basin
31.03   Al Hasa Plain (in Saudi Arabia)
31.04   Central arch with Tuwaig Mountains
31.05   Rub-al-Khali Basin
31.06   Marib and Shabwa basins in Yemen
31.07   Masila-Jeza Basin (with wadi Hadramawt)
31.08   Mountains and plains of Oman

Groundwater resources are abundant in terms of stored volumes, but very limited in terms of replenishment. Large regional aquifer systems are found in sandstones (e.g. Mukalla Aquifer System) and fissured carbonates (e.g. Umm-Er-Rhaduma Aquifer System). Macroscopically, the very thick multilayer sequence of Cambrian to Neogene sediments present in most of the region is sometimes considered as one large aquifer system with non-renewable groundwater (Arabian Aquifer System). Limestone complexes along the Mediterranean Sea and unconsolidated alluvial sediments along the main wadis form local aquifers with renewable groundwater.

4.5.13   Global Groundwater Region 32: Peninsular India and Sri Lanka (B)

This region includes the low to moderately elevated areas of the Indian craton. The region is predominantly composed of crystalline rocks. Volcanic rocks (basalts) belonging to the Deccan Trap formation cover a large area in the Western part of the peninsula. In the coastal areas, aquiferous sedimentary rocks occur (predominantly sandstones). These rocks may be covered by thick accumulations of unconsolidated sediments especially in the deltas of larger rivers. Climate is hot and varies from arid to humid.

The region is subdivided into the following groundwater provinces:

32.01   Precambrian basement areas in southern and eastern India
32.02   Precambrian basement area of Aravalli Range in Rajasthan
32.03   Precambrian basement and sediments of Sri Lanka
32.04   Deccan Trap
32.05   Coastal sedimentary areas

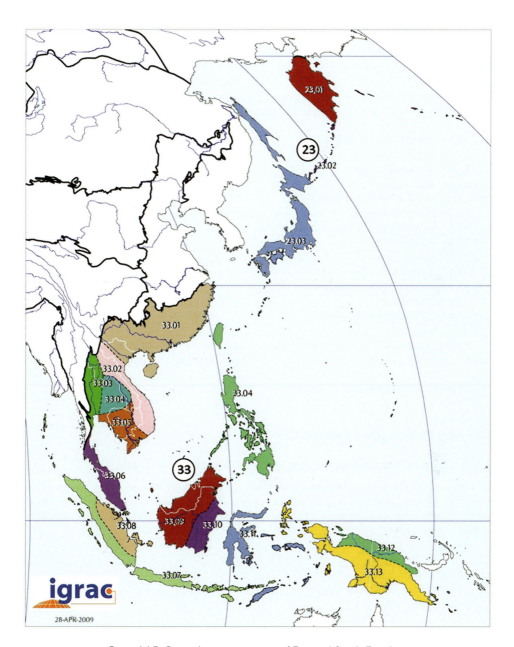

Figure A4.7 Groundwater provinces of East and South-East Asia.

Groundwater resources are limited to moderate. Groundwater in the crystalline rock areas is restricted to local pockets of fractured and weathered zones or to shallow layers of overlying alluvial sediments. Sedimentary intercalations between lava flows contain important groundwater resources in the volcanic areas. Major deltaic and coastal aquifers, particularly along the East coast, have the highest potential. The coastal zones are prone to seawater intrusion.

4.5.14 Global Groundwater Region 33: Peninsulas and Islands of South-East Asia (V)

This region includes low to moderate elevated areas of peninsulas and islands of South-East Asia associated with the Circum-Pacific Belt. Tectonic activity in this area produces a complex geological setting. The region is characterised by outcrops of old crystalline rocks, deep sedimentary basins and recent volcanic eruptions. The climate is hot and humid.

The region is subdivided into the following groundwater provinces:

33.01 South China Fold Belt
33.02 Truong Son Fold Belt
33.03 Thailand Basin
33.04 Khorat Platform
33.05 Tonle Sap-Phnom Penh Basin
33.06 Malay Peninsula
33.07 Sumatra/Java Magmatic Belt
33.08 Sumatra Basin
33.09 Sunda Platform
33.10 Barito-Kutei Basin
33.11 Sulawesi Magmatic Arc
33.12 Irian Basins
33.13 New Guinea Mobile Belt

Groundwater resources are variable. Groundwater in the zones of crystalline and volcanic rocks is restricted to local pockets of fractured and weathered rocks. Unconsolidated sediments (e.g. in the Jakarta groundwater basin) and fissured sedimentary rocks (e.g. karst zones in Vietnam) form regional aquifers.

4.6 AUSTRALIA AND THE PACIFIC

4.6.1 Global Groundwater Region 34: Western Australia (B)

This region includes low to moderately elevated, predominantly flat basement blocks of the Australian Craton, separated by deep sedimentary basins. The sedimentary basins contain thick layers of sandstones and karstified limestone and local alluvial sediments. Climate is semi-arid to arid in most of the region, but changes to tropical humid in the North.

The region is subdivided into the following groundwater provinces:

34.01 Pilbara Block
34.02 Yilgarn Basement Block
34.03 Carnavon Basin
34.04 Canning Basin
34.05 Officer Basins
34.06 Eucla Basin
34.07 Kimberly Basement Block
34.08 Musgrave Basement Block
34.09 McArthur Basin
34.10 Wiso and Georgina Basins

Groundwater resources are generally limited to moderate. Groundwater in the crystalline rock areas is restricted to local pockets of fractured and weathered zones or to shallow layers of overlying alluvial sediments. Large regional aquifer systems, however, are formed by extensive bodies of fissured sandstones (e.g. Canning Basin) and limestone (e.g. Eucla Basin). Groundwater renewal is limited compared to the total volumes of groundwater stored. Palaeochannel sands (representing former riverbeds) are prospective aquifers in the region, though the salinity of its groundwater may be high.

4.6.2   Global Groundwater Region 35: Eastern Australia (S)

This region includes low to moderately elevated, flat areas of the East Australian sedimentary basins. The older consolidated sediments are frequently overlain by large alluvial fans. Uplifted areas at the Eastern margin (Great Dividing Range), belonging to the Tasman Mobile Belt, are also included. Climate is arid to semi-arid in the continental part of the region, but rather humid within a few hundred kilometres from the coast.

The region is subdivided into the following groundwater provinces:

35.01   Gawler Ranges
35.02   Great Artesian Basin
35.03   Murray Basin
35.04   Great Dividing Range
35.05   Australian Alps
35.06   Tasmania Island

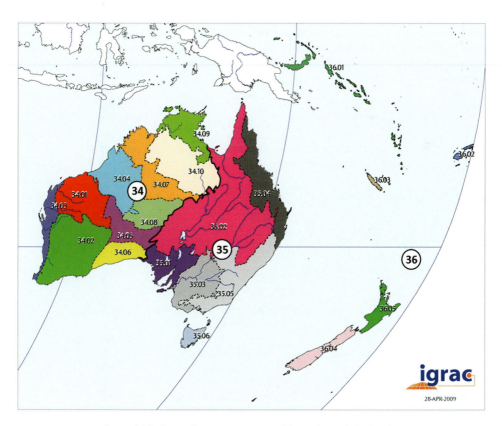

Figure A4.8 Groundwater provinces of Australia and the Pacific.

Groundwater resources are moderate to high. Thick layers of sandstones form one of the world's largest aquifer systems: the Great Artesian Basin Aquifer System. Fissured limestone aquifers also occur (e.g. Murray Group Aquifer). Extensive alluvial aquifers, associated with the large rivers draining the uplifted areas, are important sources of shallow groundwater. Uplifted areas have only local aquifers in fractured rocks.

4.6.3 Global Groundwater Region 36: Islands of the Pacific (V)

This region includes the numerous small islands of the South-Eastern Pacific and New Zealand, belonging to the Circum-Pacific Belt. Pacific islands West of the American continents are also included. The region has a large variation in elevation and relief. Volcanic rocks are found in the Northern part. The Southern part (New Zealand) includes also crystalline rocks, uplifted by the orogenesis, and thick sequences of sedimentary rocks. Climate is humid.

The region is subdivided into the following groundwater provinces:

36.01 Bismarck -New Hebrides Volcanic Arcs
36.02 Fiji Islands
36.03 Orogenic belt of New Caledonia
36.04 Axial tectonic belt of New Zealand
36.05 Sedimentary basins of New Zealand
36.06 Pacific islands West of the American continents

Groundwater resources are variable. Some of the recent volcanic rocks are highly porous and contain large volumes of groundwater. Karstified limestone and porous calcareous formations in coastal areas are also important aquifers. Shallow aquifers occur in unconsolidated alluvial sediments. Freshwater lenses are usually shallow and saline water intrusions are very common.

Groundwater abstraction estimates by country

| | Country or territory | Population 2010 (in thousands) | Estimated groundwater abstraction 2010 km³/yr | Groundwater abstraction — Breakdown by sector | | | Groundwater share in total freshwater withdrawal (excl. reservoir losses) | | | | Indicators | | Groundwater recharge per capita |
				Grw abstraction for irrigation %	Grw abstraction for domestic use %	Grw abstraction for industry %	All water use sectors %	Irrigation sector %	Domestic water sector %	Industry sector %	Groundwater development stress %	Renewable groundwater development stress %	m³/yr
1	Afghanistan	31 412	7.12	94	6	0	23	22	71		110	110	206
2	Albania	3 204	0.90	61	33	6	53	52	67	25	16	16	1 716
3	Algeria	35 468	2.87	65	32	2	42	43	66	7	46	19	175
4	American Samoa	68	0.01	25	60	15					6	6	
5	Andorra	85	0.01										
6	Angola	19 082	0.41				6				0	0	6 840
7	Anguilla	15											
8	Antarctica	0											
9	Antigua and Barbuda	89	0.00	20	60	20	60						
10	Argentina	40 412	6.66	79	12	9	14				5	5	3 642
11	Armenia	3 092	0.69	56	34	10	22	19	25	52	90	90	248
12	Aruba	107	0.01										
13	Australia	22 268	4.96	52	47	1	21	14	66	2	2	2	11 771
14	Austria	8 394	1.12	5	52	43	30				8	8	1 632
15	Azerbaijan	9 188	0.86	95	5	0	6	8	8	0	30	30	314
16	Bahamas	343	0.01								1	1	5 263
17	Bahrain	1 262	0.31	60	35	5	100	90	100	100	>1 000	>1 000	
18	Bangladesh	148 692	30.21	86	13	1	79				86	86	237
19	Barbados	273	0.09	20	80	0	100				124	124	
20	Belarus	9 595	1.20	28	52	20	28				6	6	2 058

(Continued)

	Country or territory	Population 2010 (in thousands)	Estimated groundwater abstraction 2010 km³/yr	Groundwater abstraction — Breakdown by sector			Groundwater share in total freshwater withdrawal (excl. reservoir losses)				Indicators		
				Grw abstraction for irrigation %	Grw abstraction for domestic use %	Grw abstraction for industry %	All water use sectors %	Irrigation sector %	Domestic water sector %	Industry sector %	Groundwater development stress %	Renewable groundwater development stress %	Groundwater recharge per capita m³/yr
21	Belgium	10 712	0.65	4	55	41	10		65*		8	8	784
22	Belize	312	0.03	5	90	5	2				0	0	24 070
23	Benin	8 850	0.17				32				2	2	1 098
24	Bermuda	65											
25	Bhutan	726	0.04				0				1	1	4 959
26	Bolivia	9 930	0.88	10	60	30	40				1	1	16 152
27	Bosnia and Herzegovina	3 760	0.30	33	67	0	32	17	87	0	3	3	3 077
28	Botswana	2 007	0.14	16	61	23	66	25	100	86	1	1	5 534
29	Brazil	194 946	10.06	38	38	24	14				0	0	14 225
30	British Virgin Islands	23											
31	Brunei Darussalam	399	0.02										
32	Bulgaria	7 494	0.58				10				7	7	1 135
33	Burkina Faso	16 469	0.39	84	14	2	45	43	48	100	4	4	648
34	Burundi	8 383	0.16								5	5	346
35	Cambodia	14 138	0.76								2	2	3 428
36	Cameroon	19 559	0.37								0	0	5 693
37	Canada	34 017	1.87	43	43	14	4			1*	0	0	24 459
38	Cape Verde	496	0.01								9	9	
39	Cayman Islands	56											
40	Central African Republic	4 401	0.08								0	0	20 964
41	Chad	11 227	0.45	81	19	0	32	28	98		1	1	3 454
42	Channel Islands	153											
43	Chile	17 114	0.98	73	20	7	5				2	2	3 561
44	China	1 341 335	111.95	54	20	26	18	15	29	20	17	17	492
45	Colombia	46 295	1.75	33	34	33					0	0	10 960
46	Comoros	735	0.01								1	1	
47	Congo	4 043	0.03	10	76	14	52				0	0	29 522
48	Cook Islands	20	0.00										
49	Costa Rica	4 659	0.80	12	71	17	26				4	4	3 876
50	Côte d'Ivoire	19 738	0.37								1	1	2 228
51	Croatia	4 403	1.16	0	86	14	97		100	100	9	9	3 076
52	Cuba	11 258	5.38				50				44	44	1 093

(Continued)

	Country or territory	Population 2010 (in thousands)	Estimated groundwater abstraction 2010 km³/yr	Grw abstraction for irrigation %	Grw abstraction for domestic use %	Grw abstraction for industry %	All water use sectors %	Irrigation sector %	Domestic water sector %	Industry sector %	Groundwater development stress %	Renewable groundwater development stress %	Groundwater recharge per capita m³/yr
				Groundwater abstraction — Breakdown by sector			**Groundwater share in total freshwater withdrawal** (excl. reservoir losses)				**Indicators**		
53	Cyprus	1 104	0.15	87	13	0	62		60		35	35	
54	Czech Republic	10 493	0.38				19				5	5	692
55	Dem. P. Republic of Korea	24 346	1.56								13	13	491
56	Dem. Rep. of the Congo	65 966	1.22								0	0	8 786
57	Denmark	5 550	0.65	38	40	22	98				4	4	2 815
58	Djibouti	889	0.02	11	89	0	95	67	100		26	26	85
59	Dominica	68	0.01										
60	Dominican Republic	9 927	1.23								21	21	590
61	Ecuador	14 465	0.44	10	60	30	3				1	1	5 818
62	Egypt	81 121	7.78	41	59	0	4	2	27	0	>1 000	>1 000	7
63	El Salvador	6 193	0.58								12	12	781
64	Equatorial Guinea	700	0.01								0	0	15 797
65	Eritrea	5 254	0.09	74	26	0	14	11	68	0	12	12	136
66	Estonia	1 341	0.33	0	20	80	18				4	4	5 814
67	Ethiopia	82 950	1.49				24				3	3	531
68	Faroe Islands	49	0.00										
69	Falkland Islands (Malvinas)	3	0.00								0	0	915 410
70	Fiji	861	0.08								1	1	6 124
71	Finland	5 365	0.28	24	65	11	17				1	1	8 010
72	France	62 787	5.71	14	63	23	18	20	63	41	5	5	1 738
73	French Guiana	231	0.01								0	0	106 245
74	French Polynesia	271	0.03										
75	Gabon	1 505	0.03	0	100	0					0	0	57 357
76	Gambia	1 728	0.03								3	3	662
77	Georgia	4 352	0.64	16	66	18	34	9	100	48	7	7	2 188
78	Germany	82 302	5.83	4	48	48	18	87*			8	8	871
79	Ghana	24 392	0.51				11				2	2	1 033
80	Gibraltar	29											
81	Greece	11 359	3.65	86	14	0	39	41	33	0	31	31	1 051

(Continued)

| Country or territory | Population 2010 (in thousands) | Estimated groundwater abstraction 2010 km³/yr | Groundwater abstraction — Breakdown by sector | | | Groundwater share in total freshwater withdrawal (excl. reservoir losses) | | | | Indicators | | |
			Grw abstraction for irrigation %	Grw abstraction for domestic use %	Grw abstraction for industry %	All water use sectors %	Irrigation sector %	Domestic water sector %	Industry sector %	Groundwater development stress %	Renewable groundwater development stress %	Groundwater recharge per capita m³/yr
82 Greenland	57	0.00								0	0	247 010
83 Grenada	104	0.01										
84 Guadeloupe	461	0.04										
85 Guam	180	0.02										
86 Guatemala	14 389	1.52								5	5	2 238
87 Guinea	9 982	0.09				5				0	0	5 226
88 Guinea-Bissau	1 515	0.03				18				0	0	4 878
89 Guyana	754	0.03								0	0	104 035
90 Haiti	9 993	0.95				6				34	34	276
91 Holy See	0	0.00										
92 Honduras	7 601	0.73								3	3	3 417
93 Hong Kong	7 053	0.39										
94 Hungary	9 984	0.37	18	35	47	7				5	5	679
95 Iceland	320	0.16				97				1	1	98 590
96 India	1 224 614	251.00	89	9	2	33		64	27	82	82	250
97 Indonesia	239 871	14.93	2	93	5	11				2	2	3 535
98 Iran (Islamic Republic of)	73 974	63.40	87	11	2	57	53	97	100	255	255	336
99 Iraq	31 672	2.69	50	5	45	3	2	2	9	34	34	247
100 Ireland	4 470	0.21	29	35	36	27		16		1	1	5 857
101 Isle of Man	83											
102 Israel	7 418	1.25	71	26	3	71	66	52	27	144	138	117
103 Italy	60 551	10.40	67	23	10	25	35	30	13	23	23	733
104 Jamaica	2 741	0.51	80	10	10					34	34	549
105 Japan	126 536	10.94	23	29	48				40	10	10	832
106 Jordan	6 187	0.64	42	51	7	65	38	100	100	57	39	182
107 Kazakhstan	16 026	3.23	71	21	8	7	6	85	3	11	11	1 763
108 Kenya	40 513	0.62	3	79	18					2	2	662
109 Kiribati	100	0.01										
110 Kuwait	2 737	0.62	79	8	14	6	100	8	14	>1 000	980	8
111 Kyrgyzstan	5 334	0.96	25	50	25					43	43	417
112 Lao PDR	6 201	0.34								1	1	9 020
113 Latvia	2 252	0.11				26				1	1	4 563
114 Lebanon	4 228	0.81	66	28	6	64	59	61	27	78	78	247
115 Lesotho	2 171	0.02				30				3	3	218
116 Liberia	3 994	0.07								0	0	11 692
117 Libyan Arab Jamahiriya	6 355	4.71	83	14	3	98	97	100	100	134	55	554

(Continued)

	Country or territory	Population 2010 (in thousands)	Estimated groundwater abstraction 2010	Groundwater abstraction Breakdown by sector			Groundwater share in total freshwater withdrawal (excl. reservoir losses)				Indicators		
				Grw abstraction for irrigation	Grw abstraction for domestic use	Grw abstraction for industry	All water use sectors	Irrigation sector	Domestic water sector	Industry sector	Groundwater development stress	Renewable groundwater development stress	Groundwater recharge per capita
			km³/yr	%	%	%	%	%	%	%	%	%	m³/yr
118	Liechtenstein	36	0.00										
119	Lithuania	3 324	0.17				7				2	2	2 995
120	Luxembourg	507	0.02				40				30	30	
121	Macao	554											
122	Madagascar	20 714	0.38				0				0	0	6 124
123	Malawi	14 901	0.28								1	1	1303
124	Malaysia	28 401	0.59	5	62	33	3				0	0	5 582
125	Maldives	316	0.02	0	98	2	54				57	57	
126	Mali	15 370	0.34	85	15	0	5	4	8	0	1	1	1 751
127	Malta	417	0.03	75	16	9	100	100	100	100	62	62	
128	Marshall Islands	54											
129	Martinique	406	0.04										
130	Mauritania	3 460	0.76	73	20	7	41	33	95	100	21	18	1 072
131	Mauritius	1 299	0.16	40	48	12					18	18	
132	Mayotte	204											
133	Mexico	113 423	29.45	72	22	6	35	33	44	37	30	30	878
134	Micronesia (Fed.States of)	111	0.01										
135	Monaco	35	0.00										
136	Mongolia	2 756	0.48	48	22	30	91	83	100	100	17	17	1022
137	Montenegro	631	0.05	0	100	0	100		100		3	3	2 625
138	Montserrat	6											
139	Morocco	31 951	3.06	83	14	3	34	28	47	63	40	40	242
140	Mozambique	23 391	0.44				5				1	1	3 554
141	Myanmar	47 963	4.02				9				3	3	3 190
142	Namibia	2 283	0.15	34	55	11	45	21	100	100	2	2	3 724
143	Nauru	10	0.00								0	0	1 000
144	Nepal	29 959	2.91								15	15	636
145	Netherlands	16 613	0.97	23	32	45	9		65*		7	7	887
146	Netherlands Antilles	201											
147	New Caledonia	251	0.07								3	3	11 094
148	New Zealand	4 368	0.80	60	30	10	17				1	1	20 532
149	Nicaragua	5 788	1.17								3	3	7 531
150	Niger	15 512	0.14	38	58	4	6	2	83	50	1	1	1 013
151	Nigeria	158 423	3.44								2	2	951

(Continued)

	Country or territory	Population 2010 (in thousands)	Estimated groundwater abstraction 2010 km³/yr	Groundwater abstraction Breakdown by sector			Groundwater share in total freshwater withdrawal (excl. reservoir losses)				Indicators		
				Grw abstraction for irrigation %	Grw abstraction for domestic use %	Grw abstraction for industry %	All water use sectors %	Irrigation sector %	Domestic water sector %	Industry sector %	Groundwater development stress %	Renewable groundwater development stress %	Groundwater recharge per capita m³/yr
152	Niue	1	0.00	10	85	5							
153	Norfolk Island	0											
154	Northern Mariana Islands	61	0.01										
155	Norway	4 883	0.41	0	27	73	14				1	1	14 289
156	Occ. Palestinian Territory	4 039	0.23	52	48	0	100	61	69		124	124	
157	Oman	2 782	0.84	96	2	2	100	97	100	100	152	152	199
158	Pakistan	173 593	64.82	94	6	0	32	32	100	0	661	661	57
159	Palau	20	0.00										
160	Panama	3 517	0.32								1	1	7 217
161	Papua New Guinea	6 858	0.61	0	90	10					0	0	30 890
162	Paraguay	6 455	0.25	20	50	30					1	1	7 146
163	Peru	29 077	2.89	60	25	15	10				1	1	12 372
164	Philippines	93 261	3.30	0	36	64	4				4	4	847
165	Pitcairn	0											
166	Poland	38 277	2.59	0	70	30	22		66*		6	6	1 174
167	Portugal	10 676	6.29	89	7	4	74				48	48	1 225
168	Puerto Rico	3 749	0.21	29	46	25	20						
169	Qatar	1 759	0.26	99	1	0	100	84	100	0	>1000	654	8
170	Republic of Korea	48 184	4.31	17	83	0	14				32	32	
171	Republic of Moldova	3 573	0.60				31				42	42	398
172	Réunion	846	0.02								1	1	
173	Romania	21 486	0.63	1	61	38	9				3	3	1 032
174	Russian Federation	142 985	11.62	3	79	18	18		79*	<15*	1	1	6 461
175	Rwanda	10 624	0.20								11	11	168
176	Saint Helena	4	0.02										
177	Saint Kitts and Nevis	52	0.00				0				24	24	
178	Saint Lucia	174	0.02										
179	Saint Pierre and Miquelon	6	0.00										

(Continued)

	Country or territory	Population 2010 (in thousands)	Estimated groundwater abstraction 2010	Groundwater abstraction			Groundwater share in total freshwater withdrawal				Indicators		
				Breakdown by sector			(excl. reservoir losses)						
				Grw abstraction for irrigation	Grw abstraction for domestic use	Grw abstraction for industry	All water use sectors	Irrigation sector	Domestic water sector	Industry sector	Groundwater development stress	Renewable groundwater development stress	Groundwater recharge per capita
			km³/yr	%	%	%	%	%	%	%	%	%	m³/yr
180	St. Vincent & the Grenadines	109	0.01										
181	Samoa	183	0.02										
182	San Marino	32	0.00										
183	Sao Tome and Principe	165	0.00										
184	Saudi Arabia	27 448	24.24	92	5	3	95	94	100	100	>1 000	238	64
185	Senegal	12 434	0.74	90	7	3	31	30	51	34	6	4	999
186	Serbia	9 856	0.53								6	6	928
187	Seychelles	87	0.00										
188	Sierra Leone	5 868	0.11								0	0	4 817
189	Singapore	5 086	0.28										
190	Slovakia	5 462	0.36	3	84	13	52				6	6	1 034
191	Slovenia	2 030	0.19	1	83	16	15		61	28	3	3	2 712
192	Solomon Islands	538	0.05								0	0	22 157
193	Somalia	9 331	0.28	97	3	0	0		80	0	5	5	649
194	South Africa	50 133	3.14	84	10	6	23				18	18	348
195	Spain	46 077	5.70	72	23	5	18	19	27	25	16	16	768
196	Sri Lanka	20 860	1.17								11	11	520
197	Sudan and South Sudan	43 552	0.59	94	4	2	1	1	2	4	1	1	1 260
198	Suriname	525	0.02								0	0	90 156
199	Svalbard and Jan Mayen	0											
200	Swaziland	1 186	0.04	96	2	2	4				7	7	550
201	Sweden	9 380	0.35	0	92	8	14				1	1	6 831
202	Switzerland	7 664	0.79	0	72	28	30		80*		8	8	1 229
203	Syrian Arab Republic	20 411	11.29	90	5	5	65	56	30	84	194	194	285
204	Taiwan Province of China	23 216	9.54								239	239	
205	Tajikistan	6 879	3.63	39	31	30	19				73	73	720
206	Thailand	69 122	10.74	14	60	26	17				12	12	1 327
207	The former Yugoslav Republic of Macedonia	2 061	0.16				16				10	10	757

(Continued)

	Country or territory	Population 2010 (in thousands)	Estimated groundwater abstraction 2010 km³/yr	Groundwater abstraction — Breakdown by sector — Grw abstraction for irrigation %	Grw abstraction for domestic use %	Grw abstraction for industry %	Groundwater share in total freshwater withdrawal (excl. reservoir losses) — All water use sectors %	Irrigation sector %	Domestic water sector %	Industry sector %	Indicators — Groundwater development stress %	Renewable groundwater development stress %	Groundwater recharge per capita m³/yr
208	Timor-Leste	1 124	0.06								4	4	1 239
209	Togo	6 028	0.11								2	2	1 232
210	Tokelau	1											
211	Tonga	104	0.01										
212	Trinidad and Tobago	1 341	0.09				39						
213	Tunisia	10 481	2.02	74	19	7	78	73	92	100	67	45	286
214	Turkey	72 752	13.22	60	32	8	16	13	32	28	30	30	604
215	Turkmenistan	5 042	0.54	38	53	9	2	1	50	19	65	65	165
216	Turks and Caicos Islands	38											
217	Tuvalu	10	0.00										
218	Uganda	33 425	0.62								3	3	669
219	Ukraine	45 448	4.02	52	30	18	10				13	13	667
220	United Arab Emirates	7 512	3.53	100	0	0	100	84	0	0	>1 000	790	33
221	United Kingdom	62 036	2.16	9	77	14	17		35*		3	3	1 338
222	United Republic of Tanzania	44 841	0.98				2				1	1	1 962
223	United States of America	310 384	111.70	71	23	6	23	41	38	21	13	13	2 739
224	United States Virgin Islands	109	0.01										
225	Uruguay	3 369	0.15	30	60	10	4				1	1	9 169
226	Uzbekistan	27 445	9.94	57	32	11	13	8	87	67	285	285	127
227	Vanuatu	240	0.02								0	0	18 239
228	Venezuela	28 980	1.55	10	60	30	15				1	1	8 869
229	Viet Nam	87 848	1.63				2				3	3	698
230	Wallis and Futuna Islands	14											
231	Western Sahara	531											200
232	Yemen	24 053	3.22	86	11	3	71	68	100	100	235	235	57
233	Zambia	13 089	0.30				4				0	0	6 233
234	Zimbabwe	12 571	0.43				10				3	3	991

Specification of information sources:

Column	Variable	Source
3	Population	UNDESA, 2011: http://esa.un.org/unpd/wpp/Excel-Data/population.htm
4	Groundwater abstraction 2010	Processed on the basis of data obtained from databases and scientific reports (GGIS, AQUASTAT, EUROSTAT, Plan Blue, etc.) and various national sources, usually selecting the most recent data (unless considered less reliable). Extrapolated to the year 2010 on the basis of observed growth rates in the region concerned.
5–7	Breakdown according to use sector	Processed on the basis of data obtained from databases and scientific reports (GGIS, AQUASTAT, EUROSTAT, Plan Blue, etc.) and various national sources, usually selecting the most recent reliable data. Years of reference differ (mostly closely around 2000), but percentages are assumed to remain relatively stable over time.
8–11	Groundwater share in total withdrawal of water	Processed on the basis of data obtained from databases (AQUASTAT, EUROSTAT, Plan Blue, etc.) and various national sources (compilation by Jean Margat). Data derived from less formal sources, but considered reliable, is added and marked by an asterisk (*). Years of reference differ (mostly close to 2000), but percentages are assumed to remain relatively stable over time.
12	GDS indicator	Calculated as GDS = 100 * Groundwater abstraction 2010 (km^3/yr)/Diffuse recharge (km^3/yr)
13	RGDS indicator	Calculated as RGDS = 100 * Groundwater abstraction 2010 (km^3/yr) from renewable sources/Diffuse recharge (km^3/yr)
14	Groundwater recharge per capita (GRC)	Calculated as GRC = 1000 * Diffuse recharge (km^3/yr)/Population 2010 (see column 3)
12–14	Diffuse recharge	Taken from: Döll, P., & K. Fiedler, 2008. Global-scale modelling of groundwater recharge. *Hydrol. Earth Syst. Sci.*, Vol. 12, pp. 863–885.

Notes:

(1) Few water related statistics were available for 2010, thus most of the estimates are based on extrapolation.

(2) There are some inconsistencies in the data sets on groundwater shares in total freshwater withdrawal; the available data do not allow this to be corrected. For countries where desalinated sea water is used, the share of groundwater in total freshwater withdrawal may be significantly different from its share in total water withdrawal.

Suggestions for additional reading

6.1 SELECTED GENERIC TEXTS AND AWARENESS RAISING MATERIALS RELATED TO GROUNDWATER

6.1.1 Introductory materials on the Internet[1]

UK Groundwater Forum
Groundwater Basics
  http://www.groundwateruk.org/Groundwater-Basics.aspx
Groundwater in Depth
  http://www.groundwateruk.org/Groundwater-in-depth.aspx

United States Geological Survey (USGS)
Water basics
  http://ga.water.usgs.gov/edu/mwater.html
Groundwater
  http://pubs.usgs.gov/gip/gw/
What is groundwater?
  http://pubs.usgs.gov/of/1993/ofr93-643/
Water science for schools
  http://ga.water.usgs.gov/edu/index.html

Environment Canada
Water basics
  http://www.ec.gc.ca/eau-water/default.asp?lang=En&n=A6A15B0B-1
Groundwater – Nature's hidden treasure
  http://www.ec.gc.ca/eau-water/default.asp?lang=En&n=3F93145A-1

The Groundwater Foundation
Learn the basics/Threats to groundwater/Protection efforts/Learn more
  http://www.groundwater.org/gi/gi.html

International Association of Hydrogeologists (IAH)
About groundwater
  http://www.iah.org/groundwater_about.asp

---

1 All internet portals and websites referenced in this book have been accessed during 2011 or 2012. Nevertheless, some of them may become inaccessible in the future.

6.1.2   Introductory texts on Groundwater

Baldwin, H.L., & C.L McGuinness, 1963. *A Primer on Groundwater.* Washington, USGS, 26 p.

Davis, S.N., & R.J.M. De Wiest, 1966. *Hydrogeology.* New York, John Wiley & Sons, 463 p.

Downing, R.A., 1998. *Groundwater: our hidden asset.* British Geological Survey, UK Groundwater Forum, NERC.

Fetter, C.W., 2001. *Applied hydrogeology.* Fourth edition, Upper Saddle River, New Jersey, Prentice Hall, 598 p.

Fits, C.R., 2002. *Groundwater Science.* Amsterdam, Academic Press, 450 p.

Heath, R.C., 1984. *Basic Groundwater Hydrology.* USGS Water-Supply Paper 2220, 84 p.

López-Geta, J., J. Fornés Azcoiti, G. Ramos González & F. Villaroya Gil, 2006. *Groundwater. A natural underground resource.* IGME, UNESCO and Fundación Marcelino Botín, 108 p.

Nonner, J.C., 2010. *Introduction to Hydrogeology.* Second edition, Leiden, Taylor and Francis/CRC Press/Balkema, 248 p.

Price, M., 1996. *Introducing groundwater.* London, Chapman and Hall, 278 p.

6.1.3   Supplementary texts – Thematic or more advanced

ADEME, BRGM & IFP, 2005. La *capture et le stockage géologique du CO₂.* Coll. Les enjeux des géosciences Paris, 44 p.

Anonymous, 2005. L'eau souterraine. In: *Géosciences,* No 2. BRGM, Orléans, 94 p.

Anonymous, 2006a. Les eaux souterraines. In: *Géochronique,* No 97. BRGM, Orléans & Société de Géologie Française, Paris.

Anonymous, 2006b. *Proceedings. European Groundwater Conference 2006.* Vienna, 22–23 June 2006, Umweltbundesamt of the Austrian Federal Environment Agency.

Bear, J., 1979. *Hydraulics of groundwater.* New York, McGraw-Hill Inc., 567 p.

Burke, J.J., & M. Moench, 2000. *Groundwater and society, resources, tensions and opportunities. Themes in groundwater management for the 21st century.* New York, United Nations, 170 p.

Chapelle, F.H., 2000. *The Hidden Sea: Groundwater, Springs, and Wells.* National Ground Water Association, Westerville, OH.

Clarke R., A.R. Lawrence & S.D.D. Foster, 1996. *Groundwater – a threatened resource.* UNEP Environment Library, No 15.

Collin, J.-J., 2004. *Les eaux souterraines. Connaissance et gestion.* Orléans/Paris, BRGM/Hermann, 169 p.

Collin, J.-J., J. Dubuisson, J. Margat, G. Rampon & J.-C. Roux., 1991. Eau souterraine. Quantité et sécheresse, qualité et pollution. *Géochronique,* No 37, pp. 10–13.

Cosgrove, W.J., & F. Rijsberman, 2000. *World Water Vision, Making water everybody's business.* Marseille, World Water Council, EARTHSCAN.

Custodio, E., & Llamas, M.R., 1976/1983. *Hidrología subterránea.* Barcelona, Ediciones Omega, 2 volumes, 1950 p.

Detay, M., 1997. *La gestion active des aquifères.* Masson, Paris, 416 p.

European Union, 2006. *Mediterranean Groundwater Report.* EU Water initiative, Mediterranean Groundwater Working Group, 113 p.

FCIHS, 2009. *Hidrogeología.* Fundación Centro Internacional de Hidrología Subterránea Barcelona, 768 p.

Fetter, C.W., 1993. *Contaminant hydrogeology.* Upper Saddle River, New Jersey, Prentice Hall, 458 p.

Foster, S., & D. Loucks, 2006. *Non-renewable groundwater resources.* UNESCO-IHP, IHP-VI, Series on Groundwater No. 10, 104 p.

Foster, S., A. Tuinhof, K. Kemper, H. Garduño & M. Nanni, 2003. *Characterization of Groundwater Systems: key concepts and frequent misconceptions*. G-MATE Briefing Note Series No 2, G-MATE, The World Bank, Washington, 6 p.

Foster, S., A. Tuinhof, K. Kemper, H. Garduño & M. Nanni, 2003. *Groundwater Management Strategies: facets of the integrated approach*. G-MATE Briefing Note Series No 3, G-MATE, The World Bank, Washington, 6 p.

Freeze, R.A., & J.A. Cherry, 1979. *Groundwater*. Englewood Cliffs, N.J., Prentice-Hall Inc., 604 p.

Giordano, M., 2009. Global groundwater? Issues and solutions. *Annual Review of Environment and Resources, 34*, 153–178.

Giordano, M. & K. Villholth (ed.), 2007. *The Agricultural Groundwater Revolution*, CABI, Wallingford, UK., 420 p.

Guillemin, C., & J.-C. Roux, 1994. Mystères et réalités des eaux souterraines. *La Vie des Sciences*, C.R.T.11, No 2, pp. 87–114.

Jones, J.A.A. (ed.), 2011. *Sustaining groundwater resources. A critical element in the global water crisis*. Series International Year of Planet Earth, Dordrecht, Springer, 228 p.

Kounine, W.N., 1964. Eaux souterraines, trésor méconnu. In: *Courrier de l'UNESCO*, July–August 1964, 'L'eau et la vie'.

Lerner, D.N., A. Issar, & I. Simmers (eds), 1990. *Groundwater Recharge. A Guide to Understanding and Estimating Natural Recharge*. International Contributions to Hydrogeology, Vol. 8, Heise, Hannover, 345 p.

Llamas, M.R., 1974. Motivaciones extraeconomicas en la utilitizacion de las aguas subterraneas: la hidroesquizofrenia. *Agua*, No 82, Jan–Feb 1974, Madrid, pp. 29–36.

Margat, J., & K. Saad, 1985. Les mines d'eau fossile. In: *Courrier de l'UNESCO, 1985*, pp. 14–16.

Margat, J., 1991. *Les eaux souterraines dans le monde. Similitudes et différences*. Soc. Hydrol. Fr., Journées Hydraul., Sophia-Antipolis, Conf. No 1.

Margat, J., 1993. *L'eau souterraine: une ressource fragile*. Paris, Almanach Jules Verne.

Margat, J., 1993. Les réserves secrètes. In: *Courrier de l'UNESCO*, 1993, pp. 15–18.

Morris, B., A. Lawrence, J. Chilton, B. Adams, R. Calow & B. Klinck, 2003. *Groundwater and its susceptibility to degradation*. Early Warning and Assessment Report Series, RS 03–03, UNEP, Nairobi, Kenya, 126 p.

Mukherji, A., 2006. Is intensive use of groundwater a solution to the world's water crisis? In: *Water Crisis. Myth or reality?* Balkema, Rotterdam/London, Balkema/Taylor & Francis, London, pp. 181–193.

Narasimhan, T.N., 2009. Groundwater: from mystery to management. *Environmental Research Letters*, Vol. 4, 035002 Available from: doi: 10.1088/1748–9326/4/3/035002.

Planet Earth, 2005. *Groundwater Reservoir for a thirsty planet?* Leiden/Paris, Planet Earth/IUGS/UNESCO, 14 p.

Postel, S., 1999. *Pillar of Sand: Can the Irrigation Miracle Last?* New York, Norton.

Price, M., 1991. Water from the ground. In: *New Scientist*, No 42, 4 p.

Puri, S., & A. Aureli, 2010. *Atlas of Transboundary Aquifers*, UNESCO-IHP, ISARM Programme.

Schmoll, O., G. Howard, J. Chilton & I. Chorus, 2006. *Protecting groundwater for Health – Managing the quality of drinking-water sources*. WHO Drinking-water Quality Series, World Health Organization, Geneva.

Shah, T., D. Molden, R. Sakthivadivel & D. Seckler, 2000. *The global groundwater situation: Overview of opportunities and challenges*. International Water Management Institute, Colombo, Sri Lanka, 19 p.

Treidel, H., J.L.Martin-Bordes & J.J. Gurdak (eds), 2012. *Climate change effects on groundwater resources. A Global synthesis of findings and recommendations*. IAH, International Contributions to Hydrogeology 27, CRC Press/Balkema, 401 p.

Tuinhof, A., C. Dumars, S. Foster, K. Kemper, H. Garduño & M. Nanni, 2002. *Groundwater Resource Management: an introduction to its scope and practice.* G-MATE Briefing Note Series No 1, G-MATE, The World Bank, Washington, 6 p.

UNESCO, 1992. *Ground Water.* Environment and Development Briefs No 2, 16 p.

UNICEF. *Groundwater: the invisible and endangered resource.* Geneva, UNICEF.

Van Dam, J.C. (ed.), 1997. Seawater Intrusion in Coastal Aquifers. Guidelines for study, monitoring and control. Water Reports No 11, FAO, Rome, 152 p.

Van der Gun, J., 2012. *Groundwater and global change: trends, opportunities and challenges.* United Nations World Water Assessment Programme, WWDR4 Side Publication Series No 01, UNESCO, Paris, 38 p. Available from: http://unesdoc.unesco.org/images/0021/002154/215496e.pdf.

WWAP, 2003. *The World Water Assessment Report. Water for People, Water for Life.* UNESCO-WWAP, 576 p. Available from: http://www.unesco.org/new/en/natural-sciences/environment/water/wwap/wwdr/wwdr1–2003/

WWAP, 2006. *The World Water Assessment Report 2. Water: a shared responsibility.* UNESCO-WWAP, 584 p. Available from: http://www.unesco.org/new/en/natural-sciences/environment/water/wwap/wwdr/wwdr2–2006/

WWAP, 2009. *The World Water Assessment Report 3. Water in a changing world.* UNESCO-WWAP, 318 p. Available from: http://www.unesco.org/new/en/natural-sciences/environment/water/wwap/wwdr/wwdr3–2009/

WWAP, 2012. *The World Water Assessment Report 4. Facing the Challenges.* UNESCO-WWAP, 3 volumes, 866 p. Available from: http://www.unesco.org/new/en/natural-sciences/environment/water/wwap/wwdr/wwdr4–2012/#c219661

6.1.4 Some groundwater-related internet portals

AHLSUD – Asociación Latinoamericana de Hidrología Subterránea para el Desarrollo
  Web: http://www.alhsud.com/
AQUASTAT – FAO's global information system on water and agriculture:
  Web: http://www.fao.org/nr/water/aquastat/main/index.stm
BGS – British Geological Survey – Groundwater
  Web: http://www.bgs.ac.uk/research/groundwater/home.html
BGR – Geological Survey of Germany – Groundwater
  Web: http://www.bgr.bund.de/EN/Themen/Wasser/wasser_node_en.html
  (*in German:* http://www.bgr.bund.de/DE/Themen/Wasser/grundwasser_node.html)
BRGM – Bureau de Recherches Géologiques et Minières (Geological Survey of France) – Water
  Web: http://www.brgm.fr/brgm/EN/expert_int_eau.htm
  (*in French*: http://www.brgm.fr/brgm/eau.htm)
CGWB – Central Ground Water Board, India
  Web: http://www.cgwb.gov.in/
Dutch Portal to International Hydrology and Water Resources
  Web: www.hydrology.nl
Environmental Protection Agency, United States
  Web: http://www.epa.gov/
EuroGeoSurveys – The Association of European Geological Surveys
  Web: http://www.eurogeosurveys.org/
European Environmental Agency
  Web: http://www.eea.europa.eu/
EUROSTAT – Statistics on the EU and candidate countries
  Web: http://epp.eurostat.ec.europa.eu/portal/page/portal/eurostat/home/

Groundwater Foundation, United States
    Web: http://www.groundwater.org/
Groundwater Protection Council, United States
    Web: http://www.gwpc.org/
G-WADI: Water and development information for arid lands – A global network
    Web: http://www.gwadi.org
Hydrologie (in French; mainly on surface water, but some interesting information on groundwater)
    Web: http://www.hydrologie.com
IAH – International Association of Hydrogeologists
    Web: http://www.iah.org/
IGME – Geological Survey of Spain
    Web: http://www.igme.es/internet/default.asp
IGRAC – International Groundwater Resources Assessment Centre (also: UNESCO-IGRAC)
    Web: http://www.un-igrac.org/
International Center on Qanats and Historic Hydraulic Structures (ICQHS)
    Web: http://www.icqhs.org/English/Default.aspx
International Network of Water-Environment Centres for the Balkans (INWEB)
    Web: http://www.inweb.gr
International Water Law Project
    Web: http://www.internationalwaterlaw.org/
ISARM – Internationally Shared Aquifer Resources Management
    Web: http://www.isarm.org/
IW:LEARN – International Waters Learning Exchange and Research Network
    Web: http://iwlearn.net/
IWMI – International Water Management Institute
    Web: http://www.iwmi.cgiar.org/
National Centre for Groundwater Research and Training – Australia
    Web: http://www.groundwater.com.au/research.php
National Groundwater Association
    Web: http://www.ngwa.org/Pages/default.aspx
SEMIDE/EMWIS – Euro-Mediterranean Information System on know-how in the Water sector
    http://www.semide.org/
SIAGUA – Sistema Iberoamericano de Información sobre el Agua
    Web: http://www.siagua.org/
The Water Channel
    http://www.thewaterchannel.tv/
UNESCO-IHP – UNESCO's International Hydrological Programme
    Web: http://www.unesco.org/new/en/natural-sciences/environment/water/ihp/
USGS – United States Geological Survey – Water
    Web: http://www.usgs.gov/water/
Water History
    http://www.waterhistory.org/

6.2   SELECTED BOOKS, PAPERS AND REPORTS SUMMARISING NATIONAL GROUNDWATER CONDITIONS

Only part of the publications listed below have been accessed and seen by the authors. The remaining titles, however, have been listed because they are likely to present an overall picture of the groundwater conditions in the corresponding countries. Several of the publications are wider in scope than only groundwater, but pay substantial attention to groundwater, e.g. in one or more chapters.

In addition to the selection shown below, reference is also made to the national summaries in the series on groundwater of the different continents, coordinated by R. Dijon and published by the United Nations Division of Technical Cooperation for Development (see References to Chapter 3: UN, 1976; 1982; 1983; 1986; 1987; 1993). Several other compilations by continent or region do exist, e.g. UNEP's *Africa Water Atlas* (UNEP/DEWA, 2010; 326 pages) and the European Environment Agency's *Groundwater quality and quantity in Europe* (EEA, 1999; 123 pages). Finally, many national institutions active in the field of groundwater have their portals on the Internet, where they share information on groundwater conditions in their country.

Afghanistan

Uhl, V.W., and M.Q. Tahiri, 2004(?). *Afghanistan – an overview of groundwater resources and challenges*. Uhl, Baron, Rana Associates, Inc., Washington Crossing, PA, USA., 60 p. Available from: http://www.vuawater.com/pages/Afghanistan_GW_Study.pdf.

Algeria

Kettab, A., 2001. *Water in Algeria: stakes, strategies, prospects and vision*. Hydrotop 2001, Scientific and Technical Conference, 24–26 April 2001, Marseille, France.

Argentina

World Bank, 2000. *Argentina Water Resources Management. Policy elements for Sustainable Development in the XXI Century*, Main Report.

Armenia

ERMC/AUA (no date). *State of the Environment Armenia: Groundwater/Hydrogeological map*. Available from: http://enrin.grida.no/htmls/armenia/soe_armenia/english/water/grwater.htm.

Australia

Jacobson, G., M.A. Habermehl & J.E. Lau, 1983. *Australia's groundwater resources*. Department of Water Resources and Energy, Australia, WATER 2000, Consultants Report 2, 65 p.

Lau, J.E., D.P. Commander & G. Jacobson, 1987. *Hydrogeology of Australia*. Bureau of Mineral Resources, Australia, Bulletin 227, 21 p.

Planet Water – OzH2O (no date). *Water resources use in Australia: Groundwater resources and use*. Available from: http://www.ozh2o.com/h2use3.html.

Bahamas

US Army Corps of Engineers, 2004. *Water resources assessment of the Bahamas*, 114 p. Available from: http://www.sam.usace.army.mil/en/wra/Bahamas/Bahamas.html.

Bahrain

Basheer, A., Z. Al Hashimi, A. Al Aradi, B. Altal, M. Al Noaim & M. Al Ansari, 2001. Development of water resources in Bahrain. In: *Proceedings of the WHO/UNEP First Regional Conference on Water Demand Management, Conservation and Pollution Control*, Amman, Jordan, 7–10 Oct. 2011.

Belarus

Shirokov, V.M., V.N. Pluznikov, 1995. Water resources of Belarus, their use and protection (Vodnye resursy Belarusii, ich ispol'zovanie i okhrana). *Vodnye Resursy (Water Resources)*, No 1, pp. 115–125, Moscow, Russian Academy of Sciences, Nauka. (in Russian).

Cherepansky, M.M., L. Konopelko & R. Oborotova, 1999. Water resources of the Republic of Belarus: development and problems. In: *Natural Resources*, 1999, No 1, Nat. Academy of Sciences of Belarus & Ministry of Natural Resources and Environmental Protection, Minsk, pp. 102–110.

Kudelsky, A.V., S. Gudak, V. Pashkevich, M. Kapora, V. Korkin, M. Fadeyeva & L. Shapoval, 1999. Underground waters of Belarus (resources, quality, utilization). In: *Natural Resources*, 1999, No 1, Nat. Academy of Sciences of Belarus & Ministry of Natural Resources and Environmental Protection, Minsk, pp. 48–58.

Belgium

M. Gulinck, 1966. *Atlas de Belgique: Hydrogéologie*. Comité National de Géographie, Royaume de Belgique, 68 p. + 2 maps.

Derycke, F., 1982. *Bilan des ressources en eau souterraine de la Belgique*. Commission of the European Communities, ECSC/EEC/EAEC, Brussels and Luxembourg. Schäfer GmbH, Hannover, 267 p.

Bolivia

Montes de Oca, I., 1997. *Geografía y recursos naturales de Bolivia*. 3ra Edición, La Paz, Bolivia, Edobol, 615 p.

Secretaría de Medio Ambiente, 1996. *Situación de los recursos hídricos en Bolivia*. Informe Final. Secretaría Nacional de Recursos Naturales y Medio Ambiente. Ministerio de Desarrollo Sostenible y Medio Ambiente, La Paz.

US Army Corps of Engineers, 2004. *Water resources assessment of Bolivia*, 118 p. Available from: http://www.sam.usace.army.mil/en/wra/Bolivia/Bolivia.html.

Botswana

Carlsson L., Selaolo E. and Von Hoyer M., 1993. Assessment of Groundwater Resources in Botswana. Experience from Botswana National Water Master Plan Study – 'Africa Needs Ground Water'. In: Univ. of Witwatersrand: *Convention Papers*, Geol. Soc. of South Africa, 11 p.

Brazil

Rebouças, A.C., 1988. Groundwater in Brazil. *Episodes,* Vol. 11, No 3, pp. 209–219.

Ministério do Meio Ambiente, dos Recursos Hídricos e da Amazônia Legal, Secretaria de Recursos Hídricos. 1998. *Water Resources of Brazil*. Brasilia.

Ministry of Environment, Secretariat of Water Resources, Brazilia, 2006. *Plano Nacional de Recursos Hídricos,* Cap.10: Situação Atual das Águas do Brazil.

Bulgaria

Galabov, M., 1994. *Water resources of Bulgaria*. National Water Council.

Canada

Rivera, A., 2005. *How well do we understand groundwater in Canada?* Geological Survey of Canada. Available from: http://ess.nrcan.gc.ca/2002_2006/gwp/index_e.php [Accessed in October 2008].

Central African Republic

Plesinger, V., 1990. *Les eaux souterraines de la RCA et leur exploitation*. UNDP, Report prepared for the project PNUD/DCTD/CAF/86/004.

Chad

Schneider, J.L., 1989. *Géologie et Hydrogéologie de la République du Tchad.* PhD thesis, University of Avignon, 3 vols.

Schneider, J.L., 2001. *Géologie, Archéologie, Hydrogéologie, République du Tchad.* Ministry of Water and the Environment, N'Djamena, Chad, Two volumes, 1100 pp.

Terap, M.M. and B.W. Kaibana, 1992. *Ressources en eau souterraine du Tchad.* Observatoire du Sahara et du Sahel, Project launch workshop 'Aquifères des grands bassins', Cairo, 22–25 Nov. 1992.

China

Chen, M., 1985. *Groundwater resources and development in China.* 18th Congress, Intern. Assoc. of Hydrogeologists, 8–13 Sept. 1985, Cambridge (United Kingdom), 14 p.

Chen, M., 1987. Groundwater resources and development in China. *Environmental Geology,* Vol. 10, No 3, pp. 141–147. Available from: doi: 10.1007/BF02580469.

Li, Jinchang, G. Zhengang, Z. Zhaoxiu, H. Xianjie, K. Fanwan & F. Dedi, 1990. Accounting for groundwater resource in China. In: *Natural Resource Accounting for Sustainable Development,* China Environmental Sciences Press, pp. 133–153.

Zhaoxin, W. (ed.), 1992. *Groundwater resources development and use in China.* Hohhot Neirnenggu People Publishing House.

Congo

Moukolo, N., 1992. État des connaissances actuelles sur l'hydrogéologie du Congo Brazzaville. *Hydrogéologie,* 1992, No 1–2, pp. 47–58.

Costa Rica

US Army Corps of Engineers, 1996. *Water resources assessment of Costa Rica.* Available from: http://www.sam.usace.army.mil/en/wra/CostaRica/CostaRica.html.

Democratic Republic of the Congo

UNEP, 2011. *Water issues in the Democratic Republic of the Congo. Challenges and opportunities.* Technical Report, UNEP, Nairobi, Kenya, 98 p.

Denmark

Kelstrup, N., A. Baekgaard and L. Andersen, 1982. *Grundvandsressourcer i Danmark.* Commission of the European Communities, ECSC/EEC/EAEC, Brussels and Luxembourg, Schäfer GmbH, Hannover, 557 p.

Henriksen, H.J., and B. Madsen, 1997. Water resources in Denmark. In: *Geologi – Nyt fra GEUS* 2/97, pp. 5–15.

Djibouti

Muller, W., 1982. *Les ressources en eau de la République de Djibouti, possibilités et limites du développement régional.*

Dominica, Antigua, Barbuda, St Kitts and Nevis

US Army Corps of Engineers, 2004. *Water resources assessment of Dominica, Antigua, Barbuda, St Kitts and Nevis,* 140 p. Available from: http://www.sam.usace.army.mil/en/wra/N_Caribbean/N_Caribbean.html)

Dominican Republic

US Army Corps of Engineers, 2002. *Water resources assessment of the Dominican Republic*, 143 p. Available from: http://www.sam.usace.army.mil/en/wra/DominicanRepublic/DominicanRepublic.html.

Ecuador

US Army Corps of Engineers, 1998. *Water resources assessment of Ecuador*, 83 p. Available from: http://www.sam.usace.army.mil/en/wra/Ecuador/Ecuador.html.

Egypt

Shahin, M., 1991. *Assessment of Groundwater Resources in Egypt.* IHE Rep. Ser. 23, Delft, 71 p.

Ground Water Research Institute, 2001. *The groundwater sector plan, national level.* National Water Research Center, Cairo, Egypt.

Khater, A.R., 2005. Groundwater Resources in Egypt: Development and Protection Measures. In: CIHEAM, *Internat. Conf. on Water, Land and Food Security in Arid and Semi-Arid Regions*, Valenzano-Bari, Italy, 6–11 Sept 2005.

El Salvador

US Army Corps of Engineers, 1998. *Water resources assessment of El Salvador*, 71 p. Available from: http://www.sam.usace.army.mil/en/wra/ElSalvador/ElSalvador.html.

Eritrea

Zerai, H., 1996. Groundwater and geothermal resources of Eritrea, with emphasis on their chemical quality. *Journal of African Earth Sciences*, Vol. 22, No 4.

Estonia

Andresmaa, E., 2001. *Groundwater management and protection in Estonia.* Presentation (20 p) and Abstract (2p), Workshop on the protection of waters uses as a source of drinking water supply, Budapest, November 2001.

Ethiopia

Alemayehu, T., 2006. *Groundwater occurrence in Ethiopia.* Addis Ababa University, UNESCO, 106 p.

Aytenffisu, M. 1981. *Groundwater in Ethiopia.*

France

Bodelle, J., & J. Margat, 1980. *L'Eau souterraine en France.* Paris, Masson, 208 p.

Blum, A., 2004. *L'état des eaux souterraines en France. Aspects quantitatifs et qualitatifs.* Orléans, IFEN, Etudes et travaux No. 43, 34 p.

Defosssez, M., J. Mouton & B. Chapignac, 1982. *Bilan des ressources en eau souterraine de la France.* Commission of the European Communities, ECSC/EEC/EAEC, Brussels and Luxembourg. Schäfer GmbH, Hannover, 557 p.

Ministère de l'Environnement, 1998. *L'eau souterraine, une richesse à partager et à protéger.* Ministry of the Environment, France.

Roux, J.C. (ed.), 2006. *Aquifères et eaux souterraines en France.* Orleans, BRGM, 1000 p.

Gambia

Fernandopulle, 1983. *Groundwater resources of the Gambia.* Preliminary Report Rural Water Supply Project, Min. Wat. Res. & Envir., Dep. Water Res., UNDP.

Ghana

Agyekum, W., 2004. Groundwater resources of Ghana with focus on international shared aquifer boundaries. In: B. Appelgren (ed.), 2004, *Managing Shared Aquifer Resources in Africa.* Proceedings of the International ISARM-Africa Workshop at Tripoli, Libya, 2–4 June 2002. UNESCO-IHP-VI Series on Groundwater No. 8, pp. 77–85.

Water Resources Management (WARM), 1998. *Ghana's water resources: management, challenges and opportunities.* WARM study, Ministry of Works and Housing, Accra, Ghana.

Germany

Bannick, C., E. Engelmann, R. Fendler, J. Frauenstein, H. Ginzky, C. Hornemann, O. Ilvonen, B. Kirschbaum, G. Penn-Bressel, J. Rechenberg, S. Richter, L. Roy and R. Wolter, 2008. *Grundwasser in Deutschland*, MBU, 72 p.

Büro Dr Pickel, 1982. *Grundwasserbilanz der Bundesrepublik Deutschland.* Commission of the European Communities, ECSC/EEC/EAEC, Brussels and Luxembourg. Schäfer GmbH, Hannover, 557 p.

Matthess, G., 1979. Die Grundwasserbilanz der Bundesrepublik Deutschland. *Umschl. Wis. u Technik*, B. 79, pp. 144–149.

Greece

Anonymous, no date (2007 or later). Groundwater in the Southern Member States of the European Union: an assessment of current knowledge and future prospects. Country report for Greece. European Academies Science Advisory Council, 37 p. Available from: http://www.easac.eu/fileadmin/PDF_s/reports_statements/Greece_Groundwater_country_report.pdf

Guatemala

US Army Corps of Engineers, 2000. *Water resources assessment of Guatemala*, 90 p. Available from: http://www.sam.usace.army.mil/en/wra/Guatemala/Guatemala.html.

Guyana

Négrel P., and E. Petelet-Giraud, 2001. *Les eaux souterraines de la Guyane: caractéristiques hydrogéochimiques et isotopiques.* BRGM/RP-50306-FR, 80 p.

US Army Corps of Engineers, 1998. *Water resources assessment of Guyana*, 67 p. Available from: http://www.sam.usace.army.mil/en/wra/Guyana/Guyana.html.

Haiti

US Army Corps of Engineers, 1999. *Water resources assessment of Haiti*, 93 p. Available from: http://www.sam.usace.army.mil/en/wra/Haiti/Haiti.html.

Hungary

Altreder, A., 1984. Experience and main results of groundwater resources assessment in improvement of general scheme for water management in Hungary. In: *Development of methods for optimal use and artificial groundwater recharge.* CMEA Symposium, Estergom, pp. 16–23.

VITUKI, 2006. *Groundwaters in Hungary, Guide II*. Prepared for the Ministry for Environment and Water, 76 p. Available from: http://www.kvvm.hu/szakmai/karmentes/kiadvanyok/fav2/fav2_eng.pdf.

India

Anonymous, 1995. *Ground Water Resources of India*. Faridabad, Central Ground Water Board, Ministry of Water Resources, 150 p.

CGWB, 1996. *Groundwater statistics (1996)*, Central Groundwater Board, New Delhi.

CGWB, 2011. *Dynamic Ground Water Resources of India (as on 31 March 2009)*. Faridabad, Central Groundwater Board, Ministry of Water Resources, Government of India, 243 p. Available from: http://cgwb.gov.in/documents/Dynamic%20GW%20Resources%20-2009.pdf.

CGWB, 2012. *Ground Water Year Book – India, 2011–2012*. Faridabad, Central Groundwater Board, Ministry of Water Resources, Government of India, 63 p. Available from: http://cgwb.gov.in/documents/Ground%20 Water%20Year%20Book%20-%202011–12.pdf.

Shah, T., 2009. *India's ground water irrigation economy: the challenge of balancing livelihoods and environment*. Paper presented at the 5th Asian Regional Conference of INCID, Vigyan Bhawan, New Delhi, 9–11 December 2009. Available from: http://cgwb.gov.in/documents/papers/incidpapers/Paper%203-%20Tushaar%20Shah.pdf.

Water Management Directorate, 2005. *National water resources at a glance*. CWC/MWR, India.

Indonesia

Soetrisno, S., 1998. *Groundwater resources of Indonesia and their management*.

Suprapto, A., 2002. Land and water resources development in Asia. In: *Investment in land and water*, RAP Publication 2002/09, FAO Regional Office for Asia and The Pacific, Bangkok.

Ireland

Wright, G, C. Aldwell, D. Daly & E. Daly, 1982. *Groundwater resources of the Republic of Ireland*. Commission of the European Communities, ECSC/EEC/EAEC, Brussels and Luxembourg, Schäfer GmbH, Hannover, 140 p.

Aldwell, C., 1991. *A review of groundwater and its development in Ireland*. Dublin, Geological Survey of Ireland, 13 p.

Israel

Shamir, U., 1993. *Development and management of groundwater resources: General principles and the case of Israel*. Paper presented at the Seminar on Groundwater, Inst. Techn. Geominero de ESpaña, Madrid, 1 April 1993, 15 p.

MAE, 2005. *L'Eau en Israël*. Ministry of Foreign Affairs (Ministère des Affaires Etrangères - MAE), France.

Italy

Mouton, J., & F. Mangano, 1982. *Studio sulle risorse in acque sotteranee dell'Italia*. Commission of the European Communities, ECSC/EEC/EAEC, Brussels and Luxembourg. Schäfer GmbH, Hannover, 193 p. + annexes

Civita, M., A. Massarutto & G. Seminara, no date (2008 or later). *Groundwater in Italy: a review. Contribution to the EASAG Report on Groundwater in South European Mediterranean countries*. European Academies Science Advisory Council, 48 p.

Ivory Coast

Jourda, J.P.R., 2004. Les ressources en eau souterraine de la Côte Ivoire et le cas des aquifères transfrontaliers entre la Côte Ivoire et le Ghana. In: B. Appelgren (ed.), 2004, *Managing Shared*

Aquifer Resources in Africa. Proceedings of the International ISARM-Africa Workshop at Tripoli, Libya, 2–4 June 2002, UNESCO-IHP-VI, Series on Groundwater No 8, pp. 87–92.

Jamaica

US Army Corps of Engineers, 2001. *Water resources assessment of Jamaica*, 118 p. Available from: http://www.sam.usace.army.mil/en/wra/Jamaica/Jamaica.html.

Japan

National Land Agency, Government of Japan, 1995. *Water resources in Japan*. White paper.

Jordan

Salameh, H. 1991. *Jordan's water resources and their future potential*. Proceedings of the Symposium 27–28 October 1991. Published by Friedrich Ebert Stiftung, 1992, Amman, Jordan.
El Naser, H., & Z. Elias, 1993. *Jordan's water sector*. Country paper presented at the regional symposium on water use and conservation, organized by ESCWA and CEHA.

Kenya

Ministry of Water Development. No date. *Ground water resources of Kenya (Reconnaissance study)*. Master planning section.
Mwango, F.K., Muhangù, C.O. Juma & I.T. Githae, 2004, Groundwater resources in Kenya. In: Appelgren, B. (ed), 2004. *Managing Shared Aquifer Resources in Africa*, Proceedings of the International ISARM-Africa Workshop at Tripoli, Libya, 2–4 June 2002, UNESCO IHP-VI Series on Groundwater No 8, pp. 93–100.

Kuwait

Senay, Y., 1981. Geohydrology. In: *Geology and groundwater hydrology of the State of Kuwait*. Ministry of Electricity and Water.

Latvia

UNECE, 1999. *Environmental Performance Reviews – Latvia. Chapter 7: Water management*. UN, New York and Geneva.

Lebanon

Jaber, B., 1995. *The water resources in Lebanon*. Conference on the environmental management for sustainable development in Lebanon. UNEP/LNCSR, Beirut.
Maksoud, S.W., 1998. *Towards a sustainable groundwater development in Lebanon*. International Conf. 'Water and Sustainable Development'.

Libya

Pallas, Ph., 1980. Water resources of the Socialist People's Libyan Arab Jamahiriya. In: *The Geology of Libya, II, Part 4: Hydrogeology*. 2nd Sympos. Geol. Libya, Tripoli, September 1978. Tripoli, Al Fateh University. London, Acad. Press Inc., pp. 539–594.
Pallas, Ph., & O. Salem, 1999. *Water Resources Utilisation and Management of the Socialist People Arab Jamahiriya*. Intern. Conf. 'Regional Aquifer Systems in Arid zones – Managing Non-Renewable Resources', Tripoli, Nov. 1999. Paris, UNESCO, 65 p.
Salem, O., 1992. *Hydrogeology of the major groundwater basins of Libya*. Observatoire du Sahara et du Sahel, Proiect launch wrkshop 'Aquifères des grands bassins', Cairo, 22–25 Nov. 1992, 15 p.

Miludi, H. & O.M. Salem, 2001. Water policies in the Libyan Arab Jamahiriya. In: *Proceedings of the WHO/UNEP First Regional Conference on Water Demand Management, Conservation and Pollution Control*, Amman, Jordan, 7–10 Oct. 2001.

Lithuania

UNECE, 1999. *Environmental Performance Reviews – Lithuania. Chapter 7: Water management*. UN, New York and Geneva.

Luxembourg

Bintz, J, 1982. *Bilan des ressources en eau souterraine du Grand-Duché de Luxembourg*. Commission of the European Communities, ECSC/EEC/EAEC, Brussels and Luxembourg. Schäfer GmbH, Hannover, 52 p.

Malawi

Mkandawire, O.P., 2004. Groundwater resources of Malawi. In: Appelgren, B. (ed), 2004. *Managing Shared Aquifer Resources in Africa*, Proceedings of the International ISARM-Africa Workshop at Tripoli, Libya, 2–4 June 2002, UNESCO IHP-VI Series on Groundwater No 8, pp. 101–104.

Mali

Alhousseini, M., 2004. État de la connaissance des eaux souterraines au Mali. In: Appelgren, B. (ed), 2004. *Managing Shared Aquifer Resources in Africa*, Proceedings of the International ISARM-Africa Workshop at Tripoli, Libya, 2–4 June 2002, UNESCO IHP-VI Series on Groundwater No 8, pp. 105–107.

PNUD, 1990. *Synthèse hydrogéologique du Mali*. République du Mali/PNUD-DCTD, Project report, MLI/84/005.

Malta

Guttierez, A., 1994. *Evaluation des ressources en eau souterraine de l'île de Malte*. PhD thesis, Paris, University of Paris, 326 p. + annexes.

Spiteri Stains, E., 1987. *Aspects of water problems in the Maltese Islands*. Brochure distributed at IAH Symposium Barcelona, October 1987, 16 p.

Mauritania

Mochar ould Mohamaden Fall, M., 1992. *Les basins sédimentaires de la Mauritanie. D'un aperçu hydrogéologique aux difficultés de l'évaluation de leur reserve en eau*. Observatoire du Sahara et du Sahel, project launch workshop 'Aquifères des grands bassins', Cairo, 22–25 Nov. 1992.

PNUD, 1990. *Les eaux souterraines de Mauritanie*. Projet PNUD/DCT/MAU/87/008.

Mauritius

Ministry of Public Utilities, 2003. *Water Resources of Mauritius*. Ministry of Public Utilities.

Mexico

Comisión Nacional del Agua, 1998. *Inventario de Obras en México*. México.

Conagua, 2010. *Atlas digital del agua, Mexico 2010*. Available from: http://www.conagua.gob.mx/atlas/.

Morocco

Anonymous, 1971–1977. *Ressources en eau du Maroc.* Direction Hydraulique, Notes Mém. Serv. Géol. Maroc, 3 volumes, Rabat, Morocco.

Nadifi, K., & Wahabi, R., 1993. *Water resources in Morocco: The state of art and the future prospects.* Presented at the 'Symposium régional sur l'utilisation et la conservation des ressources en eau'.

Mokhtar Bzioui, 2004. *Rapport National 2004 sur les Ressources en Eau au Maroc*, UN Water-Africa, Nov. 2004.

Mozambique

Carmo Vaz, A. 1999. *Recursos hidricos de Moçambique, potencial, problemas politicos.* Associaçao Moçambicana de Ciencias e Tecnologia. Maputo. In: Tauacale, 2002 (see below).

Direcção Nacional de Agues (DNA). 1999. *Water resources of Mozambique.* In: Tauacale, 2002 (see below).

Tauacale, F. 2002. *Water Resources of Mozambique and the situation of the shared rivers.*

Namibia

Christelis, G., & W. Struckmeier, (eds), 2001. *Groundwater in Namibia: an explanation to the Hydrogeological Map.*

The Netherlands

Van den Berg, C., S. Jelgersma, C. Meinardi, J. van der Gun & M. Damoiseaux, 1982. *Grondwater in Nederland: Behoeften en beschikbare hoeveelheden.* Commission of the European Communities, ECSC/EEC/EAAC, Brussels and Luxembourg. Schäfer GmbH, Hannover, 65 p.

NHV, 1998. *Water in The Netherlands.* Netherlands Hydrological Society and Netherlands National Committee of the IASH, NHV-special 3, Delft, The Netherlands, 186 p.

Dufour, F.C., 1998. *Grondwater in Nederland. Onzichtbaar water waarop wij lopen.* Geologie van Nederland, deel 3 (in Dutch). NITG-TNO, Delft, The Netherlands, 265 p.

Dufour, F.C., 2000. *Groundwater in The Netherlands. Facts and figures.* NITG-TNO, Delft/Utrecht, The Netherlands, 90 p.

New Zealand

Brown, R.J., 1992. Environmental hydrogeology and sustainable groundwater withdrawal. In: *Proc. Internat. Workshop "Groundwater and the Environment"*, Beijing, 1992. China National Committee for the IAH and China Institute of Hydrogeology and Engineering Geology, Seismological Press, Beijing, China, pp. 55–67.

Brown, R.J., and R.C. Gregg, 1994. Groundwater in New Zealand. *New Zealand Mining*, Vol. 14, pp. 11–14.

Nicaragua

US Army Corps of Engineers, 2001. *Water resources assessment of Nicaragua*, 90 p. Available from: http://www.sam.usace.army.mil/en/wra/Nicaragua/Nicaragua.html.

Niger

CNEDD, OSS, Ce.S.I.A. 2000. *Exploitation et état des ressources naturelles au Niger.*

Ousmane, B., 2004. La gestion et l'exploitation des eaux souterraines au Niger. In: Appelgren, B. (ed.), 2004. *Managing Shared Aquifer Resources in Africa*, Proceedings of the International ISARM-Africa Workshop at Tripoli, Libya, 2–4 June 2002, UNESCO-IHP-VI, Series on Groundwater No. 8, pp. 129–133.

Oman

Ministry of Water Resources, 1995. *Water resources of the Sultanate of Oman: an introductory guide*. MWR, Muscat, 78 p.

Pakistan

Amin, M., no date (2002 or later). *Pakistan's groundwater reservoir and its sustainability*. WAPDA, Lahore, Pakistan. Available from: http://www.watertech.cn/english/amin.pdf.

Kahlown, M.A., & A. Majeed, 2001. Water resources situation in Pakistan: challenges and future strategies. In: *Science Vision*, Vol. 7, No 3, Islamabad, Pakistan, pp. 33–45.

Kahlown, M.A., & A. Majeed, 2004. *Water-resources situation in Pakistan: challenges and future strategies*. Pakistan Council of Research in Water Resources.

Papua New Guinea

Carter, J.A., 1979. *Groundwater in Papua New Guinea*. Geological Survey of Papua New Guinea, Report 79/129.

Peru

Dirección General de Aguas y Suelos (DGSA). 1992. *Estudio Básico Situacional de los Recursos Hídricos del Perú*. Ministerio de Agricultura, Lima, 335 pp.

US Army Corps of Engineers, 2004. *Water resources assessment of Peru*, 118 p.

Portugal

INAG, 2001. *Plano Nacional da Água, Portugal*. Ministério do Ambiente e do Ordenamento do Território, Instituto da Água (INAG).

Ribeiro, L., & L. Veiga da Cunha, no date (2006 or later). *Portuguese Groundwater Report – EASAG WG on the role of groundwater in the water resources policy of Southern EU member states*. European Academies Science Advisory Council.

Russian Federation

Tcherneyev, A.M. *et al*. 1992. Water resources and water management in Russia (Vodnyye resursy i vodnoye khozaystvo Rosii). *Drainage and Water Management (Melioratsia i Vodnoye Khozaystvo)*, No 9–12, pp. 2–5, Agropromizdat/Ministry of Agriculture, Moscow. (in Russian).

Yazvin, L.S. and Zektser, I.S., 1996. *Fresh groundwater resources in Russia: modern state, perspectives of use and tasks for future investigations*. Water Resources, Vol. 23, No 1, pp. 24–30.

Saudi Arabia

Anonymous, 1987. *Groundwater Resources Evaluation in Saudi Arabia and Long Term Strategic Plan for Fresh Groundwater Use*. King Fahd University of Petroleum and Minerals, Dahran.

Senegal

PNUD, 1993. *Synthèse des ressources en eau de la République de Sénégal, cartography des systems aquifères (piézometrie-géometrie, salinité)*. PNUD/DAD/SG.

Somalia

Anonymous, 1986. *Comprehensive Groundwater Development project, End of Project Report, Draft Copy, Vol 3: Hydrogeology, Somalia*. LBI/MMWR/WDA, Mogadishu, Somalia.

South Africa

Vegter, J.R., 1995. *Groundwater resources of South Africa – an explanation of a set of national groundwater maps. Water Research Commission Technical Report TT74/95*. Pretoria.

Woodford, A., P. Rosewarne & J. Girman, no date (2005 or later). *How much groundwater does South Africa have?* Department of Water Affairs and Forestry, 6 p.

Spain

Anonymous, 1990. *Unidades hidrogeológicas de la España peninsular e Islas Baleares*. Madrid, MOPU, Serv. Geológico No 52 (text + map at 1/1 000 000), 32 p.

Anonymous, 1994. *Libro Blanco de las aguas subterráneas*. Madrid, Ministerios de Industria y Energía, Obras Públicas, Transportes y Medio Ambiente, 135 p.

De Stefano L., & M.R. Llamas, 2012. *Water and food security and care of nature in Spain*. Taylor and Francis.

Dumont, A., L. De Stefano & E. López-Gunn, 2011. *El agua subterránea en España según la Directiva Marco del Agua: una visión de conjunto*. VII Congreso Ibérico sobre Gestión y Planificación del Agua, 16–19 Feb 2011, Talavera de la Reina, 8 p.

Garrido, A., & M.R. Llamas, 2009. *Water policy in Spain*. CRC Press, 234 pp.

ITGE, 1993. *Las aguas subterráneas en Espana. Estudio de síntesis*. Second edition. Madrid, Instituto Tecnológico Geominero de Espana, 591 p. + 12 maps.

Ministerio de Medio Ambiente, 2000. *Libro Blanco del agua en España*. Madrid, Ministerio de Medio Ambiente, 637 p.

Sri Lanka

WRB (Water Resources Board), 2005. *Groundwater resources of Sri Lanka*. Available from: http://tsunami.obeysekera.net.

Sudan and South Sudan

Ibrahim, M.E. and Salih, M.K., 1996. *Groundwater resources of Sudan: development potential*. Fourth Nile 2002 Conference, Kampala, Uganda.

Suriname

US Army Corps of Engineers, 1998. *Water resources assessment of Suriname*, 111 p. Available from: http://www.sam.usace.army.mil/en/wra/Suriname/Suriname.html.

Sweden

Johansson, S., no date. Groundwater in Sweden. IAHS, Enquête, Rapport 10, Commission des Eaux Souterraines, 5 p. Available from: http://iahs.info/redbooks/a026/Sout_Q0_R10.pdf.

Pousette, J.B., 1994. Shallow groundwater in Sweden – a vulnerable resource. In: *Water Down Under 94: Groundwater papers, preprints of papers*. National conference publication, Institution of Engineers, Australia, 1994, pp. 723–726. Available from: http://search.informit.com.a

Tajikistan

Tahirov, I.G., and Kupayi, G.D. 1994. *Water resources of the Republic of Tajikistan* (in Russian). Dushanbe, Vol. 1: 181 p., Vol. 2: 119 p.

Tunisia

Ennabli, M., 2000. *Analyse des stratégies et prospective de l'eau en Tunisie.*

Hamdane, A., 1994. *La gestion de l'eau en Tunisie, Rapport de synthèse*. DG Agricultural Engineering, Min. Agric., Tunis, Tunisia, 24 p.

Uganda

Kitakarugire, J.A., 2004. Managing shared aquifer resources in Africa – Uganda case. In: Appelgren, B. (ed), 2004. *Managing Shared Aquifer Resources in Africa*, Proceedings of the International ISARM-Africa Workshop at Tripoli, Libya, 2–4 June 2002, UNESCO IHP-VI Series on Groundwater No 8, pp. 171–174.

United Arab Emirates

Brook, M.C., H. Al Houqani, T. Darawsha, M. Al Alawneh & S. Achary, 2006. *Groundwater resources: development and management in the Emirate of Abu Dhabi, United Arab Emirates.*

United Kingdom

Monkhouse, R.A. & H.J. Richards, 1982. *Groundwater resources of the United Kingdom*. Commission of the European Communities, ECSC/EEC/EAEC, Brussels and Luxembourg. Schäfer GmbH, Hannover, 252 p.

UK Groundwater Forum, no date. *Groundwater: a valuable resource*. Briefing Note. Available from: http://www.groundwateruk.org/downloads/groundwater_valuable_resource.pdf.

United States of America

Miller, J.A. ed., 2000. *Ground Water Atlas of the United States*. US Geological Survey, 404 p. On-line version available from: http://pubs.usgs.gov/ha/ha730/gwa.html.

Uruguay

Montaño, X., Gagliardi, J., S. and M. Montaño, 2006. Recursos hídricos subterráneas del Uruguay. In: IGME, España: *Boletín Geológico y Minero*, Vol. 117, No 1.

Yemen

TS/HWC-UNDP/UNDESA. 1995. *Final reports. Volume IV: Groundwater resources.*

Van der Gun, J.A.M., & A.A.Ahmed, 1995. *The water resources of Yemen – A summary and digest of available information*. Report WRAY–35. MOMR, Sana'a and TNO, Delft, 228 p. Available from: http://www.un-igrac.org/dynamics/modules/SFIL0100/view.php?fil_Id=122.

Rybakov, V., R. Tkachencko, N. Mikhailin, N. Gamal, M. G. Ali, M. Danikh, A. Al Khouri & A. Al Thary, 1995. *Republic of Yemen: Groundwater resources available for development*. GCC/IASNSR, Moscow, and MOMR, Sana'a.

6.3 SELECTED SOURCES OF COMPILED INFORMATION ON THE WORLD'S MEGA AQUIFERS

6.3.1 Africa

Nubian Aquifer System

Bakhbakhi, M., 2002. *Nubian Sandstone Aquifer System Resource Evaluation*. Paris, CEDARE/UNESCO.

Bakhbakhi, M., 2004. Hydrogeological framework of the Nubian Sandstone Aquifer System. In: Appelgren, B. (ed), 2004. *Managing Shared Aquifer Resources in Africa*, Proceedings of the International ISARM-Africa Workshop at Tripoli, Libya, 2–4 June 2002, UNESCO IHP-VI Series on Groundwater No 8, pp. 177–201.

Bakhbahki, M., 2006. Nubian Sandstone Aquifer System. In: Foster, S., & D. Loucks, 2006. *Non-renewable groundwater resources*. UNESCO-IHP, IHP-VI Series on Groundwater No 10, pp. 75–81.

CEDARE/IFAD, 2002. *Regional Strategy for the Utilisation of the Nubian Sandstone Aquifer System, Volume II*. CEDARE, Heliopolis Bahry, Cairo, Egypt.

IME, 2008. *Les aquifères fossiles au sud de la Méditerranée*. Institut Méditerranéen de l'Eau, Marseille, France, 30 p.

Thorweihe, U., & M. Heinl, 1996. *Groundwater Resources of the Nubian Aquifer System*. Berlin, OSS/Technical University of Berlin, 95 p.

UNESCO/OSS, 2005. *Ressources en eau et gestion des aquifères transfrontaliers de l'Afrique du Nord et du Sahel*. ISARM-Africa, UNESCO IHP-IV Series on Groundwater No 11, Paris.

North Western Sahara Aquifer System (NWSAS)

Besbes, M., M.Babasy, S. Kadri, D. Latrech, A. Mamou. P. Pallas & M. Zammouri., 2004. Conceptual framework of the North Western Sahara Aquifer System. In: Appelgren, B. (ed), 2004. *Managing Shared Aquifer Resources in Africa*, Proceedings of the International ISARM-Africa Workshop at Tripoli, Libya, 2–4 June 2002, UNESCO IHP-VI Series on Groundwater No 8, pp. 163–169.

IME, 2008. *Les aquifères fossiles au sud de la Méditerranée*. Institut Méditerranéen de l'Eau, Marseille, France, 30 p.

Mamou, A., M. Besbes, B. Abdous, D. Latrech & C. Fezzani, 2006. North Western Sahara Aquifer System. In: Foster, S., & D. Loucks, 2006. *Non-renewable groundwater resources*. UNESCO-IHP, IHP-VI Series on Groundwater No 10, pp. 68–74.

OSS, 2003. *Système aquifère du Sahara septentrional: Gestion commune d'un basin transfrontière*. Synthesis Report. Observatoire du Sahara et du Sahel, Tunis, 147 p.

OSS, 2008a. *Système aquifère du Sahara septentrional: Gestion concertée d'un basin transfrontalier*. Collection Synthèse No 1. Tunis, Observatoire du Sahara et du Sahel, 56 p. Available from: http://www.oss-online.org/pdf/synth-sass_Fr.pdf.

OSS, 2008b. *The North-Western Sahara Aquifer System: Concerted management of a transboundary water basin*. Synthesis Collection No 1. Tunis, Observatoire du Sahara et du Sahel, 56 p. Available from: http://www.oss-online.org/pdf/synth-sass_En.pdf.

UNESCO/OSS, 2005. *Ressources en eau et gestion des aquifères transfrontaliers de l'Afrique du Nord et du Sahel*. ISARM-Africa, UNESCO IHP-IV Series on Groundwater No 11, Paris.

Murzuk–Djado Basin

IME, 2008. *Les aquifères fossiles au sud de a Méditerranée*. Institut Méditerranéen de l'Eau, Marseille, France, 30 p.

Sola, M.A., & D. Worsley (ed.), 2000. *Geological Exploration in Murzuq Basin*. Amsterdam, Elsevier, 519 p.

UNESCO/OSS, 2005. *Ressources en eau et gestion des aquifères transfrontaliers de l'Afrique du Nord et du Sahel*. ISARM-Africa, UNESCO IHP-IV, Series on Groundwater No 11, Paris.

Taoudéni–Tanezrouft Basin

Dakoure, D., 2010. *Multi-disciplinary approach to improve the knowledge of southeastern border of Taoudeni sedimentary basin*. International Conference ISARM 2010.

Derouane, J., & D. Daukoure, 2007. Hydrogeological structure of the aquifer system in the Taoudenni sedimentary basin, Burkina Faso. In: L. Chéry & G. De Marsily (eds), *Aquifer*

Systems Management: Darcy's legacy in a world of impending water shortage. IAH Selected Papers, Vol 10, Taylor & Francis/Balkema, Leiden, The Netherlands, pp. 137–148.

Derouane, J., and D. Daukoure, 2008. Etude hydrogéologique et modélisation mathémathique du système aquifère du basin sedimentaire de Taoudeni au Burkina Faso. In: *International Symposium Darcy 2006 – Aquifer Systems Management*, Darcy-45, IAH and BRGM, May–June 2006, Dijon, France.

IME, 2008. *Les aquifères fossiles au sud del a Méditerranée.* Institut Méditerranéen de l'Eau, Marseille, France, 30 p.

UNESCO/OSS, 2005. *Ressources en eau et gestion des aquifères transfrontaliers de l'Afrique du Nord et du Sahel.* ISARM-Africa, UNESCO IHP-IV Series on Groundwater No 11, Paris.

Senegalo-Mauritanian Basin

Diagana, B., 1997. *Gestion des eaux internationales en Afrique Sub-Saharienne. Bilan diagnostic de la gestion integrée des eaux et des contraintes environnementales dans la vallée du fleuve Sénégal.* Rapport No A/PNUD/DASDG/RAF/94/01C-SAT1.

Diagana, B., & S. Thieye, 2010. Gestión intégrée des ressources en eau dans les bassins transfrontaliers – Bassin côtier sénégalo-mauritanien – Comportement du champ captant d'Idini pour l'alimentation en eau potable de la ville de Nouakchott, Mauritanie. In: *Proceedings Third Intern. Conference 'Managing Shared Resources in Africa'*, Tripoli, 25–27 May 2008, UNESCO, Paris, pp. 163–179.

IME, 2008. *Les aquifères fossiles au sud de la Méditerranée.* Institut Méditerranéen de l'Eau, Marseille, France, 30 p.

SGPRE, 1999/2000. Étude hydrogéologique de la nappe profonde du Maestrichtien.

UNESCO/OSS, 2005. *Ressources en eau et gestion des aquifères transfrontaliers de l'Afrique du Nord et du Sahel.* ISARM-Africa, UNESCO IHP-IV Series on Groundwater No 11, Paris.

Iullemeden–Irhazer Basin

Dodo, A., 1992. *Etude des circulations profondes dans le grand bassin sédimentaire du Niger: identification des aquifères et compréhension de leurs fonctionnements.* PhD thesis, Univ. of Neuchâtel, Switzerland.

Dodo, A., 2004. Caractérisation des systèmes aquifères transfrontaliers du Niger. In: Appelgren, B. (ed.), 2004. *Managing Shared Aquifer Resources in Africa*, Proceedings of the International ISARM-Africa Workshop at Tripoli, Libya, 2–4 June 2002, UNESCO IHP-VI Series on Groundwater No 8, pp. 123–128.

Dodo, A.K., M. O. Baba Sy & A. Mamou, 2010. La gestion concertée des ressources en eau partagées du Système Aquifère saharo-sahélien d'Iullemeden (Afrique de l'Ouest). In: *Proceedings Third Intern. Conference 'Managing Shared Resources in Africa'*, Tripoli, 25–27 May 2008, UNESCO, Paris, pp. 266–272.

Maduabuchi, C.M., 2004. Case studies on transboundary aquifers in Nigeria. In: Appelgren, B. (ed), 2004. *Managing Shared Aquifer Resources in Africa*, Proceedings of the International ISARM-Africa Workshop at Tripoli, Libya, 2–4 June 2002, UNESCO IHP-VI Series on Groundwater No 8, pp. 135–141.

OSS, 2007. *Gestion des Risques Hydrogéologiques dans le Système Aquifère d'Iullemeden (SAI): Analyse diagnostique transfrontalière.* OSS/GEF, March 2007, 108 p.

OSS, 2008. *Management of the Hydrogeological Risks of the Iullemeden Aquifer System (IAS): Scientific Report.* OSS/GEF, March 2008, 17 p.

UNESCO/OSS, 2005. *Ressources en eau et gestion des aquifères transfrontaliers de l'Afrique du Nord et du Sahel.* ISARM-Africa, UNESCO IHP-IV Series on Groundwater No 11, Paris.

Lake Chad Basin

Dodo, A., 2004. Caractérisation des systèmes aquifères transfrontaliers du Niger. In: Appelgren, B. (ed.), 2004. *Managing Shared Aquifer Resources in Africa*, Proceedings of the International ISARM-Africa Workshop at Tripoli, Libya, 2–4 June 2002, UNESCO IHP-VI Series on Groundwater No. 8, pp. 123–128.

Eberschweiler, C., 1992. *Suivi et gestion des ressources en eaux souterraines dans le basin du lac Tchad. Prémodélisation des systèmes aquifères, évaluation des ressources et simulation d'exploitation*. Unpublished report, BRGM, 106 p.

Maduabuchi, C.M., 2004. Case studies on transboundary aquifers in Nigeria. In: Appelgren, B. (ed), 2004. *Managing Shared Aquifer Resources in Africa*, Proceedings of the International ISARM-Africa Workshop at Tripoli, Libya, 2–4 June 2002, UNESCO IHP-VI Series on Groundwater No. 8, pp. 135–141.

Massuel, S., 2001. *Modélisation hydrodynamique de la nappe phréatique quaternaire du basin du lac Tchad*. Thesis, University of Montpellier II and University of Avignon.

Ngatcha, B.N., B. Laignel, J. Mudry and P. Genthon, 2010. Gestion des eaux souterraines dans une région sous contraintes naturelles et anthropiques sévères: le bassin du lac Tchad. In: *Proceedings Third Intern. Conference 'Managing Shared Resources in Africa'*, Tripoli, 25–27 May 2008, UNESCO, Paris, pp. 180–185.

Oguntola, J.A., 2004. Management of transboundary aquifer systems in Africa. A case study of the Lake Chad Basin Commission. In: Appelgren, B. (ed), 2004. *Managing Shared Aquifer Resources in Africa*, Proceedings of the International ISARM-Africa Workshop at Tripoli, Libya, 2–4 June 2002, UNESCO IHP-VI Series on Groundwater No. 8, pp. 203–208.

Schneider, J.L., 2001. *Carte de valorisation des eaux souterraines de la République du Tchad (à 1/1500000) et géologie-archéologie-hydrologie de la République du Tchad*. Two volumes, 1100 p.

UNESCO/OSS, 2005. *Ressources en eau et gestion des aquifères transfrontaliers de l'Afrique du Nord et du Sahel*. ISARM-Africa, UNESCO IHP-IV Series on Groundwater No 11, Paris.

Sudd Basin (Umm Ruwaba Aquifer)

ICID, 1983. *Irrigation and Drainage in the World. Vol. III, Sudan*. Third ed. 1983, New Delhi, pp. 1262–1280.

Safar-Zitoun, M., 1993. *Notice explicative de la carte hydrogéologique international de l'Afrique, Feuille 3*. Association Africaine de Cartographie, Algiers, Algeria.

UN, 1987. *Groundwater in Africa*. UN-DTCD, New York, Natural Resources Water Series No 18.

Yousif, M.A. & S.H. Abdallah, 2010. *Transboundary aquifers, Sudan Country Paper*. ISARM for IGAD Region Meeting, 23–25 February 2010, Addis Ababa, Ethiopia, 24 p.

Ogaden-Juba Basin

UN, 1987. *Groundwater in Africa*. UN-DTCD, New York, Natural Resources Water Series No 18.

Congo Basin

Moukolo, N., 1992. État des connaissances actuelles sur l'hydrogéologie du Congo Brazzaville. *Hydrogéologie*, No 1–2, 1992, pp. 47–58.

Safar-Zitoun, M., 1993. *Notice explicative de la carte hydrogéologique international de l'Afrique, Feuille 4*. Association Africaine de Cartographie, Algiers, Algeria.

Seguin, J.J., 2005. *Projet Réseau SIG-Afrique, Carte hydrogéologique de l'Afrique á l'échelle du 1/10 M*. BRGM/RP-54404-FR.

Snel, M.J., 1957. *Contribution à l'étude hydrogéologique du Congo Belge*. Congo Belge, 4ᵉ Direction Générale, Service Géologique, 36 p.

Zektser, I.S., & L.G. Everett (eds), 2004. *Groundwater resources of the world and their use.* UNESCO IHP-VI Series on Groundwater No 6, Paris, UNESCO.

Upper Kalahari-Cuvelai-Upper Zambezi Basin

Ndengu, S., 2004. International shared aquifers in Namibia. In: Appelgren, B. (ed), 2004. *Managing Shared Aquifer Resources in Africa*, Proceedings of the International ISARM-Africa Workshop at Tripoli, Libya, 2–4 June 2002, UNESCO IHP-VI Series on Groundwater No 8, pp. 117–122.

Lower Kalahari–Stampriet Basin

Ndengu, S., 2004. International shared aquifers in Namibia. In: Appelgren, B. (ed), 2004. *Managing Shared Aquifer Resources in Africa*, Proceedings of the International ISARM-Africa Workshop at Tripoli, Libya, 2–4 June 2002, UNESCO IHP-VI Series on Groundwater No 8, pp. 117–122.

Peck, H., 2000. *The preliminary study of the Stampriet Transboundary aquifer in the South East Kalahari/Karoo Basin.* MSc Thesis, University of the Western Cape, Cape Town, 93 p.

Karoo Basin

Safar-Zitoun, M., 1993. *Notice explicative de la carte hydrogéologique international de l'Afrique, Feuille 4.* Association Africaine de Cartographie, Algiers, Algeria.

Zektser, I.S. and Everett, L.G. (eds), 2004. *Groundwater resources of the world and their use.* UNESCO IHP-VI Series on Groundwater No 6. Paris, UNESCO, pp. 230–232.

6.3.2 North America

Northern Great Plains Aquifer

UN, 1976. Groundwater in the Western hemisphere. UN-DTCD, New York, Natural Resources Water Series No 4.

USGS, 2003. Principal Aquifers. In: *National Atlas of the United States of America, scale 1: 5 000 000.* USGS (revised in 2003).

USGS, 2005. *Ground Water Atlas of the United States – Montana, North Dakota, South Dakota, Wyoming: Regional Aquifer Systems.* Available from: http://pubs.usgs.gov/ha/ha730/ch_i/I-text2.html.

Vogelsberg, A., 2007. *Northern Great Plains Aquifer System.* Available from: http://academic.emporia.edu/schulmem/hydro/TERM%20PROJECTS/2007/Vogelsberg/Beryl_to_Aquifer.htm

Cambrian-Ordovician Aquifer System

Grundl, T., & L. Schmidt, 2002. *Delineation of high-salinity conditions in the Cambro-Ordovician aquifer of Eastern Wisconsin.* University of Wisconsin-Madison, Department of Natural Resources, 55 p.

USGS, 2003. Principal Aquifers. In: *National Atlas of the United States of America, scale 1: 5 000 000.* USGS (revised in 2003).

Young, H.L. and D.I. Siegel, 1992. *Hydrogeology of the Cambrian-Ordovician Aquifer System in the Northern Midwest, United States.* USGS Professional Paper 1405-B, USGS, Denver, 99 p.

Californian Central Valley Aquifer System

Famiglietti,J., S. Swenson & M. Rodell, 2009. *Water storage changes in California's Sacramento and San Joaquin river basins, including groundwater depletion in the Central Valley.*

PowerPoint presentation, American Geophysical Union Press Conference, 14 December 2009, CSR, GFZ, DLR and JPL.

Famiglietti, J. S, M. Lo, S.L. Ho, J. Bethune, K.J. Anderson, T.H. Syed, S. C. Swenson, C.R. de Linage & M. Rodell, 2011. Satellite measure recent rates of groundwater depletion in California's Central valley. *Geoph. Res. Letters,* Vol. 38, L03403, 4 p. Available from: https://webfiles.uci.edu/jfamigli/blog/CentralValleyGW.pdf.

Ireland, R.H., J.F. Poland & F.S. Riley, 1984. *Land subsidence in the San Joaquin Valley, California, as of 1980.* USGS Professional Paper 437–1.

USGS, 2003. Principal Aquifers. In: *National Atlas of the United States of America, scale 1: 5 000 000.* USGS (revised in 2003).

Willamson, A.K., D.E. Prudic and L.A. Swain, 1989. *Ground-water flow in the Central Valley, California.* USGS Professional Paper 1401-D.

Ogallala Aquifer (High Plains)

McGuire, V., 2003. *Water-level changes in the High Plains aquifer, predevelopment to 2001, 1999 to 2000, and 2000 to 2001.* USGS Fact Sheet FS-078–03, 4p, USGS, Reston, Virginia.

McGuire, V., 2009. *Water-level changes in the High Plains aquifer, predevelopment to 2007, 2005–06 and 2006–07.* Scientific Investigations Report 2009–5019, USGS, Reston, Virginia.

Sophocleus, M., 2010. Review: groundwater management practices, challenges and innovations in the High Plains aquifer, USA – lessons and recommended actions. *Hydrogeology Journal,* Vol. 18, pp. 559–575.

USGS, 2003. Principal Aquifers. In: *National Atlas of the United States of America, scale 1: 5 000 000.* USGS (revised in 2003).

Atlantic and Gulf Coastal Plains Aquifer

Sonenshein, R.S., 1997. *Delineation and extent of saltwater intrusion in the Biscayne aquifer, Eastern Dade county, Florida, 1995.* USGS Water Resources Investigations Report 96–4285.

USGS, 1990. *Ground Water Atlas of the United States – Alabama, Florida, Georgia, South Carolina: Biscayne Aquifer.* Available from: http://pubs.usgs.gov/ha/ha730/ch_g/G-text4.html.

USGS, 1990. *Ground Water Atlas of the United States – Alabama, Florida, Georgia, South Carolina: Floridan Aquifer.* Available from: http://pubs.usgs.gov/ha/ha730/ch_g/G-text6.html.

USGS, 1990. *Ground Water Atlas of the United States – Alabama, Florida, Georgia, South Carolina: Southeastern Coastal Plain Aquifer System.* Available from: http://pubs.usgs.gov/ha/ha730/ch_g/G-text7.html.

USGS, 1996. *Ground Water Atlas of the United States – Oklahoma, Texas: Coastal Lowlands Aquifer System.* Available from: http://pubs.usgs.gov/ha/ha730/ch_e/E-text6.html.

USGS, 1998. *Ground Water Atlas of the United States – Arkansas, Louisiana, Mississippi: Coastal Lowlands Aquifer System.* Available from: http://pubs.usgs.gov/ha/ha730/ch_f/F-text3.html.

USGS, 1998. *Ground Water Atlas of the United States – Arkansas, Louisiana, Mississippi: Missisippi Embayment Aquifer System.* Available from: http://pubs.usgs.gov/ha/ha730/ch_f/F-text4.html.

USGS, 2003. Principal Aquifers. In: *National Atlas of the United States of America, scale 1: 5 000 000.* USGS (revised in 2003).

6.3.3   South America

Amazon Basin

Da Franca, N., M. Miletto, M. Donoso, A. Aureli, S. Puri, J. van der Gun, O. Tujchneider & A. Rivera, 2007. *Sistemas acuíferos transfronterizos en las Américas. Evaluación preliminar.* PHI-VI Serie ISARM Américas No 1, UNESCO-IHP, Montevideo and OEA Washington, pp. 124–125.

Miotto, K., 2010. *En el subsuelo de la Amazonia, el mayor acuífero del mundo?* Available from: http://www.oecoamazonia.com/es/reportajes/brasil/61-no-subsolo-da-amazonia-eis-o-maior-aquifero-no-mundo.

Rebouças, A., 1988. Groundwater in Brazil. *Episodes*, Vol 11, No 3, pp. 209–214.

Ferreira do Rosario, F., 2011. O sistema aquifer cretáceo multicamada Tikuna, subunidade do Sistema Aquífero Amazonas. Federal University of Rio de Janeiro, 224 pp.

Maranhão Basin

Araújo dos Santos, L.C., 2010. *Reflexões sobre água subterranea do Estado de Maranhão.* XVI Congresso Brasileiro de Águas Subterráneas e XVII Encontro Nacional de Perfuradores de Poços, 17 p.

Rebouças, A., 1976. Le grand basin hydrogéologique du Maranhão, Brésil. Perspectives sur l'exploitation. In: *Proceedings XIth IAH Congress, Budapest*, pp. 448–458.

Rebouças, A., 1988. Groundwater in Brazil. *Episodes*, Vol 11, No 3, pp. 209–214.

Guaraní Aquifer System

Araújo, L.M., .A. B. França & P.E. Potter, 1999. Hydrogeology of the Mercosul aquifer system in the Paraná and Chaco-Paraná basins, South America, and comparison with the Navajo-Nugget aquifer system, USA. *Hydrogeology Journal*, Vol. 7, No 3, pp. 317–336.

Da Franca, N., M. Miletto, M. Donoso, A. Aureli, S. Puri, J. van der Gun, O. Tujchneider & A. Rivera, 2007. *Sistemas acuíferos transfronterizos en las Américas. Evaluación preliminar.* PHI-VI Serie ISARM Américas No 1, UNESCO-IHP, Montevideo and OEA Washington, pp. 140–141.

Fili, M., E. Rosa Filho, N. Auge & J. Xavier, 1998. *El acuífero Guaraní. Un recurso compartido por Argentina, Brasil, Paraguay y Uruguay (América del Sur). Hidrología Subterránea.* Bol. Géol. y Minero, Madrid, Vol. 109, No 4, pp. 389–394,

GEF, 2007. *Project for the Environmental Protection and Sustainable Development of the Guaraní System: Transboundary Diagnosyic Analyis (TDA).* Montevideo, 115 p.

Geodatos SRL, 2008. *Sistema Acuífero Guaraní: Informe final geológico-geofísico.* Geodatos SRL, Buenos Aires, Argentina, 166 p.

Perez, M., 2006.The Guarani aquifer system. State-of-the-art in Argentina. In: *International Symposium Darcy 2006 – Aquifer Systems Management*, Darcy-58, IAH and BRGM, May–June 2006, Dijon, France.

Rebouças, A., 1988. Groundwater in Brazil. *Episodes*, Vol 11, No 3, pp. 209–214.

Rebouças, A.C., 1994. *Sistema agüífero Botucatu no Brasil.* Anais 8° Congr. Bras. Águas Subterrâneas – ABAS, Recife, pp. 500–509.

6.3.4   Asia

Arabian Aquifer System

Abdurrahman, W., 2002. *Development and management of groundwater in Saudi Arabia.* GW-MATE/UNESCO Expert Group Meeting, Socially sustainable management of groundwater mining from aquifer storage, Paris, 6 p.

Abderrahman, W.A., 2006. Saudi Arabia Aquifers. In: Foster, S., & D. Loucks, 2006. *Non-renewable groundwater resources.* UNESCO IHP-VI Series on Groundwater No 10, pp. 63–67.

Bakiewicz, W., D.M. Milne & M. Noori, 1982. Hydrogeology of the Umm Er Radhuma aquifer, Saudi Arabia, with reference to fossil gradients. Q.J. Eng. Geol., 15, pp. 105–126.

UN, 1982. *Groundwater in the Eastern Mediterranean and Western Asia.* UN-DTCD, New York, Natural Resources Water Series No. 9.

Indus Basin

Archer, D.R., N. Forsythe, H.J. Fowler & S.M. Shah, 2010. Sustainability of water resources management in the Indus Basin under changing climatic and socio economic conditions. *Hydrol. Earth Syst. Sci. Discuss.*, Vol. 7, pp 1883–1912. Available from: http://www.hydrol-earth-syst-sci-discuss.net/7/1883/2010/hessd-7-1883-2010.pdf.

ICID, 2000. *Irrigation and drainage in the world.* Chapter Pakistan, 12 p., New Delhi.

Kahlown, M.A., & M. Azam, 2004. Groundwater development and management in Indus Basin: issues and challenges. In: *Proceedings Int. Conf. 'Research basins and hydrological planning'*, Heifei/Anhui, China, 22–31 March 2004, Tailor & Francis 2004, pp. 193–200. Available from: doi: 10. 1201/9781439833858.ch27.

Khan, S., T. Rana, F. Gabriel, & M.K. Ullah, 2008. Hydrogeological assessment of escalating groundwater exploitation in the Indus Basin, Pakistan. *Hydrogeology Journal*, Vol. 16, No. 8, pp. 1635–1654. Available from: doi: 10.1007/s10040-008-0336-8.

Qureshi, A.S., P.G. McCornick, M. Qadir and Z. Aslam, 2008. Managing salinity and water-logging in the Indus Basin of Pakistan. *Agricultural Water Management*, Vol. 95, pp. 1–10. Available from: http://www.icarda.org/docrep/Articles/Managing_salinity.pdf.

Qureshi, A.S., P.G. McCornick, A. Sarwar & B.R. Sharma, 2010. Challenges and prospects of sustainable groundwater management in the Indus Basin, Pakistan. *Water resources Management*, Vol. 24, No 8, pp. 1551–1569. Available from: doi: 10.1007/s11269-009-9513-3.

Rodell, M., I. Velicogna & J. Famiglietti, 2009. *Satellite-based estimates of groundwater depletion in India.* Nature, Vol. 460. Available from: doi:10.1038/nature08238.

Tiwari, V., J. Wahr & S. Swenson, 2009. Dwindling groundwater resources in northern India, from satellite gravity observations. *Geophysical Research Letters*, Vol. 36.

Indus-Ganges-Brahmaputra Basin

IWMI, no date (2009 or later). *Water resources of Indus-Gangetic Basin: Continuing threats and emerging challenges.* CPWF-IWMI Basin Focal Project for Indus Gangetic Basin. Research Brief-Series 1, IWMI, CGIAR, 4 p.

Rodell, M., I. Velicogna & J. Famiglietti, 2009. *Satellite-based estimates of groundwater depletion in India.* Nature, Vol. 460. Available from: doi:10.1038/nature08238.

Scott, C.A., & B. Sharma, 2009. Energy surplus and the expansion of groundwater irrigation in the Indus-Ganges Basin. *Intl. J. River Basin Management*, Vol. 7, No 1, pp. 1–6.

Sharma, B.R., & G. Ambili, 2009. *Water resources and hydrogeology of the Indus-Gangetic basin: Comparative analysis of issues and opportunities*, Review Paper, Annals of Arid Zone (India), 41 p.

Tiwari, V., J. Wahr & S. Swenson, 2009. Dwindling groundwater resources in northern India, from satellite gravity observations. *Geophysical Research Letters*, Vol. 36.

West Siberian Basin

UN, 1986. *Groundwater in Continental Asia.* UN-DTCD, New York, Natural Resources Water Series No 15.

Vsevolozhsky, V.A., 1973. *Groundwater resources of the southern part of the West Siberian Lowland.* Moscow, Nauka.

Zektser, I.S., & L.G. Everett (eds), 2004. *Groundwater resources of the world and their use.* UNESCO IHP-VI Series on Groundwater No 6. Paris, UNESCO, 342 p. Available from: http://unesdoc.unesco.org/images/0013/001344/134433e.pdf.

Tunguss Basin

UN, 1986. *Groundwater in Continental Asia.* UN-DTCD, New York, Natural Resources Water Series No 15.

Zektser, I.S., & L.G. Everett (eds), 2004. *Groundwater resources of the world and their use.* UNESCO IHP-VI Series on Groundwater No 6. Paris, UNESCO, 342 p. Available from: http://unesdoc.unesco.org/images/0013/001344/134433e.pdf.

Angara-Lena Basin

UN, 1986. *Groundwater in Continental Asia.* UN-DTCD, New York, Natural Resources Water Series No 15.

Zektser, I.S., & L.G. Everett (eds), 2004. *Groundwater resources of the world and their use.* UNESCO IHP-VI Series on Groundwater No 6. Paris, UNESCO, 342 p. Available from: http://unesdoc.unesco.org/images/0013/001344/134433e.pdf.

Yakut Basin

UN, 1986. *Groundwater in Continental Asia.* UN-DTCD, New York, Natural Resources Water Series No 15.

Zektser, I.S., & L.G. Everett (eds), 2004. *Groundwater resources of the world and their use.* UNESCO IHP-VI Series on Groundwater No 6. Paris, UNESCO, 342 p. Available from: http://unesdoc.unesco.org/images/0013/001344/134433e.pdf.

North China Aquifer System (Huang Huai Hai Plain)

Jia, Y., & J. You, 2010. *Sustainable groundwater management in the North China Plain: main issues, practices and foresights.* Extended abstracts XXXVIIIth IAH Congress, Krakow, 12–17 Sept 2010, No. 517, pp 855–862.

Foster, S., & H. Garduño, 2004. Towards sustainable groundwater resource use for irrigated agriculture on the North China Plain. GW.MATE Case Profile Collection Number 8, The World Bank, Washington, 16 p.

Foster, S., H. Garduño, R. Evans, D. Olson, Y. Tian, W. Zhang & Z. Han, 2004. Quaternary Aquifer of the North China Plain – assessing and achieving groundwater resource sustainability. *Hydrogeology Journal*, Vol. 12, No 1, pp. 81–93.

Kendy, L., Y. Zhang, Ch. Liu, J. Wang & T. Steenhuis, 2004. Groundwater recharge from irrigated cropland in the North China Plain: case study of Luancheng County, Hebei Province, 1949–2000. *Hydrol. Process.* Vol. 18, pp. 2289–2302.

Liu, Ch., J. Yu & E. Kendy, 2001. Groundwater exploitation and its impact on the environment in the North China Plain. *Water International*, Vol. 26, No 2, pp. 265–272.

Wang, R. H. Ren & Z. Ouyang (eds), 2000. *China Water Vision, 2000. The ecosphere of water, life, environment and development.* Beijing, China Meteorological Press, 178 p.

Zaicheng, H., 2002. Artificial recharge of groundwater in North China Plain. In: *Management of Aquifer Recharge for Sustainability*, Swets & Zeitlinger BV, Lisse, The Netherlands.

Zaicheng, H., 2006. Alluvial aquifers in North China Plain. In: *International Symposium Darcy 2006 – Aquifer Systems Management*, Darcy-01, IAH and BRGM, May–June 2006, Dijon, France.

Zektser, I.S., & L.G. Everett (eds), 2004. *Groundwater resources of the world and their use.* UNESCO IHP-VI Series on Groundwater No 6. Paris, UNESCO, 342 p. Available from: http://unesdoc.unesco.org/images/0013/001344/134433e.pdf.

Song-Liao Plain

Chen, M., 1987. Groundwater resources and development in China. *Environmental Geology*, Vol 10, No 3, pp. 141–147. Available from: doi: 10.1007/BF02580469.

Zektser, I.S., & L.G. Everett (eds), 2004. *Groundwater resources of the world and their use.* UNESCO IHP-VI Series on Groundwater No 6. Paris, UNESCO, 342 p. Available from: http://unesdoc.unesco.org/images/0013/001344/134433e.pdf.

Tarim Basin

Cui, Y., & J. Shao, 2005. The role of ground water in arid/semiarid ecosystems, Northwest China. *Ground Water*, Vol. 43, No. 4 (2005), pp 471–477.

Ma, J., & J. Li, 2001. The groundwater resources and its sustainable development in the south edge of the Tarim Basin. *Chinese Geographical Science*, Vol. 11, No 1 (2001), pp. 57–62. Available from: doi: 10.1007/s11769–001–0008–2.

Ma, L., T. Lowenstein, J.Li, P.Jiang, C. Liu, J. Zhong, J. Sheng, H. Qiu & H. Wu, 2010. Hydrochemical characteristics and brine evolution paths of Lop Nor Nasin, Xinjiang Province, Western China. *Applied Geochemistry*, Vol. 25, pp. 1770–1782.

Sun, H., 2008. The characteristics of Tarim Basin's groundwater resources and its sustainable utilization. *Ground Water*, 2008–04, Bimonthly CNKI Journal, China.

Wang, Z., T. Chen, D. Yu & F. Song, 2004. Division of groundwater system of Tarim Basin. *Xinjiang Geology*, 2004–3.

6.3.5   Europe

Paris Basin

Goncalves, J., 2006. 3D modelling of the permafrost development in the Paris basin to ascertain its hydrogeological impact. In: *International Symposium Darcy 2006 – Aquifer Systems Management*, Darcy 104, IAH and BRGM, May–June 2006, Dijon, France.

Jost, A., 2006. Are large-scale aquifer systems in equilibrium with their environmental conditions? A modelling approach on the example of the Paris basin. In: *International Symposium Darcy 2006 – Aquifer Systems Management*, Darcy-55, IAH and BRGM, May–June 2006, Dijon, France.

UN, 1990. *Groundwater in Europe*. UN-DTCD, New York, Natural Resources Water Series No 19.

Russian Platform Basins

Juodcasis & A. Klimas, 1991. *The change of groundwater quality in the Baltic artesian basin*. Vilnius, 39 p.

Lebedeva, N.A., 1972. *Natural groundwater resources of the Moscow artesian basin*. Moscow, Nedra, 148 p.

UN, 1990. *Groundwater in Europe*. UN-DTCD, New York, Natural Resources Water Series No 19.

North Caucasus Basin

UN, 1990. *Groundwater in Europe*. UN-DTCD, New York, Natural Resources Water Series No 19.

Pechora Basin

UN, 1990. *Groundwater in Europe*. UN-DTCD, New York, Natural Resources Water Series No 19.

6.3.6   AUSTRALIA

Great Artesian Basin

GABCC, 2009. *Great Artesian Basin Strategic Management Plan: Progress and achievements to 2008*. GABCC Secretariat, Manuka, Australia.

Habermehl, M.A., 1980. The Great Artesian Basin, Australia, in *BMR Journal of Australian Geology & Geophysics*, 5, pp 9–38.

Habermehl, M.A., 1982. *Springs in the Great Artesian Basin, Australia, – their origin and nature*. Bureau of Mineral Resources, Australia, Report 235.

Habermehl, M., 2002. *Groundwater development in the Great Artesian Basin, Australia*. GW-MATE UNESCO Expert Group meeting, Socially Sustainable Management of Groundwater Mining from Aquifer Storage, 35 p., Paris.

Habermehl, M., 2006. The Great Artesian Basin, Australia. In: Foster, S., & D. Loucks, 2006. *Non-renewable groundwater resources*. UNESCO-IHP, IHPVI, Series on Groundwater No. 10, pp 82–88.

Herczeg, A., & A. Love, 2007. *Review of recharge mechanisms for the Great Artesian Basin*. Report to the Great Artesian Basin Coordinating Committee, CSIRO, Australia.

Queensland Government, 2012. *Great Artesian Basin*. Internet portal. Available from: http://www.derm.qld.gov.au/water/gab/ [Accessed in 2012].

Sinclair Knight Merz, 2008. *Great Artesian Basin sustainability initiative. Mid-term review of Phase 2*. Report prepared for Australian Government, Department of the Environment and Water Resources.

Welsh, W., 2006. *Great Artesian Basin transient groundwater model*. Australian Government, Bureau of Rural Sciences, Canberra.

Canning Basin

Bestow, T.T., 1991. *The geothermal energy development potential of large basins in Western Australia*. Proc. Int. Conf. 'Groundwater in large sedimentary basins', Perth, 1990. AWRC, Conference Series, 20, pp 319–328.

Ghassemi, F., H. Ethminan and J. Ferguson, 1992. A reconnaissance investigation of the major Palaeozoic aquifers in the Canning Basin, Western Australia, in relation to Zn-Pb mineralisation. *BMR Journal of Australian Geology and Geophysics*, Vol. 13, pp. 37–54.

Laws, A.T., 1991. *Outline of the groundwater resource potential of the Canning Basin, Western Australia*. In: Proceedings Int. Conf. 'Groundwater in large sedimentary basins', Perth, 1990. AWRC, Conference Series No 20, pp. 47–58.

Lau, J.E., D.P. Commander & G. Jacobson, 1987. *Hydrogeology of Australia*. Bureau of Mineral Resources, Geology and Geophysics, Bull. 227, Canberra, Australian Gov. Publ. Service.

Leech, R.E.J., 1979. Geology and groundwater resources of the southwestern Canning Basin, Western Australia. In: *Annual Report 1978 of the Geological Survey of Western Australia*, pp. 66–74.

Zektser, I.S., & L.G. Everett (eds), 2004. *Groundwater resources of the world and their use*. UNESCO IHP-VI Series on Groundwater No 6. Paris, UNESCO, 342 p. Available from: http://unesdoc.unesco.org/images/0013/001344/134433e.pdf.

6.4 SELECTED GLOBAL AND REGIONAL HYDROGEOLOGICAL MAPS[2]

6.4.1 Global maps

Dzhamalov, D.G., & I.S. Zektser (eds), 1999. *World Map of Hydrogeological Conditions and Groundwater Flow, scale 1: 10 M*. Water Problems Institute, Russian Academy of Sciences, under UNESCO supervision.

IGRAC, 2009. *Transboundary aquifers of the world, 1: 50 M, update*. IGRAC, UNESCO and WMO, Special edition for the 5th World Water Forum, Istanbul.

---

2 Meta-information on a large number of national hydrogeological maps can be found in WHYMAP's Web Map Application: http://www.bgr.de/app/fishy/whymap/ (using the WHYMIS button).

IGRAC, 2012. *Transboundary aquifers of the world, 1: 50 M, update.* IGRAC, UNESCO and WMO, Special edition for the 6th World Water Forum, Marseille.

UNESCO, CGMW, IAH, IAEA & BGR, 2004. *Groundwater Resources of the World, 1: 50 M.* WHYMAP, Special edition for the International Geological Congress, Florence, Italy, August 2004. UNESCO & BGR, Paris, Hannover.

UNESCO, CGMW, IAH, IAEA & BGR, 2006. *Groundwater Resources of the World – Transboundary Aquifer Systems, 1: 50 M.* WHYMAP, Special edition for the Fourth World Water Forum, Mexico, March 2006. UNESCO & BGR, Paris, Hannover.

UNESCO, CGMW, IAH, IAEA & BGR, 2008a. *Groundwater Resources of the World, 1: 25 M.* WHYMAP. UNESCO & BGR, Paris, Hannover.

UNESCO, CGMW, IAH, IAEA & BGR, 2008b. *Groundwater Resources of the World, 1: 25 M.* WHYMAP, Special edition reduced to the scale 1: 40 M. UNESCO & BGR, Paris, Hannover.

UNESCO, CGMW, IAH, IAEA & BGR, 2012. *River and Groundwater Basins of the World, 1: 50 M.* WHYMAP, Special edition for the 6th World Water Forum, Marseille, March 2012. UNESCO & BGR, Paris, Hannover.

The Americas

Heath, R.C., 1988. *Hydrogeological Map of North America – Showing the Major Units that Comprise the Surficial Layer. 1:13.3 M.* The Geological Society of America, Inc.

Heath, R.C., 1988. *Hydrogeological Map of North America – Showing the Major Rock Units that Underlie the Surficial Layer. 1:13.3 M.* The Geological Society of America, Inc.

Da Franca R dos Anjos, N., A. Mente, A.A. Frota Mont'Alverne, G. Ruy Derze & E. Godoy, 1996. *Mapa Hidrogeológico de America del Sur, escala 1:5 M.* UNESCO-PHI, Departamento Nacional de Producão Mineral (DNPM) and Servicio Geológico de Brasil (CPRM).

Europe

Planet Earth, 2008. *International Hydrogeological Map of Europe, reduced to the scale of 1: 50 M.* Special map mosaic printed on the occasion of the International Year of Planet Earth 2008, for the International Geological Congress in Oslo (2008), EuroGeoSurveys, UNESCO-IHP, IAH, CGMW and BGR.

UNESCO and BGR, 1960 – present. *Hydrogeological Map of Europe 1:1.5 M.* In production since 1960, UNESCO, Paris, and BGR, Hannover.

Africa

Margat, J., 1987. *Ressources en eaux souterraines 1:20 M.* BRGM and United Nations.

Safar Zitoun, M., & A.C. Nouiouat, 1992. *International Hydrogeological Map of Africa 1:5 M.* Organization of African Unity (OAU) and Organisation Africaine de Cartographie et Télédétection (OACT), Algiers.

MacDonald, A.M., H.C. Bonsor, B.E.O. Dochartaigh & R.G. Taylor, 2012. Quantitative maps of groundwater resources in Africa. *Envir. Res. Lett. 7 (2012),* 024009, 7 p. Available from: doi: 10.1088/1748–9326/7/2/024009.

Seguin, J.J., 2005. *Carte hydrogéologique de l'Afrique: une maquette à l'échelle du 1:10 M. (Hydrogeological Map of Africa: a prototype at 1: 10 M scale).* SIGAfrique, BRGM, Orleans.

Asia

Anonymous, 1987. *International Hydrogeological Map of South and East Asia 1:5 M.*

Nanzhang, T., T. Huijing & J. Xiaoqing, 1997. *Hydrogeological Map of Asia 1:8 M.* Compiled by the Institute of Hydrogeology and Engineering Geology, Chinese Academy of Geological Sciences, China.

Australia

D'Addario, G.W. (ed.), G. Jacobson & J.E. Lau (comp.), 1987. *Hydrogeology of Australia 1:5 M.* Bureau of Mineral Resources, Geology and Geophysics, Department of Resources and Energy, in association with the Australian Water Resources Council, Canberra, Australia.

Arab Region

UNESCO/ACSAD, 1988. Hydrogeological Map of the Arab Region and Adjacent Areas 1:5 M. Damascus.

Ex-Soviet Union

Anonymous, 1963. *Hydrogeological Map of the USSR 1:5 M.* Leningrad, USSR.

GUGK, 1975. *Map of groundwater runoff of the USSR, Scale 1:2.5M.* GUGK, Moscow, USSR.

Kudelin, B.I., & O.V. Popov (eds), 1964. *Map of the USSR Ground Water Discharge, Scale 1:5 M.* The RSFSR Ministry of Higher and Special Secondary Education, and the Moscow State University, USSR.

Marinov, N.A. (ed.), 1962, 1969. *Hydrogeological Map of the USSR 1:2.5 M.* Moscow, USSR.

Saizev, I.K. *et al.*, 1966. *Gidrogeologiczeskaja Karta SSST.* Ministerswa Geologii USSR& Geologiczeski Institut, Moscow.

Subject index

Note: page numbers in normal font refer to text; those in **bold** and *bold italics* to figures and tables, respectively.